WATER RESOURCES MANAGEMENT IN THE FACE OF CLIMATIC/HYDROLOGIC UNCERTAINTIES

Water Science and Technology Library

VOLUME 18

8500
Geol

#37

WATER RESOURCES MANAGEMENT IN THE FACE OF CLIMATIC/HYDROLOGIC UNCERTAINTIES

edited by

ZDZISLAW KACZMAREK
Institute of Geophysics,
Warsaw, Poland (IIASA)

KENNETH M. STRZEPEK
University of Colorado,
Boulder, Colorado, U.S.A. (IIASA)

LÁSZLÓ SOMLYÓDY
Technical University of Budapest,
Budapest, Hungary (IIASA)

and

VALENTINA PRIAZHINSKAYA
Institute of Water Problems,
Moscow, Russia

KLUWER ACADEMIC PUBLISHERS
DORDRECHT / BOSTON / LONDON

INTERNATIONAL INSTITUTE FOR APPLIED SYSTEMS ANALYSIS
LAXENBURG, AUSTRIA

Library of Congress Cataloging-in-Publication Data

Water resources monagement in the face of climatic/hydrologic
uncertainties / edited by Zdzislaw Kaczmarek ... [et al.].
p. cm. -- (Water science and technology library ; v. 18)
Includes index.
ISBN 0-7923-3927-4 (hardbound : alk. paper)
1. Water-supply. 2. Water quality. 3. Climatic changes.
I. Kaczmarek, Zdzisław. II. Series.
TD353.W354 1996
628.1'1--dc20 95-48402

ISBN 0-7923-3927-4

Published by Kluwer Academic Publishers,
P.O. Box 17, 3300 AA Dordrecht, The Netherlands.

Kluwer Academic Publishers incorporates
the publishing programmes of
D. Reidel, Martinus Nijhoff, Dr W. Junk and MTP Press.

Sold and distributed in the U.S.A. and Canada
by Kluwer Academic Publishers,
101 Philip Drive, Norwell, MA 02061, U.S.A.

In all other countries, sold and distributed
by Kluwer Academic Publishers Group,
P.O. Box 322, 3300 AH Dordrecht, The Netherlands.

Printed on acid-free paper

Printed in the Netherlands

The International Institute for Applied Systems Analysis

is an interdisciplinary, nongovernmental research institution founded in 1972 by leading scientific organizations in 12 countries. Situated near Vienna, in the center of Europe, IIASA has been for more than two decades producing valuable scientific research on economic, technological, and environmental issues.

IIASA was one of the first international institutes to systematically study global issues of environment, technology, and development. IIASA's Governing Council states that the Institute's goal is: *to conduct international and interdisciplinary scientific studies to provide timely and relevant information and options, addressing critical issues of global environmental, economic, and social change, for the benefit of the public, the scientific community, and national and international institutions.* Research is organized around three central themes:

- Global Environmental Change;
- Global Economic and Technological Change;
- Systems Methods for the Analysis of Global Issues.

The Institute now has national member organizations in the following countries:

Austria
The Austrian Academy of Sciences

Bulgaria
The National Committee for Applied Systems Analysis and Management

Canada
The Canadian Committee for IIASA

Czech Republic
The Czech Committee for IIASA

Finland
The Finnish Committee for IIASA

Germany
The Association for the Advancement of IIASA

Hungary
The Hungarian Committee for Applied Systems Analysis

Italy
The Italian National Member Organization is in the process of reorganization

Japan
The Japan Committee for IIASA

Kazakhstan
The National Academy of Sciences

Netherlands
The Netherlands Organization for Scientific Research (NWO)

Poland
The Polish Academy of Sciences

Russian Federation
The Russian Academy of Sciences

Slovak Republic
The Center for Strategic Studies

Sweden
The Swedish Council for Planning and Coordination of Research (FRN)

Ukraine
The Ukrainian Academy of Sciences

United States of America
The American Academy of Arts and Sciences

Contents

Preface

This volume was prepared by the Water Resources Project at the International Institute for Applied Systems Analysis (IIASA), Laxenburg, Austria, to provide a comprehensive, international analysis of the impact of climate change and variability on hydrologic resources and water resources management. This study is the first international effort of its kind and the first to examine a broad range of impacts, including not only hydrologic impacts, but also impacts on water use and demand, water quality, and water management. The volume includes studies of river basins in Europe, Asia, Africa, and North America. The analyses were undertaken by leading scientists from the many disciplines related to water resources research and range in scope between the watershed scale and the global scale.

This book presents the results of an international study; therefore, we feel that it will be of global interest. It is intended for government policy makers and staff members; government, private, and university researchers; and university scholars. It can be used as a main or supplementary text for upper-level undergraduate and graduate students in courses on such topics as climate change, global change, and environmental change.

This work is of an international and comprehensive nature: it reports, not on the latest finding, but on the state of the art of climate change impact assessment in late 1993, when the first draft of the manuscript was prepared for reviewers. The manuscript was externally reviewed by a panel of international water resource experts; each of its chapters should be considered equivalent to a peer-reviewed journal article.

The editors would like to thank Dr. Peter de Jánosi, Director of IIASA, for his support of this project. Robert McInnes, Ellen Bergschneider, Lilo Roggenland, and Ewa Delpos are due warm thanks; their hard work and professionalism made this volume possible.

List of Contributors

Nigel W. Arnell
Department of Geography
University of Southampton
Southampton, UK

Stewart Cohen
Canadian Climate Centre
Toronto, Ontario, Canada

Daniel Gellens
Royal Meterological Institute
of Belgium
Brussels, Belgium

Christian Jokiel
Institute for Hydraulic Engineering
and Water Resources Management
Aachen, Germany; also
Department of Civil and
Environmental Engineering
University of California, Davis
Davis, CA, USA

Danuta Jurak
Institute of Meteorology and
Water Management
Warsaw, Poland

Zdzislaw Kaczmarek
Water Resources Division
Institute of Geophysics
Warsaw, Poland; also
IIASA, Laxenburg, Austria

Janusz Kindler
Institute of Environment Engineering
Warsaw Technical University
Warsaw, Poland; also
World Bank
Washington, DC, USA

Zdenek Kos
Faculty for Civil Engineering
Technical University of Prague
Prague, Czech Republic

Surendra N. Kulshreshtha
Department of Agricultural Economics
University of Saskatchewan
Saskatoon, Canada; also
IIASA, Laxenburg, Austria

Zbigniew Kundzewicz
Research Center of Agricultural and
Forest Environment Studies
Poznan, Poland; also
World Meteorological Organization
Geneva, Switzerland

Ilya Masliev
IIASA, Laxenburg, Austria

Gregory J. McCabe, Jr.
United States Geological Survey
Denver, CO, USA

Gabriela K. Meyer
Department of Civil and
Environmental Engineering
University of California, Davis
Davis, CA, USA; also
IIASA, Laxenburg, Austria

Marshall E. Moss
United States Geological Survey
Tucson, AZ, USA

Jeffrey D. Niemann
Department of Civil and
Environmental Engineering
Massachusetts Institute of Technology
Cambridge, MA, USA; also
IIASA, Laxenburg, Austria

Gerald T. Orlob
Department of Civil and
Environmental Engineering
University of Californa, Davis
Davis, CA, USA

Valentina Priazhinskaya
Institute of Water Problems
Moscow, Russia

Kazimierz A. Salewicz
Imes Strobl Gruppe
Vösendorf, Austria; also
IIASA, Laxenburg, Austria

Vladimir M. Shnaydman
Advanced Visual Data, Inc.
Waltham, MA, USA; also
Institute of Water Problems
Moscow, Russia

László Somlyódy
Technical University of Budapest
Budapest, Hungary; also
IIASA, Laxenburg, Austria

Leszek Starkel
Institute of Geography and Spatial
Organization
Krakow, Poland

Kenneth M. Strzepek
Department of Civil, Environmental,
and Architectural Engineering
University of Colorado
Boulder, CO, USA; also
IIASA, Laxenburg, Austria

Karolý Szesztay
Constultant
Budapest, Hungary

Gary D. Tasker
United States Geological Survey
Reston, VA, USA

Olli Varis
Laboratory of Water Resources
Helsinki University of Technology
Helsinki, Finland; also
IIASA, Laxenburg, Austria

David M. Wolock
United States Geological Survey
Lawrence, KS, USA

David N. Yates
Department of Civil, Environmental,
and Architectural Engineering
University of Colorado
Boulder, CO, USA; also
IIASA, Laxenburg, Austria

Part I

Overview

Chapter 1

Climate, Hydrology, and Water Resources

Zdzislaw Kaczmarek, Nigel W. Arnell, and Leszek Starkel

1.1 The Issue

Scientists, and more recently politicians, worry about the possibility of global climate change due to increased levels of carbon dioxide (CO_2) and other anthropogenic greenhouse gases. Although there is no general consensus in the scientific community on the possible scale of global warming, the problem has been highly politicized by a number of international activities, such as the evaluation process initiated by the Intergovernmental Panel on Climate Change (IPCC) and by the Rio Convention on Climate. In spite of all the uncertainties, it is widely acknowledged that if climate changes due to various emissions of human origin, then the impact of such change on an increasing human population may create dramatic social and economic problems, at least in some vulnerable regions of the world.

Policy makers should be guided by Science, even in those unfortunate situations where scientific findings are far from being final and complete. According to the report of the World Commission on Environment and Development (Brundtland, 1987), environmental concerns must be integrated at all levels of national and international planning. When the earth's long-term future is considered the issue of climate change cannot be neglected, and various scenarios of possible trends in the environment and in society should be investigated. Science is responsible for developing a methodology for the assessment of climate impact and for providing a rational basis

Z. Kaczmarek et al. (eds.),
Water Resources Management in the Face of Climatic/Hydrologic Uncertainties 3–29.

for undertaking necessary adaptive measures. Science also should inform decision makers and the public about possible global threats and their accompanying uncertainties. The complexity of the atmosphere–ocean–land surface system means that we cannot eliminate surprises, and the world's inhabitants should be prepared to cope with them. This concerns regional water supply and demand to a great extent.

A statement unanimously accepted by the scientists gathered at the Second World Climate Conference, November 1990, declares, "Among the most important impacts of climate change will be its effects on the hydrological cycle and water management systems, and through these, on socioeconomic systems." Thus, this is the subject of our book, initiated by the water resources group at the International Institute for Applied Systems Analysis (IIASA) in Laxenburg, Austria, and written by authors representing research institutes from eight countries.

At least five areas of research related to the climate–water resources interface can be identified.

- Detection of changes in atmospheric and hydrological variables using various indicators, including paleohydrological studies.
- Sound, increasingly accurate parameterization of the hydrologic components of global and mesoscale atmospheric models.
- Assessment of the sensitivity of land surface processes to changes in climate characteristics.
- Assessment of possible implications of climate change and variability for regional water supply and demand, and consequently for management of water resources systems.
- Study of the impact of climate change on physical, chemical, and biological processes in rivers, lakes, and water reservoirs.

The progress made in all of these areas during the last decade is evident, but most of the relevant theories and models still need to be improved in order to meet the requirements of water resources practice. Our book concentrates on sensitivity and impact analysis, and is based on experience gained in a number of developed countries. Theoretical investigations are accompanied by a set of case studies showing the great diversity of problems facing the water resources community as we approach the twenty-first century.

The complex relations between climate, hydrology, various non-climatic factors, and water resources are illustrated in *Figure 1.1*. Carter *et al.* (1992) distinguish three methodological approaches to assessing climate impact. The simplest one is a direct "if–then" approach in which the effects

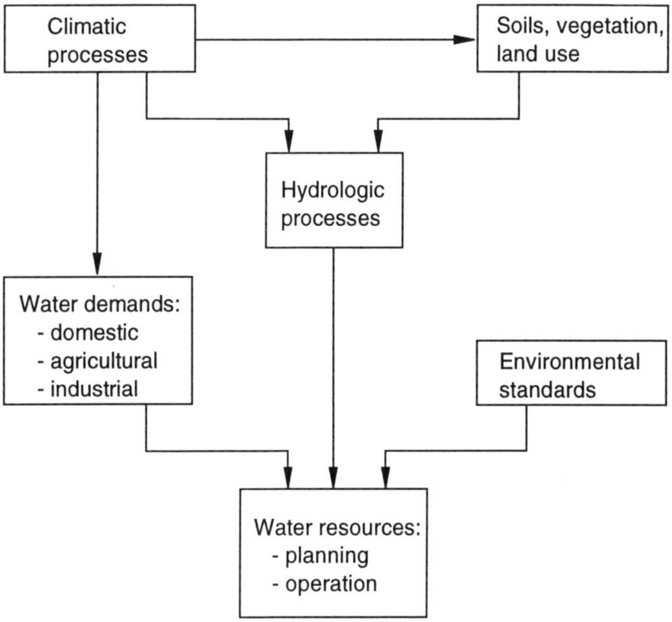

Figure 1.1. Climate, hydrology, and water resources.

of climatic events on an exposure unit (e.g., the water resources system) are studied in isolation from other non-climatic factors, which are assumed to be constant. In this direct-impact approach, the possible feedback effects usually are also neglected. Carter *et al.* (1992) stress that the "if–then" approach also may be applied to sectoral studies where a chain of direct impacts may extend to higher-order effects.

The second methodological approach takes into account that climate is only one of many causal factors influencing an environmental or economic system. For example, regional water demands will change due to fluctuations in temperature and precipitation patterns, but also due to population growth, economic development, and technological progress. In this case, the sensitivity of an exposure unit to climate disturbances should be distinguished in some way from the influence of other factors: in general, this is a rather difficult task.

The third methodological approach, an integrated regional impact assessment, combines all possible interactions, feedback, and adjustments to determine the impact of climate on society and the economy. The primary effects of global warming are likely to be outweighed by a number of secondary effects, which at present are difficult to assess and predict. To date, very few studies have used this third approach.

The first approach, the direct-impact approach, may be particularly useful for analyzing the possible effects of climatic shifts on physical, biological, and chemical characteristics of the natural environment, including hydrological systems. Study results on the impact of climate on river runoff, soil moisture index, temperature, and energy budget of water bodies for various regions of the world are reported in literature. Some of these results will be discussed in later chapters of this book. The second and the third of the above-mentioned approaches to treating the climate issue are more suitable for analyzing complex water resources systems, which are usually highly influenced by climate-independent (i.e., social and economic) factors, such as population growth.

This book does not propose a single, preferred methodology for solving water resources problems in the face of climatic uncertainties. First, such a generally accepted methodology does not exist; second, water problems are highly site specific. We would therefore like to offer a systematic view on the state of the art of a range of possible methods, models, and approaches, based on existing knowledge and on international experience. A number of regional case studies are presented to illustrate the variety of problems that may arise due to changes in regional climatic characteristics.

1.2 Is the Stationarity Concept of Climate Still Valid?

Atmospheric and land surface processes are governed by physical laws. However, by changing scales, the explanations for variability of natural phenomena may switch from being deterministic to being stochastic. Both the inputs to environmental systems and the systems' internal structure may include stochastic components, which result in random variability of output characteristics. The key questions are whether random properties of a natural system are time invariant or undergo changes; what the reasons are for non-stationarity of environmental processes; and what the scale of their variations is.

At the First World Climate Conference (Schaake and Kaczmarek, 1979), it was stressed that during the preceding decades hydrologists had recognized the need to create a scientific basis for analyzing the effects of climatic variability in order to plan and operate water resources systems. Emerging from this recognition was a science of stochastic hydrology. At the same time, a philosophy of stationarity of climatic and hydrological processes was widely accepted as the basis for water resources planning and for calculating performance criteria of water systems, such as risk, reliability, resilience, and robustness. A great number of statistical techniques of various levels of sophistication are used in hydrology (Kaczmarek, 1977); these tests usually assume that climatic and hydrologic time series are single realizations of stationary periodic-stochastic processes.

The stationarity concept of earth's environment has long been questioned by paleoclimatologists who have documented large changes in climate over diverse scales of time. Nevertheless, as has been observed by Amorocho (1970), long-term changes are "disregarded in many applications, under the assumption that the natural processes of watershed evolution are so slow that the changes they induce are of a negligible scale during the period of interest." What is most worrisome now is the superimposition of rapid human-made climate alterations on the usually less dynamic changes caused by natural factors.

Climate can be defined as a periodic-stochastic process, realizations of which are states of the atmosphere (called weather) at a point or over a given region; these atmospheric states are described by means of a set of quantifiable attributes $w_1, w_2, \ldots, w_r$. Realizations of a hydrologic stochastic process are, in turn, states of the hydrosphere, which are described by quantifiable attributes $h_1, h_2, \ldots, h_s$. The two processes are linked by a number of functional relations, called *transfer functions.* Due to seasonality, both processes are periodic, i.e., their probabilistic properties change within the year. They are, however, called stationary if the joint probability distribution, $F(w_1 \ldots w_r)$ or $F(h_1 \ldots h_s)$, is unaffected by shifting the observation times by any number of years. In other words, change or non-stationarity of climate means that probability distributions of climatic events or their parameters differ significantly for succeeding time intervals, for example, from one decade to another.

From the point of view of engineering practice, stationarity of climate and hydrology means that observations made in the past can be used in planning for the future determined by the expected useful life of the engineering structure. Although conclusions based on historical data may be affected by estimation errors, in the case of stationarity it is at least possible to evaluate these errors using standard mathematical methods. The question arises whether it is possible to detect non-stationarity of climatic and hydrological processes using statistical tests based on data routinely collected by weather services. According to our present experience, such expectations seem to be unrealistic, because of the high natural variability

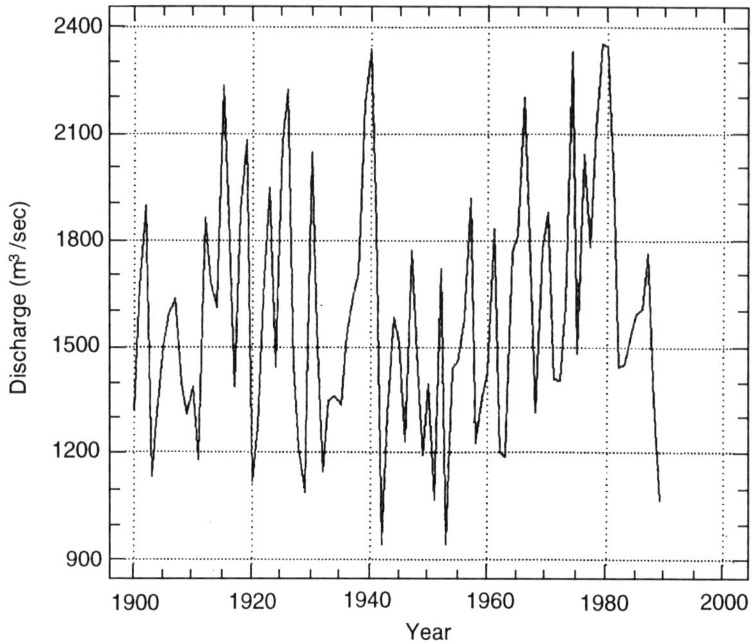

Figure 1.2. Yearly combined discharges of Odra and Vistula Rivers.

of these processes. For instance, *Figure 1.2* shows a 90-year time series of the combined average yearly runoff of two large Polish rivers, the Odra and the Vistula. With the standard deviations of this series being about 22% of the multi-annual average flow, it is practically impossible to prove any significant tendency in this series by means of standard statistical tests.

Statistical analysis of more than 150 long-term hydrological time series was implemented by IIASA in cooperation with the Institute of Geophysics of the Polish Academy of Sciences in Warsaw, within the framework of the World Climate Program. The Global Runoff Data Center in Koblenz, Germany, has provided the bulk of data – the amounts of water discharged monthly for several rivers in Europe, Africa, Asia, Australia, and North America. The study results, based on the application of some nonparametric tests with a significance level of 5%, show that for most cases the stationarity hypothesis could not be rejected. Various tests give various results; for example, the Mann's test seems to be more sensitive to changes than is the Kruskal-Wallis test. No firm conclusion can be drawn about the regional pattern of trends in runoff characteristics: in some cases, the neighboring river catchments have shown trends in the opposite direction, as can be seen

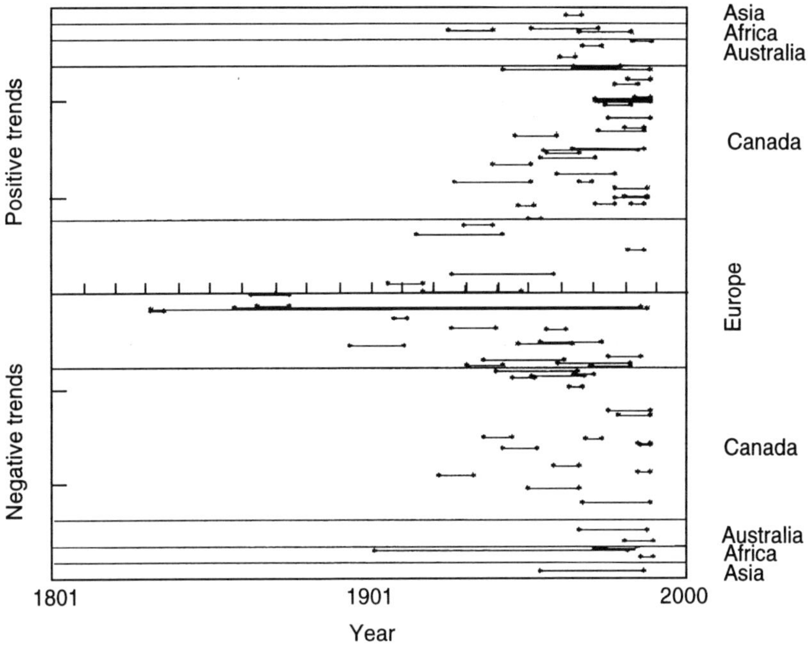

Figure 1.3. Trends in runoff values of selected world rivers: Mann's test (based on Mitosek, 1992).

in *Figure 1.3.* Further details on the results of this study can be found in a report by Mitosek (1992).

If the stationarity concept must be rejected, then from a practical point of view it is extremely important to agree on a methodology of defining future climate scenarios, which may then be used for studies of climate impact. Five types of methods are advocated in the literature.

- Prescribed changes in climatic elements.
- Temporal climatic analogues.
- Spatial climatic analogues.
- Scenarios based on paleoclimatic and paleohydrologic data.
- Scenarios based on atmospheric models.

Each of these methods has its advantages and disadvantages. The prescribed changes in temperature, precipitation, and other climatic characteristics are used as a tool for sensitivity analysis, and in that respect are essentially different from other types of scenarios.

In spite of all the difficulties of applying instrumental data for analysis of climate impact, the temporal analogue approach has been applied in many hydrological studies because it is straightforward and intuitively appealing. For example, Lough *et al.* (1983) created scenarios for changes in rainfall and temperature for northern Europe using the years from 1901 to 1920 as the cool period and those from 1934 to 1955 as the warm period (based on Northern Hemisphere temperature data). Palutikof (1987) applied rainfall data from these periods to an empirical river flow simulation model to estimate changes in river flow regimes in the UK.

The Palutikof analogue method is based on two major assumptions. First, it is assumed that the rainfall differences between a warm and cool period are independent of the causes of the difference in temperature. It does not matter, for example, if the selected cool period comes after the identified warm period; the difference in rainfall between the two periods can still be taken as an analogue for the effect of an increase in temperature. Second, it is assumed that the local climate system has responded to the changed global or hemispherical climatic circumstances. The longer the period selected, the more likely this is to be true. One problem with this approach is that the difference in temperature between the "cool" and "warm" periods may be rather smaller than the increase in temperature anticipated due to global warming. The approach does, however, indicate the qualitative characteristics of change. In principle, the method can be used to produce climate change scenarios of any time scale, although most studies have concentrated on monthly climatic and hydrological characteristics.

The stochastic properties of climatic and hydrological events vary from one region to another. This leads to the spatial analogue approach in constructing climate scenarios for impact studies, which assumes that the climate that will be experienced in the future in a particular region is the same as the climate currently experienced by another region. Information is transferred across space rather than across time, as is the case with the temporal analogue.

Some doubts arise, however, about using this method for water resources studies. Local and regional climate are a result of the combination of large-scale atmospheric circulation features with local topography and land–ocean interactions, and these local features are not transportable. Spatial analogues are expected to work best where large-scale controls on climate dominate – as is the case in the center of the North American continent – and to be less appropriate where topography and other local conditions have a significant effect on local climate – such as in large parts of Europe. It is difficult to transfer hydrological data from one region to another. Not only are there differences between the climate in the "source" and "destination" regions, but there are also differences between the catchment characteristics that control flow regimes and hydrological behavior. Even if it were appropriate to transfer climate data from one region to another, it would be dangerous to attempt to transfer hydrological data if, for example, one region were underlain by impermeable rocks and the other by a major aquifer. Arnell *et al.* (1990) attempted to compare river flow regimes in southern England and southwest France, but major differences in topography and catchment geology meant that any comparisons were simply meaningless. The spatial analogue approach can only be recommended where large-scale processes dominate and where there is little difference between the catchment characteristics in the source and destination regions. These two criteria are unlikely to be widely met.

The practical difficulties of evaluating possible changes in climatic and hydrological phenomena by means of routinely collected data have lead climatologists and water resources analysts to seek alternative methods of constructing future scenarios of geophysical processes. Two approaches are possible: paleoclimatic and paleohydrological analogues, and a modeling approach in which hydrologic scenarios are based on the combined application of macroscale atmospheric models and regional semi-empirical transfer functions, by which water balance components are related to climatic forcings. Both methods, which are widely used in water resources impact studies, are presented in the next sections.

1.3 What Can Be Learned from Paleodata?

Searching for analogues from the past is one way to construct hydrological scenarios for the future. Similarly, the recognition of changes during the Upper Quaternary helps to identify long-term tendencies. For retrodiction, the mechanism of processes, including hydrological processes, is reconstructed based on various long-term paleorecords. When looking for the analogues of the expected greenhouse effects, scientists usually concentrate on the recognition of periods and areas, when and where it was warmer, higher evaporation rates, a reduction in the permafrost, ice sheets, and winter snow cover, and a higher sea level (Budyko and Izrael, 1987; Gleick, 1989). This is not enough: the expectation of a rapid anthropogenic transformation of geosystems superimposed on the long-term trends of extraterrestial origin should lead to a careful examination of the past periods of rapid climatic and hydrological changes (Baker 1991; Starkel, 1983). An interesting problem is how to distinguish the role of anthropogenic factors, which have been superimposed on natural fluctuations from at least the Neolithic time (Starkel, 1987).

The information preserved in the deposits, landforms, or vegetation and faunal remains can often be related to one of the components of the water balance equation. The type of vegetation and its annual mass production reflect the amount of available water. Evaporation and evapotranspiration may be reconstructed from the past vegetation, soils, and chemistry and biology of lakes. Runoff characteristic of the past is mainly based on paleochannel parameters, facial differentiation of alluvial deposits, and reconstructed fluctuations of closed lake basins. The information on the subsurface runoff and ground water reservoirs is stored in dated ground waters, calcareous tufa, lake-level fluctuations, and soil features. Various types of water storage can be reconstructed on the basis of lake and marine coastal features, glacial forms and deposits, and periglacial indicators of the permafrost.

The reconstruction of the past environment, including past water resources, is based on the examination of the deposits and forms created continuously over a long time, and dated by precise radiometric and other methods. The retrodiction of water balance requires the parameterization of geological phenomena and their relation to hydrological parameters. This procedure is based on several assumptions. The first is that of the uniformitarian approach, which is that observed elements of the past (sediments, vegetation, etc.) reflect interrelations with paleoclimate and paleohydrology similar to those in the present. The second, and more problematic, assumption is that the components of the past that are analyzed reflect the equilibrium stage of the environment. In reality, at present as well as in the past, especially during phases of rapid changes and in ecotonal zones, the geoecosystems are far from the equilibrium stage between climate and vegetation, and between soils and water balance (Gleick, 1989; Starkel, 1991). The application of atmospheric and climatic models may lead to another way of reconstructing the past water cycle that requires a double testing: correlation of the simulated image of the present with the instrumental records, and comparison of the modeled picture of the past with the geological and other data.

A comprehensive review of methodology and results of paleohydrologic reconstructions can be found in the literature (Budyko and Izrael, 1987; IPCC Scientific Assessment, 1990; Starkel, 1989; Starkel *et al.*, 1991). Shiklomanov (1990) has defined three types of paleoclimatic analogues:

- The Holocene climatic optimum (5,300 to 6,200 years ago), representing an increase in temperature of 1°C.
- The last interglacial period (125,000 years ago), as an analogue for a 2°C increase in temperature.
- The Pliocene (3,000 to 4,000 years ago), representing a 4°C increase in temperature.

According to some authors, a remarkable increase of annual precipitation could be observed in the middle Holocene and in the last interglacial period in some parts of Eurasia and North America (Frenzel *et al.*, 1992).

Changes in the global water budget during the last 18,000 years were relatively small, with distinct regional differences, as can be seen in *Figure 1.4*. The probable causes of these disturbances are very complex. Only the geological and geomorphological conditions of water storage in the depressions remain relatively unchanged. Long-term variations are related to the cycles of earth rotation and solar radiation, which were about 23,000, 40,000, and 100,000 years long. These variations manifested in the seasonal distribution of insolation in the various latitudinal belts, which in turn caused changes in the circulation patterns of air masses. The general warming trend between 6,000 and 18,000 years ago in the higher latitudes caused the northward shift of climatic and hydrological zones, the melting of ice sheets and retreat of permafrost, and a change in runoff. Additional factors influencing the warming of the higher latitudes were the impact of the Gulf Stream and the land–ocean water exchange in the North Atlantic region.

Short rhythmical variations of climate characteristics are superimposed on these general trends. In the middle latitudes, the fluctuations of the snow line and upper tree line in the mountains indicate this type of rhythmic oscillation. Short fluctuations of 1,500 to 2,300 years' and 200 to 300 years' duration have been observed in North American lakes and in some tropical lakes. Usually, these cooler and wetter phases were only 300 to 500 years long and were separated by more stable phases of 1,000 to 1,500 years' duration each (Starkel, 1983; Magny, 1993).

It is interesting to note that various versions of the general circulation models (GCMs) were used for paleoclimatic reconstructions. For instance, the National Center for Atmospheric Research (NCAR) Community Climatic Model was applied for the simulation of every 3,000-year transect during the last 18,000 years (COHMAP, 1988). Data produced by the model were verified by correlating them with paleoenvironmental records. The main sources of information, covering most of the climatic zones, are

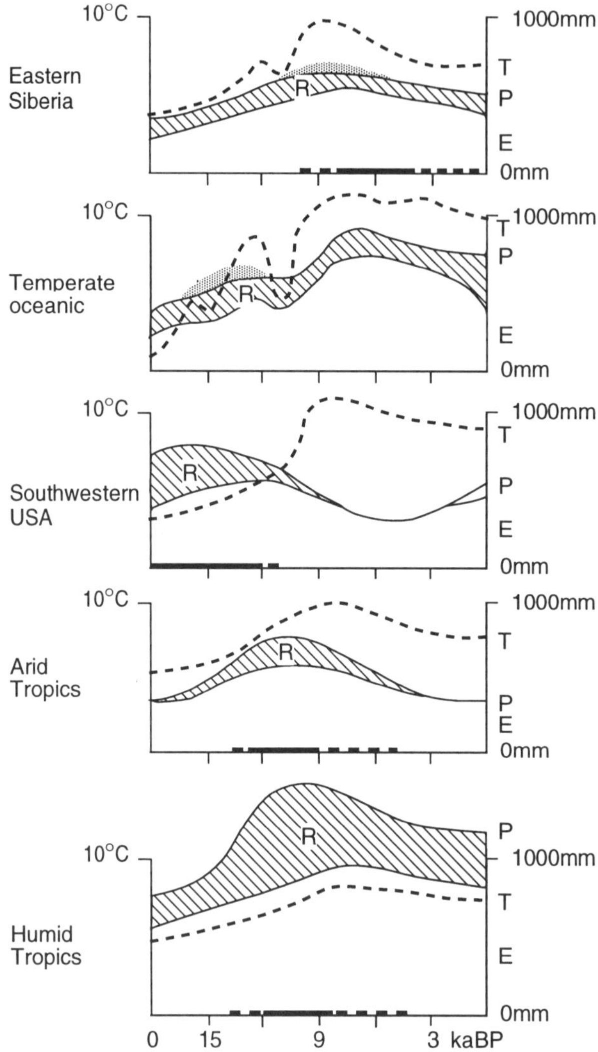

Figure 1.4. Regional water balance elements in the last 18,000 years. T = temperature; P = precipitation; E = evaporation; R = runoff; broad line at base indicates phases of high lake levels (based on Starkel, 1989).

the palynological records and fluctuations of the lake levels. The question considered above is how far the reconstructed vegetation is from the equilibrium stage with the climate. In the case of the Holocene transect, the coincidence of models and geological data is very high; but before 9,000 years ago deviations were relatively distinct. Another GCM was used by Kutzbach and Otto-Bliesner (1982) to reconstruct the temperature and precipitation in the transitional phase, 9,000 years ago, when the rise of radiation in the summer and its decrease in winter was accompanied by a general warming and the shift of climatic zones toward the poles.

The crucial question is to what extent the paleodata can be used for the current climate impact studies. The answer is difficult, and climatologists disagree about the practical value of using past analogues for prediction. For example, the Russian school considers the Holocene climatic optimum – during which precipitation and runoff show zonal deviations with the highest positive values in the arid zones and in the polar regions – as an analogue for an anthropogenic 1°C warming (Budyko and Izrael, 1987). As the analogue of 2°C global warming, there is the suggestion of using the optimum of the last interglacial period, about 125,000 years ago, during which precipitation throughout Europe was high. Using these analogues, some experts (IPCC Impacts Assessment, 1990) tried to forecast the changes in the extension of permafrost.

It should be remembered, however, that this approach must be used with caution. Dates are not exact, reconstructions of temperature and precipitation for a given region differ substantially according to the various methods of analysis used, and above of all, the causes of changes in the natural environment were different from the anthropogenic influences currently anticipated (Gleick, 1989). With all these reservations, the paleo-analogues may still serve as a useful tool for evaluating various climatic and hydrologic scenarios. It would be particularly useful to consider past periods of rapid warming, for example, the early Holocene between 9,000 and 10,000 years ago. Rapid warming around 10,000 years ago, documented in the varved lacustrine sediments, caused a simultaneous change in biomass production (and evapotranspiration), runoff, and sediment load in the higher latitudes, which was reflected in the rapid decline of parameters of the paleo-channels (Starkel, 1991). Other analogues that may be followed are those of rhythmic variations of different duration, observed in the temperate and the tropical zones. In all cases, it is necessary to distinguish the role of geophysical factors from trends directly connected with human activities, such as forest clearing, overgrazing by livestock, and irrigation practices. Good examples of direct human influences are found in the excavated archeological sites of the Mediterranean region and the Near East, and in observed changes, such as the aridization of North Africa, the salinization of the Aral Sea basin, and the impact of urbanization in some parts of Europe and the USA.

1.4 Modeling of Climate Scenarios

To be useful for water resources impact studies, a climate change scenario should present a feasible and internally consistent quantitative estimate of possible future climate. The GCMs are currently the most sophisticated tools for evaluating the likely effects of increasing concentrations of CO_2 and other greenhouse gases on the global climate (Carter *et al.*, 1992). A comprehensive, state-of-the-art review of existing GCMs and results obtained from running them can be found in the reports of the Intergovernmental Panel on Climate Change (IPCC Scientific Assessment, 1990, 1992) and in a great number of papers published in scientific journals devoted to meteorology and climatology. Two main types of GCM results are usually identified:

- The equilibrium response to a certain level of changes (usually a doubling) in greenhouse gas concentration in the atmosphere.
- The transient reaction of the climate system to time-dependent changes in this concentration.

Most of the water resources impact studies reported in the literature were implemented based on the equilibrium climate scenarios. The usual procedure is to add to the measured temperature data the temperature differences (ΔT) of model results for $2\times CO_2$ and $1\times CO_2$ conditions. Similarly, observed precipitation values are multiplied by the ratios (rP) of GCM precipitation for the above conditions. Because of simplified model parameterization and the still very coarse spatial resolution, the GCM results are not very accurate, particularly at the regional scale.

There are two main problems with using GCM data as inputs into hydrological assessment studies at present. First, the spatial scale of current global atmospheric models is too coarse for water resources analysis. Second, regional climate estimates, particularly of rainfall, are currently very uncertain – the finer the time and spatial resolution, the greater the uncertainty. Giorgi and Mearns (1991) suggest a number of approaches to the simulation of regional climate change scenarios and conclude, "The problem of projecting regional climate changes can be identified as that of representing effects of atmospheric forcings on two different spatial scales: large-scale forcings ... and mesoscale forcings, i.e., forcings which modify local circulations, thereby regulating the local distribution of climatic variables ...."

Methods of projecting regional climate scenarios based on GCM output can be divided into following groups:

- Straightforward statistical interpolation of temperature "deltas" and precipitation "ratios" for 2×CO_2 and 1×CO_2 conditions, calculated for the GCM's grid boxes. The interpolated values can than be superimposed on the instrumental records of climatic characteristics.
- "Guided" interpolation of ΔT and rP with regard, for example, to regional topography or current spatial patterns of climate.
- Application of empirical relationships between large-scale free atmosphere processes, e.g., circulation patterns, and regional surface climatic characteristics (Bárdossy and Plate, 1992; Wigley *et al.*, 1990).
- Nesting a mesoscale meteorological model in a coarse-resolution GCM over the region, e.g., a river catchment under consideration (Dickinson *et al.*, 1989; Giorgi and Mearns, 1991).

At present, none of these approaches give perfect results, but the inability to accurately predict future climate should not be used "as an argument to negate the application of GCMs to river basin level assessment" (IPCC Impacts Assessment, 1992).

The regional climate change scenarios based on global atmospheric models should not be understood as a forecast or a prediction of future climatic conditions. They represent a plausible future climate, based on a specific parameterization of a given model and on a certain methodology of projecting global warming to regional climate change. For the water resources impact study, several different feasible scenarios should be used to assess the possible range of hydrologic implications and to evaluate the performance of water resources schemes under changed climatic conditions. At present it is not feasible to assign numerical probabilities to different climate scenarios. There are two ways in which a number of scenarios can be used in an impact assessment. First, each different scenario can be treated equally, with no one scenario assumed to be a more "likely" future than any other. Second, it is possible to subjectively choose a so-called best scenario, and to express other scenarios as a range centered at this core. The question of how to interpret the results of impact studies based on various scenarios is discussed in the following sections.

1.5 From Climate to Hydrology

The idea that climate plays a decisive role in shaping hydrological processes was suggested more than 100 years ago by the Russian climatologist A.I. Voejkov (1886), who said that "rivers are the product of climate." Sixty years later, in the pioneering paper on runoff processes, Langbein (1949) wrote that "runoff, like soil moisture, evaporation and other components of the hydrological cycle, may be regarded as the manifestation of climate." The problem of interrelation between atmospheric and hydrologic processes is of great concern, both from a theoretical and a practical point of view. There are at least five main areas of research related to the climate–water resources interface:

- Research to improve parameterization of land surface processes in global and mesoscale atmospheric models.
- Analysis of the sensitivity of water balance components – runoff, evaporation, and catchment storage – to changes in precipitation, air temperature, potential evapotranspiration, and other climatic variables.
- Studies on the impact of climate change on thermal regime, and on physical, chemical, and biological processes in rivers, lakes, and reservoirs.
- Detection of changes in climate and environment by means of hydrologic indicators, including paleohydrological data.
- Assessment of possible implications of climatic fluctuations on water supply and demand, and consequently on planning and operation of water resources systems.

A detailed presentation of some of the above subjects follows in Chapters 2 to 6 of this book. Here we will investigate some general questions concerning the role of climatic variables in forming hydrologic processes.

To assess the impact of climate on water resources, some kind of model of the hydrological process is required if scenarios are expressed in terms of changes in certain climatic variables. There is a range of possible model types, from simple empirical relationships to complicated conceptual models with a great number of parameters representing different processes. Empirical models are "black box" representations of the hydrological process, and are usually based on regression functions linking climate inputs with some indices of hydrological response. Good examples are the Langbein relationship (Langbein, 1949) and the Turc formula (Turc, 1954), which allow the calculation of annual runoff from average annual precipitation and average annual temperature representative of a given catchment.

There are two major problems in applying empirical models. First, the implied sensitivity to climate change is dependent on the form of the model and the values of model parameters (Kaczmarek, 1990). Second, existing empirical relationships were developed for current climate conditions and it is not clear how they will work for alternative climates. Empirical models tend to be used to estimate changes in average annual or seasonal runoff. They are easy to use, but their results should be used with extreme caution.

In contrast, the conceptual hydrological models are simplified representations of the processes involved in the hydrological cycle. They usually consist of a series of storage components, each defined and linked by functional relationships containing a number of parameters. Conceptual models are based on a particular idea of how a river catchment works, and all include some parameters that need to be calibrated or eventually estimated from empirical relationships with measurable catchment properties.

Authors of most climate change impact assessment studies have used conceptual hydrological models, for which the most important assumption is that the calibration made under current climatic conditions will remain appropriate under a changed climate. The validity of this assumption can be evaluated by examining the performance of the model during extreme periods in the currently available data sets. It may be possible to estimate how particular model parameters might change (particularly those reflecting vegetation cover), but very few studies have attempted this so far. The physical background of conceptual models (e.g., the mass conservation law) means that the implied sensitivity to climate change is not as dependent on current climate, as is the case with purely empirical methods.

Another group of hydrological models is strictly based on physical laws of water movement and transformation of chemicals. These models differ from conceptual models in that their parameters do not need to be calibrated and can be calculated from various field data. In principle, such models should be very good at simulating soil moisture characteristics, recharge, river flow, and water quality, and should give credible estimates of future changes. Their usefulness to climate impact studies is, however, very limited at present, because they require large amounts of high-resolution input data and typically work on a sub-daily time scale.

As can be seen from the above discussion, methodological approaches to transferring climate forcings to water balance characteristics vary widely, from simple empirical relations to sophisticated hydrological models. Because there currently are serious discrepancies among different climate change scenarios, it is not appropriate to select a "best" hydrological model, as long as the model simulates current hydrological processes reasonably well. The variations caused by the application of different climate scenarios are usually considerably larger than are the eventual differences in simulation results obtained using different models for the same scenario.

The key input variables to most of the hydrologic models are catchment precipitation and potential evapotranspiration (PET). The latter can be seen as a "secondary order" climate characteristic, because it is itself calculated from other variables, such as temperature, radiation balance, wind speed, and air humidity. Therefore, it seems reasonable to briefly discuss ways of estimating PET, and possible changes in this variable. As temperature increases, the rate of PET can be expected to rise. For example, the Budyko-Zubenok model (Kaczmarek and Krasuski, 1991) gives a 4.1% PET increase per 1°C for a location in central Poland. For a site typical of southeast England, calculations of PET made by the Penman model lead to a 2.8% PET increase per 1°C, and the Thornthwaite model gives a 4.3% PET increase per 1°C (Arnell *et al.*, 1990). Unfortunately, each formula implies a different sensitivity of PET to climate change. This means that selection of the method used for estimating PET, to be used as the input variable to the hydrological model, may greatly influence conclusions concerning the implications of global warming for runoff and other water balance characteristics.

We will close this section with some important comments concerning terminology. Throughout the book the terms *sensitivity analysis* and *impact analysis* are used. In the literature they are often used in an interchangeable way, which may lead to some confusion and misunderstandings. To avoid this, these two concepts should be clearly distinguished. Sensitivity is defined as the reaction of a physical system to any disturbances in the input variables. In mathematical formulation, if

$$y = f(x_1, x_2, \ldots, x_n) \ , \tag{1.1}$$

then sensitivity of y to change of x_i will be represented by the partial derivative (Kaczmarek, 1990):

$$\text{sensitivity } (y/x_i) = s_{y/x_i} = \left[\frac{\partial y}{\partial x_i}\right]_{x_i = x_{io}} \ , \quad i = 1, \ldots, n \ , \tag{1.2}$$

if all other forcings are assumed to be constant. A similar concept, elasticity, defined as

$$\text{elasticity } (y/x_i) = e_{y/x_i} = s_{y/x_i}\left(\frac{x_i}{y}\right) \ , \quad i = 1, \ldots, n \ , \tag{1.3}$$

has been proposed by Schaake (Schaake, 1990).

Let us assume, for example, that the average annual runoff is related to annual precipitation (P) and mean annual temperature (T) by the Turc equation:

$$R = (P)\left[\frac{1}{1 - \sqrt{k + P^2/L^2}}\right] \ , \tag{1.4}$$

where k is a coefficient usually taken as 0.9, and

$$L = 300 + 25T + 0.05T^3 \quad . \tag{1.5}$$

The sensitivity and elasticity parameters of runoff to precipitation are therefore equal:

$$s_{R/P} = \left[1 - \frac{1}{\sqrt{k + P^2/L^2}}\right]\left[1 - \frac{P^2}{L^2(k + P^2/L^2)}\right] \tag{1.6}$$

$$e_{R/P} = s_{R/P}\left(\frac{P}{R}\right) \quad . \tag{1.7}$$

For climatic conditions typical of Central Europe, $P = 600$ mm, and $T = 8°$C. From the relationships seen in equations (1.4), (1.6), and (1.7), one can calculate that

- annual runoff $R = 196$ mm,
- sensitivity $s_{R/P} = 0.725$, and
- elasticity $e_{R/P} = 2.22$.

The elasticity value shows that, using the Turc formula for those climatic conditions, the 10% change in precipitation will give a 22% change in runoff, assuming that the air temperature remains unchanged. Similar calculations can be done to obtain the sensitivity values of runoff to the change of temperature.

Such simple relations can be obtained only for direct empirical relationships between hydrological and climatological variables. For complex conceptual models, partial derivatives cannot be easily calculated, and only approximate values of sensitivity parameters can be calculated by means of the following formula:

$$s_{y/x_i} \cong \frac{y[(x_{io} + \varepsilon), \Omega] - y[(x_{io} - \varepsilon), \Omega]}{2\varepsilon} \quad , \tag{1.8}$$

where Ω is the set of forcing factors other than x_i, and ε is a small disturbance of x_i. Consequently an approximate value of the elasticity parameters may by obtained from equation (1.3).

To find sensitivity and elasticity parameters of water balance, one need not assume alternative climate scenarios. On the other hand, a climate change impact on a hydrological variable is defined as the difference of numerical values of that variable, obtained for the alternative climate scenario, versus present climatic conditions. As before, it should simultaneously be assumed that the effect of non-climatic factors can be held constant. An impact characteristic is then represented by

$$\Delta y = y(\Omega_{\text{alternative}}) - y(\Omega_{\text{present}}) \quad , \tag{1.9}$$

where Ω is again a set of climatic forcings. This means that to perform an impact analysis, one or more alternative climate scenarios must be specified and applied to a model that serves as the transfer function between climatic forcings and hydrological phenomena. It may be that in the context of limited knowledge of possible regional climate change, the sensitivity analysis as defined above is more informative for decision making than are impact studies that are based on questionable assumptions concerning climate scenarios.

1.6 Decision Making in an Uncertain Environment

Water resources management has been undergoing changes due to population growth, improved sanitation, economic development, technological revolution, and changing legislative and administrative conditions. These changes will continue in the future, with their intensity being dependent on demographic and economic processes in various parts of the world. The issue of climate change adds a new dimension to the ongoing dynamics of water supply and demand. Kundzewicz and Somlyódy (1993) state, "Nations have adapted themselves to definite and stationary climate conditions. Any change would destroy the fragile balance resulting from the long-lasting adaptation." The key question is how water resources managers will react to the enhanced greenhouse effect on the basis of imprecise information of its scale and timing.

The fundamental problem in responding to possible consequences of anthropogenic global warming is deciding what kinds of adaptive measures should be undertaken today in the face of the uncertainties of predicting future climate on a regional scale. In the last century, water resources planning was based on the assumptions that variability of hydrological phenomena is governed by a stationary stochastic process and that certain trends can be expected only with regard to demographic and socioeconomic factors. The extensive experience of various water resources agencies teaches that these non-climatic factors of water planning cannot be accurately predicted. According to current practice, hydrological statistics based on instrumental records are assumed to be constant values for the expected lifetime of the water systems schemes. In this era of concern about global change, these assumptions have become questionable.

Another chapter in our book is devoted to a detailed analysis of possible consequences of climate change on water resources management. In this chapter, only general questions will be discussed regarding the policy implications of the expected global warming. Three different approaches are possible for dealing with the adaptation of water systems to cope with global warming. First, there is the "wait and see" or "business as usual" approach, which means that decisions on water systems' adaptation to possible disturbances are postponed until new and more reliable information on global climatic processes is available. According to such a policy, the existing water schemes remain unchanged and new ones are planned and implemented according to standard methodological procedures. For instance, the Commission for Hydrology of the World Meteorological Organization has published a "CHy Statement on the Hydrological and Water Resources Impacts of Global Climate Change" (Commission for Hydrology, 1988). The statement reads: "... given the added burden of uncertainty about climate change it is certainly inappropriate at this time to discard available analytical procedures or to engage in expensive alterations to build facilities." Taking into account that in the case of large water schemes the amount of time needed for planning and implementation is usually very long, this approach may lead to decisions being made too late to effectively cope with negative consequences of extreme hydrological situations.

Second, there is the "minimum regret" approach, in which decisions are made to solve current problems in the best possible way, while simultaneously preparing the water resources system for possible changes and shocks. Waggoner and Revelle (1990) expressed this policy by saying, "So long as the future remains unsure, and that seems a long time ahead, rational people will make decisions that solve present problems and make water supply robust, resilient and flexible for any future." The question of how to measure performance criteria and how to calculate these criteria based on "historical" data needs further discussion. This method does not require any *a priori* knowledge about future climatic and hydrologic scenarios.

Finally, the third approach assumes that a certain optimality rule is applied to a range of climatic and water resources scenarios. Final decisions may be made by comparing costs, benefits, and losses for each scenario, and will usually be based on some subjective interpretation of the results of analysis. As mentioned before, it is not feasible to assign numerical probabilities to various climate scenarios. Nevertheless, some analysts suggest the possible application of Bayesian theory in the decision-making process in an uncertain environment. However, this requires an assignment of prior probabilities, which by the present circumstances may only reflect different degrees of belief about the future world emission policy and the accuracy of climatic and hydrologic models.

The effects of climate variability and climate change on water resources are highly site specific. In arid and semiarid regions, a relatively small decrease in runoff may have dramatic consequences for the provision of water for domestic, agricultural, and industrial use. On the other hand, in some regions of the world, water availability greatly exceeds expected water demands, and the possible impact of climate change on regional water management will be negligible. Consequently, the methodology for assessing the effects of assumed climate scenarios on water resources should differ from one region to another. There is no reason to apply sophisticated methods in the decision-making process for those river catchments where there is an abundance of water and where the results of analysis will be trivial, independent of all the uncertainties accompanying the issue of climate change.

In analyzing conditions for sustainable regional development, three types of possible changes should be examined: population growth, economic development, and global climate change (Kulshreshtha, 1993). In some countries, the impact of demographic trends will clearly surpass the effects of changes in water balance due to climatic disturbances; in other countries the latter will dominate. According to Kulshreshtha (1993), about two-thirds of the world population "will become more vulnerable to availability and use of water under current projections of population growth and climate change." For a country with a population of 2 billion, climate change itself would be a greater contributing factor to vulnerability than would the population growth alone. These figures clearly show the scale of possible water resources problems facing the world's population in the coming decades.

Until now, we have focused our attention on quantitative aspects of the impact of climate change on water resources. It is obvious, however, that changes in the temperature and precipitation patterns may also influence the quality of inland waters, which consequently may affect water resources management. As has been shown by Varis and Somlyódy (1993), many water quality problems "are at risk of becoming worsened by the climatic change: particularly eutrophication, oxygen depletion, salinization, acidification, and thermal problems." The quality of rivers, lakes, and ground water depends on catchment processes, on pollution control, and on physical and biochemical processes in water. All of these factors may be affected by changes in climatic characteristics. Here again, the uncertainty about these changes makes the prediction of future water quality regime very difficult.

Rational water resources management decisions cannot be made without due consideration of the quality of water used for domestic, industrial, or agricultural purposes, and of environmental requirements. The possibility for efficient adaptation to changes in water quality due to climate transformation depends on the ability to formulate adequate climate scenarios and on the development of a methodology allowing use of climatic information in modeling water quality processes.

1.7 Organization of the Book

The purpose of this book is to summarize the experience gained during the last decade regarding the possible effect of global warming on water management. It is a joint effort of scientists working at IIASA in Laxenburg, Austria, and those working in various national institutions in Belgium, Canada, the Czech Republic, Hungary, Poland, the Russian Federation, the UK, and the USA. The authors met in Laxenburg in October 1991 to discuss the structure of the book and the content of the various chapters. The book contains six chapters presenting a review of key methodological problems facing water resources management, which has been affected by changed climatic conditions. The book also contains eight case studies describing specific water resources situations in different regions of the world.

This chapter gives a general outline of the problem and a short presentation of the methods used for creating alternative climatic scenarios – from prescribed methods, through temporal analogues based on instrumental records and paleodata, to various modeling approaches. Finally, it provides an overview of concepts and methods for assessing possible effects of climate change on hydrology and water resources. The second chapter deals with the conceptualization of relations between climate and water balance components, and presents a range of hydrologic models used for evaluating the possible effect of changes in regional temperature, precipitation, and other climatic variables on runoff, evaporation, and catchment storage. In the next two chapters, authors' views on the possible impact of climate change on water quality of rivers, lakes, and reservoirs are presented. It should be stressed that this aspect of the climate issue has often been underestimated in previous studies and discussions, and deserves much more attention in the future. Key variables, boundary conditions, and parameters needed for evaluating effects of climate variability and variations are identified and discussed. In the fifth chapter, implications of climate change for future water demand are examined. Because trends in population growth and economic development also have an impact on water use, the climatic factors affecting regional demand for water are separated and identified. Also, criteria for selecting the most vulnerable world regions are discussed. Chapter 6, the last chapter of the methodological part of the book, presents a systems approach to water resources planning for future uncertainties. Natural and human-made changes in climatic and hydrologic processes add a new dimension to the reliability of water systems. Therefore, discussion on how the current assumptions, methods, and criteria in water resources planning should be altered for non-stationary conditions is offered.

In the second part of the book, the following case studies are presented:

- Assessment of the potential impact of climate change on water resources of the Delaware River basin.
- Global and regional impacts of climate change on lake stratification.
- Effects of climate change on water resources in the Central Valley of California.
- Possible consequences of climate change on the management of the Kariba hydropower system in the Zambezi River basin.
- Regional water resources management under climate change in a region of North Caucasia.
- Repercussions of a doubling of CO_2 on reservoir design and operation in Belgium.
- Repercussions of a doubling of CO_2 on national water resources assessment in Poland.

All of the case studies include a short description of the country or region under consideration: geography, climate, hydrology, and main water resources problems. This is followed by a presentation of basic assumptions, models and techniques applied, information on data and climate scenarios used for the analysis, and main results and conclusions.

The book's concluding chapter summarizes the views, approaches, and methods presented, as well as lessons derived from the case studies. Recommendations on how economic and technical criteria and procedures for water resources planning should be altered or reevaluated in light of the likely climate change are discussed. The book concludes with proposals concerning research needs aimed at closing existing gaps in our knowledge on water resources management in the face of an uncertain environment.

References

Amorocho, J., 1970, Uncertainty and Determinism in Hydrology, Water Science and Engineering Papers, No. 1043, University of California, Davis, CA, USA.

Arnell, N.W., Brown, R.P.C., and Reynard, N.S., 1990, Impact of Climatic Variability and Change on River Flow Regimes in the UK, Report No. 107, Institute of Hydrology, Wallingford, UK.

Baker, V.R., 1991, A bright future for old flows, in L. Starkel *et al.*, eds., *Temperate Paleohydrology*, John Wiley & Sons, Chichester, UK.

Bárdossy, A., and Plate, E.J., 1992, Space-time model for daily rainfall using atmospheric circulation patterns, *Water Resources Research*, **28**(5):1247–1259.

Brundtland, G.H., 1987, *Our Common Future*, Report of the World Commission on Environment and Development, Oxford University Press, Oxford, UK.

Budyko, M.I., and Izrael, Y.A., eds., 1987, *Anthropogenic Climatic Change*, Gidrometeoizdat, Leningrad, Russia [in Russian].

Carter, T.R., Parry, M.L., Nishioka, S., and Harasawa, H., 1992, *Preliminary Guidelines for Assessing Impacts of Climate Change*, Intergovernmental Panel on Climate Change (IPCC), Geneva, Switzerland.

COHMAP members, 1988, Climatic changes of the last 18000 years: Observations and simulations, *Science*, **241**(Aug. 26):1043–1052.

Commission for Hydrology, 1988, Abridged Final Report of the Eighth Session, World Meteorological Organization (WMO), No. 175, Geneva, Switzerland.

Dickinson, R.E., Errico, R.M., Giorgi, F., and Bates, G.T., 1989, A regional climate model for the western US, *Climatic Change*, **15**(3):383–422.

Frenzel, B., Pecsi, M., and Velichko, A.A., 1992, *Paleogeographical Atlas of the Northern Hemisphere*, Hungarian Academy of Sciences, Budapest, Hungary.

Giorgi, F., and Mearns, L.O., 1991, Approaches to the simulation of regional climate change: A review, *Reviews of Geophysics*, **29**(2):191–216.

Gleick, P.H., 1989, Climate change, hydrology and water resources, *Reviews of Geophysics*, **27**:329–344.

IPCC (Intergovernmental Panel on Climate Change), 1990a, *Climate Change: The IPCC Impacts Assessment*, W.J.McG. Tegart, G.W. Sheldon, and D.C. Griffiths, eds., Australian Government Publishing Service, Canberra, Australia.

IPCC (Intergovernmental Panel on Climate Change), 1990b, *Climate Change: The IPCC Scientific Assessment*, J.T. Houghton, G.J. Jenkins, and J.J. Ephraums, eds., Cambridge University Press, Cambridge, UK.

IPCC (Intergovernmental Panel on Climate Change), 1992a, *Climate Change, The IPCC Impacts Assessment*, W.J.McG. Tegart, G.W. Sheldon, and J.H. Hellyer, eds., Australian Government Publishing Service, Canberra, Australia.

IPCC (Intergovernmental Panel on Climate Change), 1992b, *Climate Change: The IPCC Scientific Assessment*, J.T. Houghton, B.A. Callander, and S.K. Varney, eds., Cambridge University Press, Cambridge, UK.

Kaczmarek, Z., 1977, *Statistical Methods in Hydrology and Meteorology*, National Center for Scientific, Technical and Economic Information, Warsaw, Poland.

Kaczmarek, Z., 1990, On the Sensitivity of Runoff to Climate Change, WP-90-058, International Institute for Applied Systems Analysis, Laxenburg, Austria.

Kaczmarek, Z., and Krasuski, D., 1991, Sensitivity of Water Balance to Climate Change and Variability, WP-91-047, International Institute for Applied Systems Analysis, Laxenburg, Austria.

Kulshreshtha, S.N., 1993, *World Water Resources and Regional Vulnerability: Impact of Future Changes*, RR-93-10, International Institute for Applied Systems Analysis, Laxenburg, Austria.

Kundzewicz, Z., and Somlyódy, L., 1993, Climatic Change Impact on Water Resources: A Systems View, WP-93-30, International Institute for Applied Systems Analysis, Laxenburg, Austria.

Kutzbach, J.E., and Otto-Bliesner, B.L., 1982, The sensitivity of the African monsoonal climate to orbital parameter changes for 9,000 years BP in a low resolution general circulation model, *Journal of Atmospheric Research*, **39**(6):1177–1188.

Langbein, W.B., 1949, Annual runoff in the United States, U.S. G.S. Circular 5, US Department of Interior, Washington, DC, USA.

Lough, J.M., Wigley, T.M.L., and Palutikof, J.P., 1983, Climate and climate impact scenarios for Europe in a warmer world, *Journal of Climate and Applied Meteorology*, **22**:1673–1684.

Magny, M., 1993, Solar influences on Holocene climatic changes illustrated by correlations between past lake-level fluctuations and the atmospheric C^{14} record, *Quaternary Research*, **40**:1–9.

Mitosek, H.T., 1992, *Occurrence of Climate Variability and Change Within the Hydrological Time Series: A Statistical Approach*, CP-92-05, International Institute for Applied Systems Analysis, Laxenburg, Austria.

Palutikof, J.P., 1987, Some Possible Impacts of Greenhouse Gas Induced Climatic Change on Water Resources in England and Wales, International Association of Hydrological Sciences (IAHS) Publication No. 168, IAHS Press, Wallingford, UK.

Schaake, J.C., 1990, From climate to flow, in P.E. Waggoner, ed., *Climate Change and U.S. Water Resources*, John Wiley & Sons, New York, NY, USA.

Schaake, J.C., and Kaczmarek, Z., 1979, Climate variability and the design and operation of water resource systems, in *Proceedings of the World Climate Conference*, WMO, Geneva, Switzerland.

Shiklomanov, I.A., 1990, Climate and water resources, *Hydrological Sciences Journal*, **34**(5):495–529.

Starkel, L., 1983, The reflection of hydrological changes in the fluvial environment of the temperate zone during the last 15,000 years, in J. Gregory, ed., *Background to Paleohydrology*, John Wiley & Sons, Chichester, UK.

Starkel, L., 1987, Man as a cause of sedimentologic changes in the Holocene, *Striae*, **26**:5–12.

Starkel, L., 1989, Global paleohydrology, *Quaternary International*, **2**:25–3.

Starkel, L., 1991, Environmental changes at the Younger Dryas Preboreal transition and during the early Holocene: Some distinctive aspects in Central Europe, *The Holocene*, **1**(3):234–242.

Starkel, L., Gregory, K., and Thornes, J., eds., 1991, *Temperate Paleohydrology*, John Wiley & Sons, Chichester, UK.

Turc, L., 1954, Water balance of soils: Relationship between precipitation, evapotranspiration and runoff, *Annales Agronomiques*, **5**:491–595 and **6**:5–131 [in French].

Varis, O., and Somlyódy, L., 1993, Potential Impacts of Climatic Change on Lake and Reservoir Water Quality, WP-93-25, International Institute for Applied Systems Analysis, Laxenburg, Austria.

Voejkov, A.I., 1886, Climates of the earth and in particular of Russia, in *Selected Works*, 1948, **1**:163–750, Leningrad, Russia [in Russian].

Waggoner, P.E., ed., 1990, *Climate Change and U.S. Water Resources*, John Wiley & Sons, New York, NY, USA.

Waggoner, P.E., and Revelle, R.R., 1990, Summary, in P.E. Waggoner, ed., *Climate Change and U.S. Water Resources*, John Wiley & Sons, New York, NY, USA.

Wigley, T.M.L., Jones, P.D., Briffa, K.R., and Smith, G., 1990, Obtaining subgrid-scale information from coarse-resolution general circulation model output, *Journal of Geophysical Research*, **95**:1943–1953.

Chapter 2

Climate Change and Water Balance Components

Stewart Cohen, Kenneth M. Strzepek, and David N. Yates

2.1 Introduction

There is growing concern that climate change will have significant impacts on water in the hydrologic cycles of many regions of the world (Skiles and Hanson, 1994). The magnitude of this concern has been expressed by the interagency Committee on Earth and Environmental Sciences (CEES) in the USA, which identified the hydrologic cycle as being the highest priority for scientific research on global change (Rind *et al.*, 1992). Concurrently, hydrologists have been busy investigating the responses of river basins to possible climatic variability. Because they are primarily concerned with water availability, hydrologists naturally focus on the hydrologic response of the physical basin and not on the atmospheric components of the evapotranspiration processes that are driving these changes. In studies of this nature a unique hydrologic model of a river catchment is often applied and then temperature and precipitation are altered to assess basin response (Riebsame *et al.*, 1994; Gleick, 1987a, 1987b; etc.). Yet Rind *et al.* (1992) point out that different formulae of physical processes and different conceptualizations of hydrologic components will likely respond differently under climate change scenarios. This suggests a need to look at the myriad approaches that have been, and continue to be, used to describe the responses of river basins to hydrologic processes. Two main modeling components have been identified with regard to the water balance of a river basin: the modeling

Z. Kaczmarek et al. (eds.),
Water Resources Management in the Face of Climatic/Hydrologic Uncertainties 30–45.

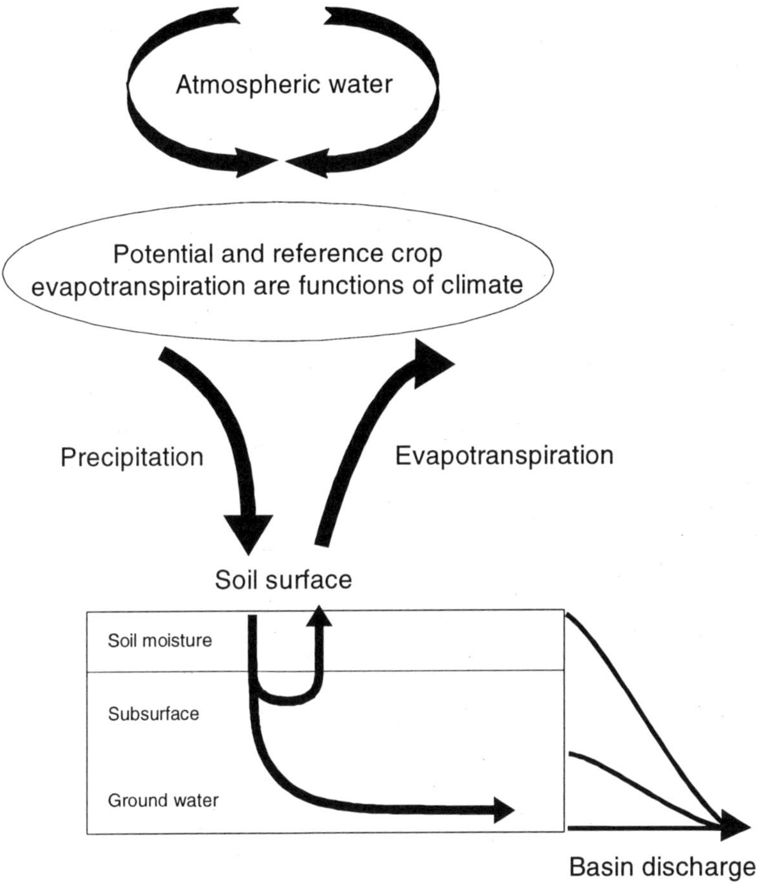

Figure 2.1. Simplified water balance showing potential and reference crop evapotranspiration (PET and RCE, respectively) as climate variables and actual evapotranspiration (AE) dependent on soil moisture, plant canopy, and PET or RCE. AE $= f$ (available moisture, PET, RCE) and PET, RCE $= f$ (climate).

of atmospheric processes to remove water vapor from the land surface and the modeling of water transport processes within the soil domain.

A key component of the hydrologic cycle (*Figure 2.1*) is evapotranspiration (EV), the conversion of liquid water to vapor at the earth–atmosphere boundary and the subsequent mixing of this vapor with the atmosphere (Hagan *et al.*, 1967). The evaluation of EV is important in the study of the impact of climate change on water resources, because EV can be considered a key link between the atmosphere and the soil matrix within the hydrologic cycle (*Figure 2.1*). The importance of this link has been observed by Dooge (1992), who states that any estimate of climate change impacts on water resources depends on the ability to relate changes in actual evapotranspiration (AE) to predicted changes in precipitation and potential evapotranspiration (PET). To correctly predict changes in EV it is obviously important to begin with good estimates of the driving mechanisms of that change.

2.2 Potential Evapotranspiration: An Imaginary Construct

Runoff from a river basin can be regarded as being the by-product of (i.e., dependent on) two larger (independent) processes, precipitation and evapotranspiration. The terms PET and reference crop evapotranspiration (RCE) have often been used for a hypothetical measurement of a climate's capacity for the removal of water vapor from the land surface. Potential and reference crop evapotranspiration can be regarded as imaginary concepts used in the estimation of AE; their relationship is depicted in *Figure 2.2*. In this idealized form, evapotranspiration is assumed to achieve the maximum level of PET when the soil moisture content achieves the value $q*$, which is smaller than the field capacity (FC) of the soil. A simpler approximation assumes a linear ratio of EV:PET between the permanent wilting point (PWP), which is a function of the soil and crop, and the FC, which is a function of only the type of soil. Hydrologists, agronomists, climatologists, and agricultural engineers often use PET and RCE as the climatological links between the dynamic processes of vapor diffusion from soil (evaporation) and crop surfaces (transpiration) to the atmosphere. However, the concepts of AE, PET, and RCE can be confused, because they are often used interchangeably within the literature (Shuttleworth, 1993; Mimikou *et al.*, 1991). For this reason, brief definitions are given below (from Shuttleworth, 1993).

- *Evaporation* (E): the rate of the transformation of liquid water to vapor from open water, bare soil, or vegetation with soil beneath it.
- *Transpiration:* the part of the total evaporation that enters the atmosphere from the soil through the plants.
- *Evapotranspiration* (EV): the combination of evaporation from bare soils and transpiration from plants.
- *Potential evapotranspiration* (PET): the idealized quantity of water evaporated per unit area, per unit time, from an idealized extensive free water surface under existing atmospheric conditions.

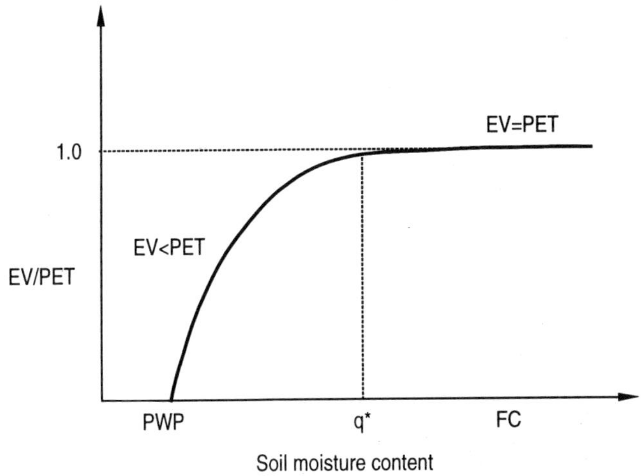

Figure 2.2. The ratio of evapotranspiration (EV) to potential evapotranspiration (PET) in an idealized form.

- *Reference crop evapotranspiration* (*RCE*): the rate of evaporation from an idealized grass crop with a fixed crop height of 0.12 m, an albedo of 0.23, and a surface resistance of 69 sm^{-1}.

It is worthwhile to highlight the difference between PET and RCE. Potential evapotranspiration can be idealized as *potential evaporation* using the strict definition, because it uses the term "extensive free water surface," which has nothing to do with the transpiration process of plants. Reference crop evapotranspiration can be thought of as a potential evapotranspiration, because it incorporates the transpiration process by assuming an idealized plant. Both PET and RCE are "potential" estimators of evapotranspiration; however, they use different assumptions. For the remainder of this chapter, PET and RCE will be referred to as PET when general methods are discussed.

A number of approaches have been used to assess EV and runoff changes within the context of an altered climate. Gleick (1987a) used the Thornthwaite method and a monthly water balance model for estimating PET in the Sacramento–San Joaquin River Basin; McCabe and Ayers (1989) did the same for the Delaware River basin. Nemec and Schaake (1982) performed an impact assessment on two basins in the USA, one a humid basin and the other arid. They used the Budyko radiation method to compute PET and a daily hydrologic model to estimate EV changes and basin discharge. Riebsame *et al.* (1994) used a mass balance approach to estimate EV (thereby eliminating the need for PET estimates) for the Zambezi River basin; the Penman method was used in this study to calculate PET for the Uruguay River basin. Mimikou *et al.* (1991) used the Blaney-Criddle method and monthly water balance to estimate PET for three river basins in mountainous regions in central Greece. Schaake (1990) applied uniform changes in PET of $\pm 10\%$ to river basins in the eastern USA, suggesting that this change was primarily caused by increased temperatures. Rind *et al.* (1990) report the use of the Thornthwaite method for computing PET within the Palmer drought severity index (PDSI) model of the USA; this was then used to create scenarios under climate change.

2.3 Methods for Estimating Actual, Potential, and Reference Crop Evapotranspiration

Methods for evaluating EV can be divided into three categories: hydrologic or water balance methods, analytical methods based on climate variables, and empirical estimates. The water balance method to determine EV consists of catchment hydrology, soil water depletion sampling, and lysimetric testing. Because this is primarily a physically based approach, its use in climate change assessment is limited to the laboratory. The second approach, referred to as a micrometeorological method, uses the scientific understanding of the physics of evaporation and transpiration. Mathematical relationships are developed that describe these processes via two key climatological components, energy balance and mass transport. The third method centers on the development of empirical relationships that are often site specific and based on local climatological conditions, and are frequently cast in regression analysis. These methods are often "calibrated" by relating the empirical estimates to observed measurements (Hagan *et al.*, 1967).

2.4 Climate Change Impacts on Components of the Water Balance

Figure 2.1 is a schematic of the hydrologic cycle. It illustrates major components of the climatic water balance as defined by Thornthwaite and Mather (1955). The water balance approach to assessing climate change impacts on water resources has been employed frequently. This chapter will describe basic processes of each water balance component and the influence of climate change on each component. Results from the literature on regional changes in the water balance due to climate change are presented. This includes case studies of small and large watersheds for which various climatological and hydrological models have been applied, as well as cases that used general circulation model (GCM) simulations of runoff and changes in soil moisture storage directly.

2.4.1 Potential and actual evapotranspiration

Potential Evapotranspiration

Potential evapotranspiration is determined solely from temperature and radiation. Actual evapotranspiration accounts for moisture availability from precipitation and soil moisture storage, as well as evaporation due to advection and possible changes in stomatal resistance due to CO_2 enrichment. Actual evapotranspiration has been estimated using methods from Thornthwaite (Thornthwaite and Mather, 1955; Mather, 1978), Morton (Morton, 1978, 1983a, 1983b), Priestly and Taylor (Priestly and Taylor, 1972), Penman and Monteith (Rosenberg *et al.*, 1990), and Budyko and Zubenok (Kaczmarek and Krasuski, 1991; Dooge, 1992).

Some studies have assumed increased stomatal resistance when calculating AE as part of a simulation of runoff (Aston, 1984; Idso and Brazel, 1984). It has been noted, however, that plant growth and yield would increase under conditions of increased CO_2, thereby leading to increased leaf area and greater transpiration (Rosenberg *et al.*, 1990). The recent MINK study, which focused on an agricultural region in the USA, assumed that the net effect of CO_2 enrichment on water supplies would be zero, but that there would be a decreased requirement for irrigation (Frederick, 1991). Bultot *et al.* (1988b) also assumed no net effect for their case study of watersheds in Belgium.

Estimates of open water evaporation (E) depend on obtaining data on water temperatures for water bodies that are not explicitly included in the GCMs. Open water evaporation has been estimated using mass transfer models, which require data on air and dewpoint temperatures, wind speed, and cloud cover (Croley, 1990). For large lakes (e.g., North American Great Lakes), there are examples of studies that have determined changes in water temperature so that the lakes' heat storage can be accounted for when lake effects on temperature and humidity are considered (Croley, 1990).

Table 2.1 lists results of various case studies.

Table 2.1. Summary of studies on impacts of climate change on evaporation (E) and evapotranspiration (AE, PET).

Basin	Area (km^2)	Model	Scenario	Results[a]	Reference
Cold-humid					
Great Lakes	770	Mass transfer	GISS	E: +33	Cohen, 1986
			GFDL80	E: +14	Cohen, 1986
			+wind	E: +10	Cohen, 1986
		Water balance	GISS	AE: +18	Cohen, 1986
			GFDL80	AE: +7	Cohen, 1986
		Mass transfer	GISS	E: +27	Croley, 1990
			GFDL86	E: +44	Croley, 1990
			OSU	E: +26	Croley, 1990
		LBRM	GISS	AE: +21	Croley, 1990
			GFDL86	AE: +19	Croley, 1990
			OSU	AE: +19	Croley, 1990
Warta	.06	Budyko-Zubenok	GFDL86	AE: +11	Kacz.&Kras., 1991
Temperate					
Delaware River		Water balance	GISS	P/PET: –73	McCa.&Wol., 1992
			GFDL86	P/PET: –76	McCa.&Wol., 1992
			OSU	P/PET: –6	McCa.&Wol., 1992
UK catchments					Arnell, 1992
Belgian catchments		IBRM	GFDL80	PET: +12	Bultot *et al.*, 1988a
				AE: +10	Bultot *et al.*, 1988a
Semiarid, arid					
Colorado River	625	4%/deg.			Gleick&Nash, 1991
Sevier River	.9	Empirical	+2C,–10%P	AE: –5	Flaschka *et al.*, 1987
Bear River	.47	Empirical	+2C,–10%P	AE: +3	Flaschka *et al.*, 1987
Arizona basins		CO_2 adj.			Idso&Brazel, 1984
Tropical wet					
Zambezi River			GFDL	AE: –3	Salewicz, Chapter 11
			GISS	AE: +11	Salewicz, Chapter 11
			GFDL	E: +7	Salewicz, Chapter 11
			GISS	E: +8	Salewicz, Chapter 11

[a] All results are expressed as the percentage change from present or control run simulations.
Water balance = derived from Thornthwaite.
LBRM = large basin runoff model.
IBRM = integrated basin runoff model.
4%/deg. = PET assumed to change by 4% per °C.
CO_2 adj. = AE reduced by an assumed rate to account for direct effects of CO_2 on stomatal resistance; does not include changes in leaf area.
GISS = simulation obtained directly from the Goddard Institute for Space Studies GCM.

Table 2.2. Summary of studies on impacts of climate change on soil moisture.

Basin	Area (km^2)	Model	Scenario	Results[a]	Reference
Cold-humid					
Great Lakes	–	Water balance	GISS	–16**	Cohen, 1987
			GFDL80	–66**	Cohen, 1987
Great Lakes sub-basins	–	LBRM	GISS	–57 to –3	Croley, 1990
			GFDL86	–43 to –19	Croley, 1990
			OSU	–33 to –8	Croley, 1990
Temperate					
Delaware River	–	Water balance	GISS	–27 to –22*	McC.&Ayers, 1989
			GFDL86	–70 to –55*	McC.&Ayers, 1989
			OSU	–25to+5*	McC.&Ayers, 1989
Semiarid, arid					
Sacramento River	–	Water balance	+2C,–10%P	DJF: –11	Gleick, 1987a
			GFDL86	DJF: –4	Gleick, 1987a
			GISS	DJF: +4	Gleick, 1987a
			GFDL86	JJA: –36	Gleick, 1987a
			GISS	JJA: –24	Gleick, 1987a
Saskatchewan River	–	Water balance	+2C,–10%P	–93*	Cohen, 1991
			+4C,–10%P	–197*	Cohen, 1991
			+4C,+20%P	–96*	Cohen, 1991
			GISS	–124*	Cohen, 1991
			GFDL80	+25*	Cohen, 1991
			GFDL86	–307*	Cohen, 1991
			OSU	–70*	Cohen, 1991

[a]Results are expressed as follows: * = change in soil moisture deficit (mm); ** = change in soil moisture deficit (%).
Water balance = derived from Thornthwaite.
LBRM = large basin runoff model.
GISS = simulation obtained directly from the Goddard Institute for Space Studies GCM.
DJF = December, January, and February; JJA = June, July, and August.

Soil Moisture

Changes in soil moisture (*Table 2.2*) depend on storage capacity, which is dictated by soil type, rooting depth, and other site factors (e.g., slope, aspect). In water balance models derived from the Thornthwaite model, soil moisture represents available water capacity for plants, which can be drawn from storage capacities varying from less than 150 mm of water per meter of soil to over 300 mm per meter (Thornthwaite and Mather, 1955).

For regions with shallow soils (less than 1 meter), capacities of 100 mm or less have been assumed for the soil column (e.g., Cohen, 1986, 1991).

Some studies have partitioned watersheds specifically to account for soil variations (e.g., Gleick, 1987b). Others include soil characteristics as part of conceptual hydrologic models (e.g., Bultot *et al.*, 1988b).

Soil moisture can also be determined directly by GCMs. Most of the older (i.e., pre-1988) GCMs include a one-layer soil with 15 cm field capacity, whereas the Goddard Institute for Space Studies (GISS) has a two-layer model with capacities that vary with surface cover (Kellogg and Zhao, 1988). Studies by Kellogg and Zhao (1988) and Zhao and Kellogg (1988) compared outputs of five GCMs for North America and eastern Asia, respectively; however, they did not provide estimates for watersheds. They found general agreement among 3 of 4 of the GCMs on direction of change at the regional scale, but unanimous agreement was rare. Rind (1988) described a number of factors that would affect soil moisture simulations from the GISS general circulation model for various grid boxes over North America. These include model resolution, sea surface temperature gradient, and the realism of the ground hydrology parameterization.

Like the older models, the newer (post-1988) high-resolution GCMs use a "bucket" method for one-layer soils and a "force-restore" method for two soil layers, in which the thicker lower layer provides a reservoir for EV. Only the GISS and Canadian Climate Center (CCC) models make allowances for soil type and vegetation. None of the GCMs consider possible direct effects of CO_2 on vegetation (IPCC, 1990).

Some of the results in *Table 2.2* are expressed as change in soil moisture deficit (in millimeters or percentage). The remainder indicate percentage changes in soil moisture, as is the case with the other tables.

Snow Regime

Impacts of climate change on snow are often determined from assumptions about the occurrence of rain or snow based on temperature thresholds. This can significantly influence the scenario estimate of runoff. Studies that use basin models with daily climatological data include snowpack accumulation and snowmelt (e.g., Bultot *et al.*, 1988a; Croley, 1990). Byrne and McNaughton (1991) used a snow-water equivalent model based on snow accumulation and net vapor transfer to and from the snowpack. The latter was derived from Morton's evaporation model (Morton, 1978). For a 15 km^2 basin with a decrease in temperature of 3°C, the model shows a 3% decrease in the annual snow water equivalent. Thornthwaite water balance models using monthly data have also been used, because scenarios of climate change have generally been based on single monthly values obtained from GCM simulations. There are obvious disadvantages to this, but for simulations of monthly and seasonal runoff for large watersheds under scenarios of climate change, additional data (e.g., daily climatological data) do not necessarily improve model accuracy (Gleick, 1987a).

Runoff

Runoff models have been used on their own and as part of integrated basin models. The literature includes studies of various regions using water balance models derived from the Thornthwaite model (e.g., Arnell, 1992; Cohen, 1986, 1987, 1991; Gleick, 1987a); assessments with customized, physically based integrated basin models that include runoff components [e.g., for the Great Lakes (Croley, 1990), for the Colorado River (Nash and Gleick, 1991; Gleick and Nash, 1991), and for Belgium (Bultot, 1988a, 1988b)]; and stochastic methods based on statistical techniques (Revelle and Waggoner, 1983).

Results from case studies are compared with runoff simulations generated directly by the GISS general circulation model (*Table 2.3*). These simulations depend on the model's precipitation, AE, soil moisture storage and oversimplified assumptions about land–atmosphere interactions. Annual runoff for major river basins was calculated from the GISS simulations by Miller and Russell (1992). It should be noted, however, that the GISS control run ($1\times CO_2$) generally showed precipitation and runoff that was higher than observed conditions. One notable exception was the underestimate of runoff for the Amazon River.

2.5 Streamflow: Seasonal Cycle and Annual Net Basin Supply

Table 2.4 summarizes case study results for various watersheds. The results come from either integrated basin models that include the above parameters as components within the models, or from sums of outputs from models of individual elements of the hydrologic cycle.

Under climate change, many watersheds show changes in the timing and magnitude of seasonal peak and low flows. In cold climate zones, peaks tend to occur earlier due to earlier snowmelt. Changes in net basin supply tend to parallel changes in runoff.

Table 2.3. Summary of studies on impacts of climate change on runoff.

Basin	Area (km^2)	Model	Scenario	Results[a]	Reference
Cold-humid					
St. Lawrence	1030	GISS	GISS	+6	Mill.&Rus., 1992
Great Lakes	770	LBRM	GISS	-24	Croley, 1990
			GFDL86	-23	Croley, 1990
			OSU	-11	Croley, 1990
		Water balance	GISS	-11	Cohen, 1987
			GFDL80	-8	Cohen, 1987
Mackenzie River	1810	GISS	GISS	+24	Mill.&Rus., 1992
Yukon River	840	GISS	GISS	+15	Mill.&Rus., 1992
Yenesei River	2580	GISS	GISS	+42	Mill.&Rus., 1992
Temperate					
Delaware River		Water balance	GISS	-39 to -25	McC.&Ayers, 1989
			GFDL86	-7	McC.&Ayers, 1989
			OSU	+2 to +9	McC.&Ayers, 1989
UK catchments	.05-1	Water balance	+4C,-10%P	-40 to -20	Arnell, 1992
			+4C,+20%P	+20 to +50	Arnell, 1992
			0C,+20%P	+30 to +70	Arnell, 1992
Belgian catchments	.1-1.2	IBRM	GFDL80	+12	Bultot *et al.*, 1988a
Mississippi River	3270	GISS	GISS	+20	Mill.&Rus., 1992
Yangtze River	1940	GISS	GISS	+36	Mill.&Rus., 1992
Semiarid, arid					
Sevier River	.9	Water balance	+2C,-10%P	-25	Flaschka *et al.*, 1987
Bear River	.47	Water balance	+2C,-10%P	-22	Flaschka *et al.*, 1987
Colorado River	640	GISS	GISS	+13	Mill.&Rus., 1992
Murray River	1060	GISS	GISS	+25	Mill.&Rus., 1992
Nile River	2960	GISS	GISS	-13	Mill.&Rus., 1992
Orange River	1060	GISS	GISS	-6	Mill.&Rus., 1992
Yellow River	740	GISS	GISS	+30	Mill.&Rus., 1992
Sacramento River		Water balance	+2C,-10%P	JJA: -32	Gleick, 1987a
				DJF: -9	Gleick, 1987a
			GFDL80	JJA: -48	Gleick, 1987a
				DJF: +34	Gleick, 1987a
			GISS	JJA: -53	Gleick, 1987a
				DJF: +81	Gleick, 1987a
Saskatchewan River	364	Water balance	+2C,-10%P	-38	Cohen, 1991
			GISS	+38	Cohen, 1991
			GFDL86	-19	Cohen, 1991
			OSU	+4	Cohen, 1991
Tropical wet					
Hsi Chiang River	440	GISS	GISS	+20	Mill.&Rus., 1992
Mekong River	790	GISS	GISS	+19	Mill.&Rus., 1992
Amazon River	6150	GISS	GISS	+3	Mill.&Rus., 1992
Zaire River	3820	GISS	GISS	+27	Mill.&Rus., 1992
Zambezi River	1200	GISS	GISS	+4	Mill.&Rus., 1992
			GFDL86	-7	Salewicz, Chapter 11
			GISS	+7	Salewicz, Chapter 11
Niger River	1210	GISS	GISS	-31	Mill.&Rus., 1992

[a]All results are expressed as the percentage change from present or control run simulations.
Water balance = derived from Thornthwaite.
LBRM = large basin runoff model.
IBRM = integrated basin runoff model.
GISS = simulation obtained directly from the Goddard Institute for Space Studies GCM.
JJA = June, July, and August; DJF = December, January, and February.

Table 2.4. Summary of studies on impacts of climate change on streamflow and net basin supply (NBS).

Basin	Area (km^2)	Model	Scenario	Results[a]	Reference
Cold-humid					
Great Lakes	770	Separate outputs	GISS	–24	Cohen, 1987
			GFDL80	–17	Cohen, 1987
			+2C,–10%P	–36	Cohen, 1987
			+4C,–10%P	–50	Cohen, 1987
			+4C,+20%P	+20	Cohen, 1987
		LBRM	GISS	–37	Croley, 1990
			GFDL86	–51	Croley, 1990
			OSU	–23	Croley, 1990
		Morton	GISS	–15	Marchand *et al.*, 1988
Warta River	.06	Budyko-Zubenok	GFDL86	+11	Kacz.&Kras., 1991
Temperate					
Delaware River					McC.&Wol., 1992
UK catchments					Arnell, 1992
Belgian catchments					Bultot *et al.*, 1988a
Semiarid, arid					
Colorado River	625	NWSRFS	GISS	–14	Gleick&Nash, 1991
			GFDL86	–24	Gleick&Nash, 1991
			UKMO	–21 to +0	Gleick&Nash, 1991
			+2C,–10%P	–23	Gleick&Nash, 1991
			+4C,–10%P	–31	Gleick&Nash, 1991
			+4C,+20%P	+2	Gleick&Nash, 1991
Saskatchewan River	364	Separate outputs	GISS	+29 to +40	Cohen, 1991
			GFDL80	–70 to –65	Cohen, 1991
			GFDL86	–36 to –27	Cohen, 1991
			OSU	–4 to +2	Cohen, 1991
			+2C,–10%P	–51 to –45	Cohen, 1991
			+4C,–10%P	–59 to –52	Cohen, 1991
			+4C,+20%P	+15 to +21	Cohen, 1991
Sacramento River					Gleick, 1987a
Bow-Oldman River					Byrne&McN., 1991
Sevier River	.9	Empirical	+2C,–10%P	AE: –5	Flaschka *et al.*, 1987
Bear River	.47	Empirical	+2C,–10%P	AE: +3	Flaschka *et al.*, 1987
Arizona basins		CO_2 adj.			Idso&Brazel, 1984
Tropical wet					
Zambezi River			GFDL86	AE: –3 33	Salewicz, Chapter 11
			GISS	AE: +11	Salewicz, Chapter 11
			GFDL86	E: +7	Salewicz, Chapter 11
			GISS	E: +8	Salewicz, Chapter 11

[a]All results are expressed as the percentage change from present or control run simulations.
Separate outputs = output from separate models is combined.
LBRM = large basin runoff model. Morton = Morton's E and AE models.
NWSRFS = US National Weather Service River Forecasting Service.
CO_2 adj. = AE reduced by an assumed rate to account for direct effects of CO_2 on stomatal resistance; does not include changes in leaf area.
GISS = simulation obtained directly from the Goddard Institute for Space Studies GCM.
AE = actual evapotranspiration. E = evapotranspiration.

2.6 Summary

Climate change, as defined in the hydrologic studies discussed in this chapter, is limited to changes in temperature and precipitation. These changes in climatic parameters impact the components of water balance to different extents.

Tables 2.1 to *2.4* present results of studies examining the impact of climate change on EV, PET, soil moisture, runoff, and streamflow. As expected, EV and PET increase under warmer temperatures. Actual evapotranspiration will increase under warmer temperatures as long as precipitation remains the same or increases. However, should precipitation decrease significantly the available soil water may be sufficiently reduced to result in a decrease in AE. Whether AE decreases depends on the catchment slope, soil depth, vegetation, and the magnitude of the temperature increases and precipitation decrease. No general statement can be made.

Across climatic zones soil moisture is generally reduced due to increases in PET, even with increased precipitation. However, as is the case for AE, soil moisture depends on the catchment slope, soil depth, vegetation, and the magnitude and direction of the temperature and precipitation changes.

The results of the studies on runoff and streamflow provide an unclear message. Because runoff is the by-product of precipitation minus AE, the direction and magnitude of changes in runoff are quite specific to each basin. The actual changes in temperature and precipitation can have different indications.

Under GCM climate change scenarios, there are seasonal variations in the temperature and precipitation changes that make it even more difficult to make a general statement about runoff and streamflow changes. In some snow regions, climate changes lead to little or no change in annual runoff, but peak flows and the temporal distribution of runoff are changed.

Although the impact of climate change upon the components of the water balance generally can be forecast, the results of this chapter show that no categorical statement can be made about the global impact of climate change and the impact on water resources. This uncertainty about runoff impacts leads to two certain conclusions: climate change impacts on runoff and streamflow must be studied at the local basin level, and the definition of climate change scenarios is essential in determining the nature of the changes.

References

Arnell, N.W., 1992, Factors controlling the effects of climate change on river flow regimes in a humid temperate environment, *Journal of Hydrology*, **132**:321–342.

Aston, A.R., 1984, The effect of doubling atmospheric CO_2 on streamflow: A simulation, *Journal of Hydrology*, **67**:273–280.

Bultot, F., Coppens, A., Dupriez, G.L., Gellens, D., and Meulenberghs, F., 1988a, Repercussions of a CO_2 doubling on the water cycle and on the water balance: A case study for Belgium, *Journal of Hydrology*, **99**:319–347.

Bultot, F., Dupriez, G.L., and Gellens, D., 1988b, Estimated annual regime of energy-balance components, evapotranspiration and soil moisture for a drainage basin in the case of CO_2 doubling, *Climatic Change*, **12**(1):39–56.

Byrne, J.M., and McNaughton, R.B., 1991, Predicting temporal and volumetric changes in runoff regimes under climate warming scenarios, *Canadian Water Resources Journal*, **16**:129–141.

Cohen, S.J., 1986, Impacts of CO_2-induced climate change on water resources in the Great Lakes Basin, *Climatic Change*, **8**(2):135–153.

Cohen, S.J., 1987, Sensitivity of water resources in the Great Lakes region to changes in temperature, precipitation, humidity and wind speed, in S.I. Solomon, M. Beran, and W. Hogg, eds., *The Influence of Climate Change and Climatic Variability on the Hydrologic Regime and Water Resources*, International Association of Hydrological Sciences (IAHS) Publication No. 168, IAHS Press, Wallingford, UK.

Cohen, S.J., 1991, Possible impacts of climatic warming scenarios on water resources in the Saskatchewan River Sub-basin, Canada, *Climatic Change*, **19**(3):291–317.

Croley, T.E., II, 1990, Laurentian Great Lakes double-CO_2 climate change: Hydrological impacts, *Climatic Change*, **17**(1):27–47.

Dooge, J.C.I., 1992, Hydrologic models and climate change, *Journal of Geophysical Research – Atmospheres*, **97**(ND3):2677–2686.

Flaschka, I., Stockton, C.W., and Boggess, W.R., 1987, Climatic variation and surface water resources in the Great Basin region, *Water Resources Bulletin*, **23**:47–57.

Frederick, K.D., 1991, *Processes for Identifying Regional Influences of and Responses to Increasing Atmospheric CO_2 and Climate Change: The MINK Project, Report IV: Water Resources*, US Department of Energy, Office of Energy Research, TR052F, Available from National Technical Information Service, US Department of Commerce, Springfield, VA, USA.

Gleick, P.H., 1987a, The development and testing of a water balance model for climate impact assessment: Modelling the Sacramento Basin, *Water Resources Research*, **23**:1049–1061.

Gleick, P.H., 1987b, Regional hydrologic consequences of increases in atmospheric CO_2 and other trace gases, *Climatic Change*, **10**(2):137–161.

Gleick, P.H., 1989, Climate change, hydrology, and water resources, *Reviews of Geophysics*, **27**:329–344.

Gleick, P.H., and Nash, L.L., 1991, The Sensitivity of Colorado River Runoff to Climate Changes, Pacific Institute, Berkeley, CA, USA.

Hagan, R., Haise, H., and Edminster, T., 1967, *Irrigation of Agricultural Lands*, American Society of Agronomy, Madison, WI, USA.

Idso, S.B., and Brazel, A.J., 1984, Rising atmospheric carbon dioxide concentrations may increase streamflow, *Nature*, **312**:51–53.

IPCC (Intergovernmental Panel on Climate Change), 1990, Report of Working Group I, IPCC, Geneva, Switzerland.

Kaczmarek, Z., and Krasuski, D., 1991, Sensitivity of Water Balance to Climate Change and Variability, International Institute for Applied Systems Analysis, WP-91-047, International Institute for Applied Systems Analysis, Laxenburg, Austria.

Kellogg, W.W., and Zhao, Z.-c., 1988, Sensitivity of soil moisture to doubling of carbon dioxide in climate model experiments: Part I: North America, *Journal of Climate*, **1**:348–366.

Marchand, D., Sanderson, M., Howe, D., and Alpaugh, C., 1988, Climatic change and Great Lakes levels: The impacts on shipping, *Climatic Change*, **12**:107–133.

Mather, J., 1978, *The Climatic Water Balance in Environmental Analysis*, Lexington Books, Lexington, MA, USA.

McCabe, G.J., Jr., and Ayers, M.A., 1989, Hydrologic effects of climate change in the Delaware River Basin, *Water Resources Bulletin*, **25**(6):1231–1242.

McCabe, G.J., Jr., and Wolock, D.M., 1992, Effects of climate change and climatic variability on the Thornthwaite moisture index in the Delaware River Basin, *Climatic Change*, **20**:143–153.

Miller, J.R., and Russell, G.L., 1992, The impact of global warming on river runoff, *Journal of Geophysical Research*, **97**(ND3):2757–2764.

Mimikou, M., Kouvopoulos, Y., Cavias, G., and Vayainos, N., 1991, Regional hydrological effects of climate change, *Journal of Hydrology*, **123**:119–146.

Morton, F.I., 1978, Estimating evapotranspiration from potential evaporation: Practicality of an iconoclastic approach, *Journal of Hydrology*, **38**:1–32.

Morton, F.I., 1983a, Operational estimates of areal evapotranspiration and their significance to the science and practice of hydrology, *Journal of Hydrology*, **66**(1/4):1–76.

Morton, F.I., 1983b, Operational estimates of lake evaporation, *Journal of Hydrology*, **66**:77–100.

Nash, L.L., and Gleick, P.H., 1991, Sensitivity of streamflow in the Colorado Basin to climatic changes, *Journal of Hydrology*, **125**:221–241.

Nemec, J., and Schaake, J.C., 1982, Sensitivity of water resources systems to climate variations, *Hydrological Sciences Journal*, **27**(3):327–343.

Priestly, C.H.B., and Taylor, R.J., 1972, On the assessment of surface heat flux and evaporation using large-scale parameters, *Monthly Weather Review*, **100**:81–92.

Revelle, R.R., and Waggoner, P.E., 1983, Effects of a carbon dioxide-induced climate change on water supplies in the western United States, in US National Research Council, *Changing Climate: Report of the Carbon Dioxide Assessment Committee*, National Academy Press, Washington, DC, USA.

Riebsame *et al.*, 1994, Rivers, in K. Strzepek and J. Smith, eds., *As Climate Changes: International Impacts and Implications*, Cambridge University Press, Cambridge, UK.

Rind, D., 1988, The doubled CO_2 climate and the sensitivity of the modeled hydrologic cycle, *Journal of Geophysical Research*, **93**(D5):5385–5412.

Rind, D., Goldberg, R., Hansen, J., Rosenzweig, C., and Ruedy, R., 1990, Potential evapotranspiration and the likelihood of future drought, *Journal of Geophysical Research*, **95**(D7):9983–10004.

Rind, D., Rosenzweig, C., and Goldberg, R., 1992, Modelling the hydrological cycle in assessments of climate change, *Nature*, **358**(July):119–122.

Rosenberg, N.J., Kimball, B.A., Martin, P., and Cooper, C.F., 1990, From climate and CO_2 enrichment to evapotranspiration, in P.E. Waggoner, ed., *Climate Change and U.S. Water Resources*, John Wiley & Sons, New York, NY, USA.

Schaake, J.C., 1990, From climate to flow, in P.E. Waggoner, ed., *Climate Change and U.S. Water Resources*, John Wiley & Sons, New York, NY, USA.

Shuttleworth, W.J., 1993, Evaporation, in D.R. Maidment, ed., *Handbook of Hydrology*, McGraw-Hill Book Company, New York, NY, USA.

Skiles, J.W., and Hanson, J.D., 1994, Responses of arid and semi-arid watersheds to increasing carbon dioxide and climate change as shown by simulation studies, *Climatic Change*, **26**:377–397.

Thornthwaite, C.W., and Mather, J.R., 1955, The water balance, *Climatology*, **8**(1):1–104.

Zhao, Z.-c., and Kellogg, W.W., 1988, Sensitivity of soil moisture to doubling of carbon dioxide in climate model experiments, Part II: The Asian monsoon region, *Journal of Climate*, **1**:367–378.

Chapter 3

Potential Impacts of Climate Change on Lake and Reservoir Water Quality

Olli Varis and László Somlyódy

3.1 Introduction

Climate change may have diverse impacts on lakes and their water quality; these impacts can be hydrologic, thermal, hydraulic, chemical, biochemical, or ecological. In this chapter, their interrelation and potential changes and their contributions to lake water quality problems are reviewed and discussed, and checklists for planning, management, and impact assessment purposes are provided. Water quality problems are grouped into the following categories: eutrophication, oxygen depletion, hygienic problems, salinization, acidification, toxic and cumulative substances, turbidity and suspended matter, and thermal pollution. Many existing problems may worsen due to the climate change, yet extreme uncertainty exists as to the extent of the change at the level of any specific case. Therefore, it has been suggested that shifts in water quality and quantity due to climate change should be integrated into contemporary planning and management in an adaptive manner, and that the research and development of the impact assessment methodology should focus on approaches that can handle extreme uncertainty (Fiering and Matalas, 1990).

As far as the water quality of lakes and reservoirs is concerned, climate change currently seems to have few implications for water resources

Z. Kaczmarek et al. (eds.),
Water Resources Management in the Face of Climatic/Hydrologic Uncertainties 46–69.

management. The implications are marginal in comparison with those of many other human activities that introduce changes into our environment. The research, impact assessment, administration, remediation, and restoration efforts concerning water quality of lakes are chiefly directed at conventional changes due to land use, waste-management practices in industry and in settlements, policies in agriculture and forestry, construction projects, water allocation, and water supply.

The issue of climate change, although still highly uncertain with respect to rate and magnitude, is currently receiving increasing attention in water resources research and management (see Kundzewicz and Somlyódy, 1993). Accordingly, the number of published studies concerning water quality of lakes is increasing. Climate change can affect water quality through many mechanisms, and often the direction of change and quantitative estimates are difficult to specify. Moreover, the assessment of water quality uses several different use-specific criteria. From the water resources management point of view climate change introduces an additional source of uncertainty whose relative importance increases rapidly with increasing planning horizon.

Climate change research that includes diverse impact assessment tasks requires the teamwork of experts from various disciplines. No education provides expertise in all of the fields of natural, engineering, social, and management sciences that are required in the assessment of the impacts of climate change on the scales encountered in water resources management. Holistic reviews and analyses are needed from the different components of the ecological and social systems to improve the communication between specialists in different, but often related, fields and to assist in focusing on the most essential impacts and problems.

Our goals are to offer a systematic view on potential changes in lakes and reservoirs; to discuss how these changes may affect water quality problems, water quality management, and water use, with special attention given to the elucidation of the complex interrelationships of the elements of lake water quality; and to create a checklist for researchers, planners, and managers. In this chapter, the most typical uses of lake water and the most frequent water quality problems are identified and the possible impacts of climate change on the natural phenomena and processes in lakes are considered. The impacts on water quality problems are also discussed and conclusions are presented.[1]

[1]It is not the objective of this chapter to analyze any water quality problems in detail, for example, by using models and case studies. For further analysis of water quality issues, see Chapters 4 and 9.

3.2 Water Quality Problems and Lake Management

The term *water quality problem* does not refer to a single problem found in all lakes; water quality problems comprise several different issues because of the diversity of natural and man-made lakes, the diversity of societal needs and preferences associated with the use of surface waters, and the complexity of lake ecosystems. The most common water quality problems are the following:

- *Eutrophication* caused by the abundance of nutrients and other prerequisites of enhanced primary production.
- *Oxygen depletion* caused by degradation of organic matter in water.
- *Hygienic problems* due to pathogenic organisms such as viruses (causing hepatitis, polio, etc.), bacteria (causing cholera, typhoid fever, etc.), or protozoa (causing schistosomiasis, malaria, amebiasis, etc.).
- *Salinization* due to high concentrations of ions such as calcium, sodium, chloride, and sulfate. Hardness is caused by high concentrations of calcium, magnesium, and carbonates (HCO_3^- and $CO_3^=$).
- *Acidification* of lakes caused globally mainly by atmospheric deposition of SO_2 and NO_x. Some industrial and mining activities also cause severe regional watershed acidification problems. The natural acidity and buffer capacity of waters differ greatly.
- *Toxic* or *cumulative compounds* such as heavy metals (mercury, chromium, zinc, cobalt, etc.), pesticides (DDT, Lindane, etc.), industrial wastes (PCB, organochlorides, etc.), or water ecosystem-borne toxins (hepatotoxins and neurotoxins produced by blue-green algae, etc.).
- *Suspended material and turbidity* caused by inorganic or organic matter.
- *Changed thermal conditions* due to thermal pollution, flow control, or changed climate.

From the water resources management perspective, the greatest motivation for water quality studies derives from the demands set by different uses and the conflicts between those uses. Water quality requirements are case specific, and human activities affect them in many ways. *Tables 3.1* and *3.2* summarize the most frequent uses of lake water and their requirements and impacts on water quality. Many uses suffer from poor water quality without contributing to it (*Table 3.1*); other uses cause water quality problems but are not harmed by them (*Table 3.2*).

Regarding the issue of climate change and management of lake water quality, the following questions are raised: Will the potential impacts of

Table 3.1. Typical influences of water quality problems on the most frequent uses of lake water.

	Water quality problems							
Water use	E	O	H	S	A	X	M	T
Conservation	•	•	•	•	•	•	•	•
Recreation	•	•	•	o	o	•	•	o
Fisheries	o	•	–	o	o	•	o	o
Aquaculture	o	•	–	o	o	•	o	o
Water withdrawal								
– Households	•	•	•	o	o	•	o	–
– Municipalities	•	o	o	o	o	•	o	–
– Irrigation, etc.	o	o	•	•	o	•	o	–
– Industry	o	o	o	o	o	o	o	–
Transport	o	–	–	–	–	–	o	–
Flood control	–	–	–	–	–	–	o	–
Hydropower	–	–	–	–	–	–	o	–
Cooling basin	–	–	–	–	–	–	–	o
Waste transport and disposal	o	–	–	–	–	–	–	–

E = eutrophication; O = oxygen depletion; H = hygienic problems; S = salinization; A = acidification; X = toxic and cumulative substances; M = turbidity and suspended matter; and T = thermal pollution.
o stands for low or occasional influence; • stands for high influence.

Table 3.2. Typical impacts of the most frequent uses of lake water on water quality of lakes.

	Water quality problems							
Water use	E	O	H	S	A	X	M	T
Conservation	–	–	–	–	–	–	–	–
Recreation	–	–	o	–	–	–	–	–
Fisheries	–	–	–	–	–	–	–	–
Aquaculture	•	o	–	–	–	–	o	–
Water withdrawal								
– Households	–	–	–	–	–	–	–	–
– Municipalities	–	–	–	–	–	–	–	–
– Irrigation, etc.	o	–	–	o	–	–	–	–
– Industry	–	–	–	–	–	–	–	–
Transport	–	–	o	–	–	o	o	–
Flood control	o	o	–	–	–	–	o	o
Hydropower	o	o	–	o	–	–	•	o
Cooling basin	o	o	–	–	–	o	–	•
Waste transport and disposal	•	•	•	–	o	•	•	–

For explanation of abbreviations and symbols see *Table 3.1*.

Table 3.3. Alternatives for management of lake water quality.

	Water quality problems							
Water use	E	O	H	S	A	X	M	T
External action								
Treatment of wastewater and sewage	•	•	•	–	–	•	•	–
Manipulation of inflow	o	o	–	–	–	–	•	–
Land use in the catchment	o	o	o	–	o	o	o	o
Non-point source pollution control	•	o	o	–	–	•	o	–
Liming of lakes and catchments	–	–	–	–	•	–	–	–
Shifts in lake water volume or level	o	o	–	o	–	–	o	–
Flow regulation	o	o	–	–	–	–	o	–
Increased throughflow	o	o	–	o	–	–	o	–
Internal action								
Ecotechnology, biomanipulation	•	o	–	–	–	–	o	–
Chemical treatment of water (P inactivation, poisoning of algae, etc.)	o	o	–	–	–	–	o	–
Removal, treatment, or covering of sediment	•	•	–	–	–	o	o	–
Macroplant harvesting	o	o	–	–	–	–	–	–
Aeration and mixing	•	•	–	–	–	–	o	–
Outflow from hypolimnion	o	o	–	o	–	–	o	–

E = eutrophication; O = oxygen depletion; H = hygienic problems; S = salinization; A = acidification; X = toxic and cumulative substances; M = turbidity and suspended matter; and T = thermal pollution.
o stands for low or occasional influence; • stands for high influence.

climate change affect decision making and water resources management? On what scale would that happen, on the river-basin scale or on a smaller scale? What technical means are available? The first two questions are rather case specific and should be considered for each case separately. The technical means can either be directed at the lake itself or at external factors affecting the quality of the inflow. One can influence the hydrology, thermal balance, chemistry, biology, or ecology of the system. A variety of alternatives for management of lake water quality have been presented (*Table 3.3*), most of which are in wide use globally (see Cooke *et al.*, 1986; Moore and Thornton, 1988; Gulati *et al.*, 1990). Many of them, particularly the internal action alternatives, can best be applied, and in some cases can only be applied, at small lakes or on a local scale.

In the following section, we discuss the potential impacts of climate change on factors driving water quality, such as water and nutrient inputs to the lake and thermal distribution in the lake water body, and on internal factors, such as the chemistry, biochemistry, and ecology of the lake. Water quality problems are also discussed.

3.3 Impacts on Factors Driving Lake Water Quality

3.3.1 Climate

Information on climate change, with variables such as air temperature, precipitation, humidity, cloudiness, solar radiation, windiness, and air pressure, and information on their temporal and spatial variability is taken here as given information; therefore, the topic is not discussed in detail. However, some notes are made on their contribution to lake water quality.

Solar radiation penetrating the water body is among the significant energy input to a lake. It warms the water and, together with wind, inflow, and air temperature, is a major contributor to the formation of thermal stratification. It is also the prerequisite for assimilation (primary production) by plants, including phytoplankton, and therefore controls many chemical cycles in a lake. Changes in solar radiation caused by changes in humidity, cloudiness, and air temperature are likely to alter the thermal balance and primary production. Humidity also controls the thermal balance through the evaporation process. Most of the climatic variables also have an impact on the hydrology of the watershed and the lake.

3.3.2 Hydrology and watershed

The hydrologic cycle is closely connected to the climate. General circulation models (GCMs) provide information from the climate change scenario on the most important components of the hydrologic cycle. However, obtaining information on the regional and local scale is usually a problem (see Smith, 1990). According to Kundzewicz and Somlyódy (1993), hydrological models are needed to estimate components of the hydrological cycle on the basis of GCMs. Problems associated with this are scale and uncertainty. Mesoscale hydrological models – such as the rainfall pattern prediction model by Bárdossy and Plate (1992), which starts from the air pressure estimate of GCMs – are a promising way to derive mesoscale climate change scenarios for water resources management. Improved integration of hydrological models and GCMs is obviously needed.

In areas where the climate becomes more humid, the increased rainfall–evaporation ratio causes a decrease in lake retention time. It is also likely to increase erosion and nutrient leaching from the catchment, and thus contribute to the non-point source pollution entering lakes. Erosion typically counteracts and reduces many lake water quality problems, such as eutrophication, oxygen depletion, and salinity. Nutrient leaching enhances these problems, as well as enhancing the turbidity and hygiene problems. In areas where the climate becomes more arid the impacts are reversed: lake retention time increases, and erosion and nutrient leaching are likely to decrease if the vegetation is not destroyed. Increased retention and decreased throughflow lead to the concentration of pollutants and salts. The risk of salinization is noteworthy in areas where evaporation is expected to become greater than precipitation. Annual runoff has a nonlinear relationship to annual precipitation. Moreover, seasonal changes are often more important than annual ones (see Williams, 1989).

The $2\times CO_2$ scenarios by Williams (1989) showed that the frequency of wet years in California's Central Valley increases from 26% to 47%. This, among other implications, increases the incentive for protecting soils and vegetation. In contrast, the mean annual total streamflow decreases 31% in the Missouri River basin, decreases 29% in the upper Mississippi River, and decreases 7% in the Arkansas River (Frederick, 1993). In Britain, where the climate is expected to get warmer and more humid, Whitehead and Jenkins (1989) predict an increase in wash off of nutrients and herbicides from soil, and a consequent increase of non-point source loading. They stress the roles of changing water pathways, increasing soil temperature, and increasing use of agricultural chemicals due to climate warming as major reasons for these changes.

The climate change may alter the water level and extension of a lake (Hartmann, 1990; Jacoby, 1990), even reducing it to a lake without outlet or drying it up. Such changes would have serious consequences for the whole ecosystem and would drastically affect water use. The Laurentian Great Lakes, for instance, are at risk of being lowered substantially. Depending on the GCM used, $2\times CO_2$ simulations have resulted in a drop in water level of 0.5 m to 2.5 m (Quinn and Croley, 1983; Hartmann, 1990; Smith, 1991; Changnon, 1993; see also critical comments by Balling, 1991).

The expected changes in the hydrology of a single river basin or a lake are still very obscure and case specific. It may not even be clear whether the attention should be paid to the mean behavior of the system – changes in precipitation, humidity, snow cover, erosion, mass flows, water levels, and lake volumes – or to the extremes – frequency and intensity of floods, storms, and droughts. Studies of climate change should, perhaps, respect the water engineering tradition more, in which extreme design conditions play a major role.

Changes in hydrology have direct effects on the thermal and hydraulic balances by affecting the loads of nutrients and suspended matter to the lake. Variability and timing of quantity and quality of the incoming water must be considered. Even small changes in the hydrology and mass flows may influence lake water quality greatly, as was the case in the simulation study by Varis (1988, 1989) at Lake Kuortaneenjärvi, Finland, where the late summer blue-green algae blooms were found to be very sensitive to variations in quantity and quality of nutrient-rich spring floodwater.

Most of the case studies addressing hydrological changes due to climate change present the implications to water quantity management and deal only marginally with water quality issues (e.g., Croley, 1990; Lettenmaier and Gan, 1990; Thomas, 1990; Cohen, 1991; Dooge, 1992; Kaczmarek and Krasuski, 1991; Lettenmaier and Rind, 1992; Miller and Russel, 1992). There has been considerable interest in the evaluation of the possibilities of adapting water allocation policies to meet the problems raised by the changing amounts of water in the respective watersheds.

3.3.3 Thermal balance of a lake

The thermal balance and thermal stratification pattern, especially the depth and stability of the thermocline (the layer with a high temperature gradient), are the result of several hydrometeorological factors. The most important of these include thermal flux from the atmosphere, evaporation and precipitation, wind-induced turbulence, and energy input by solar radiation (see Chapter 4 for details). Typically they all are subject to substantial natural variability.

Altered water temperature influences almost every hydraulic, chemical, or biological process in a lake. Considering thermal stratification, which is caused by density differences in water layers of different temperatures, the climate change will have the most notable impacts where the stratification pattern is essentially altered. The characteristic thermal stratification patterns are as follows (see also Chapter 9):

- *Dimictic:* the lake stratifies biannually. In summer the upper water layers are warmer than the lower ones; in winter, during ice cover, the upper layers are cooler than the lower ones. Typical for deep lakes that freeze in the winter.

- *Monomictic:* the lake stratifies only once a year, in the winter or in the summer. Typical for deep lakes that do not freeze in winter and for shallow lakes that freeze.
- *Amictic and polymictic:* the lake develops no stable stratification, and the lake stratifies several times in a year, respectively. Typical for tropical lakes and for shallow lakes that do not freeze.

A shift, such as from dimictic-pattern dominance to monomictic-pattern dominance, or from monomictic- to amictic- or polymictic-pattern dominance, or vice versa, is likely to change the lake ecosystem markedly. The balances of oxygen, nutrients, sulfur, iron, and many other substances may be considerably altered. Substantial changes may also be induced by the shift in the depth of the thermocline, which alters the volumes and balances of different water layers. The issue grows in complexity if the lake is also chemically stratified, i.e., meromictic. This is often the case in deep, brackish water basins. In addition to changes in air temperature, important factors contributing to stratification are altered radiation, quantity and temperature of inflow, and windiness.

McCormick (1990) has studied the potential thermal and hydraulic impacts on Lake Michigan; Croley (1990) studied these impacts on the Laurentian Great Lakes; and Orlob *et al.* (1990) studied them on the Shasta-Trinity system in California. Hondzo and Stefan (1991) modeled and compared the heat balance and stratification of a particularly warm year to those of a year with average temperatures in three lakes in north-central USA. A regional simulation study in Minnesota (Hondzo and Stefan, 1993), covering 3,002 lakes from 27 classifications, suggested that climate change will increase the epilimnetic temperatures; however, this increase will be less than the projected air-temperature increase. Hypolimnetic temperatures in seasonally stratified lakes are expected to remain largely unchanged or to become even lower than at present. Annual evaporation will increase by 300 mm, onset of stratification will take place earlier and it will overturn later, and stratification will become more stable in spring and summer. In Chapter 9, a number of hypothetical lakes, from shallow to deep and located around the globe, are defined in order to find sensitive latitudes where the risk for the change in the stratification pattern is great. Eight real lakes then were selected from these sensitive latitudes and $2\times CO_2$ scenarios were calculated for the heat balances and thermal stratification patterns of those lakes.

The length of ice cover and other characteristics of the ice-cover period have many physical, chemical, and biological influences. The most important among these are thermal balance, control of input solar radiation, and control of mass flux between the atmosphere and the water body. The most drastic impact would be the total lack of ice cover in lakes that previously froze.

Several studies (Assel *et al.*, 1985; Kuusisto, 1987; Schindler *et al.*, 1990; Assel, 1991; Kauppi *et al.*, 1992) have shown that climate change may produce remarkable changes in ice cover of lakes. For instance, in Lake Superior and Lake Erie the annual ice-cover duration from 1951 to 1980 was 13 to 16 weeks. The $2\times CO_2$ scenarios by Smith (1991) and Assel (1991) resulted in a reduction in the ice-cover period from 1.0 to 2.5 months, and winters without ice cover are to be expected. These results vary greatly depending on the GCM used. In six Finnish lakes that are currently frozen from five to six months a year, $2\times CO_2$ results show a decrease of roughly two months in the ice-cover period (Kauppi *et al.*, 1992).

Because lake ice records are sensitive to climate changes, they have been used to detect changes in the climate and to project future ice cover at both Red River, Manitoba, Canada (Rannie, 1983), using records from the nineteenth century on, and at Lake Mendota, Wisconsin, USA (Robertson *et al.*, 1992), with data from 1855 to the present. The ice-cover period of Lake Mendota will decrease by 11 days with a 1°C warming; with a warming of 4°C to 5°C, ice-free winters will occur in approximately 1 out of every 15 to 30 years. Both analyses showed clear-cut climatic periods with relatively stable conditions and some periods with rapid changes. The years prior to 1888–1890 were notably cooler than were subsequent years. Another shift to warmer temperatures occurred in 1979.

3.4 Impacts on Water Quality

3.4.1 Chemistry and biochemistry

Climate, thermal, and hydrologic changes in a lake affect the chemical and biochemical cycles and balances in many ways. The most important elements among them with regard to water quality studies include dissolved oxygen (DO), inorganic carbon, phosphorus, nitrogen, sulfur, and iron. Additionally, ions responsible for water salinity deserve attention.

Dissolved oxygen balance, including processes influenced by oxygen, is closely related to climatic, thermal, and hydraulic impacts (ice cover, water temperature, solar radiation, and stratification). It is also closely associated with nutrient cycles, biotic impacts, and the eutrophication problem. Primary production in the illuminated water layer consumes CO_2 and produces oxygen. Degradation of organic material has an opposite impact: CO_2 is produced and oxygen is consumed. Dissolved oxygen concentration influences many biochemical processes, for example in sulfur and iron dynamics. A major negative consequence of eutrophication is the depletion of DO from lower water layers; this is caused by the degradation of organic material in layers not in contact with the atmosphere. Many fish species are very sensitive to low DO concentrations in terms of both population maintenance and reproduction. Often, short extremes become critical.

The role of the inorganic carbon system has been emphasized in many textbooks (Ruttner, 1940; Hutchinson, 1957; Stumm and Morgan, 1981). Increased partial pressure of CO_2 in the atmosphere causes increased CO_2 concentration in water. Increased temperature decreases solubility of gases. The CO_2–H_2CO_3 equilibrium is sensitive to pH, and it buffers the pH. Increasing CO_2 concentration in water also moves the equilibrium in the acidic direction. Additionally, the free inorganic carbon increases the aggressivity of water and releases calcium (Ca) and magnesium (Mg) from the rock and sediments. If Ca and Mg are available, the hardness of the water increases; if not, the acidification continues until other buffering cations become available. The expression "hard water" refers to water with high alkalinity. Soft-water lakes are typically located in areas with igneous, acidic rock, and are especially prone to acidification.

For the Ca- and Mg-buffered Lake Balaton, Hungary, Szilágyi and Somlyódy (1991) have calculated that the increased dissociation of inorganic carbon from water will probably be a more important contributor to acidification and to ionic composition than will acid deposition. Doubling the atmospheric CO_2 concentration would induce a lowering of the pH value by 0.2 units, and would lead to a significant increase in the salt content and hardness of the lake water. Laboratory experiments and calculations for another set of lakes are under way.

Increased inorganic carbon adds to the availability of carbon to primary producers. The carbon availability has been suggested as a major limiting factor in eutrophic lakes (King, 1970; Shapiro, 1973, 1990). During the hours when primary production is most intensive, the concentration of inorganic carbon in water declines locally and biomass growth is limited accordingly. Therefore, increased availability of inorganic carbon may enhance primary production and contribute to eutrophication (Wetzel and Grace, 1983; Byron and Goldman, 1991). In the global carbon cycle, the fixation of carbon by phytoplankton is an important link and balancing feedback mechanism (Lashof, 1989; Williamson and Gribbin, 1991). Also, the leaching of organic carbon, such as humic compounds, from the catchment may alter a lake's carbon cycle and water quality (Arvola *et al.*, 1992).

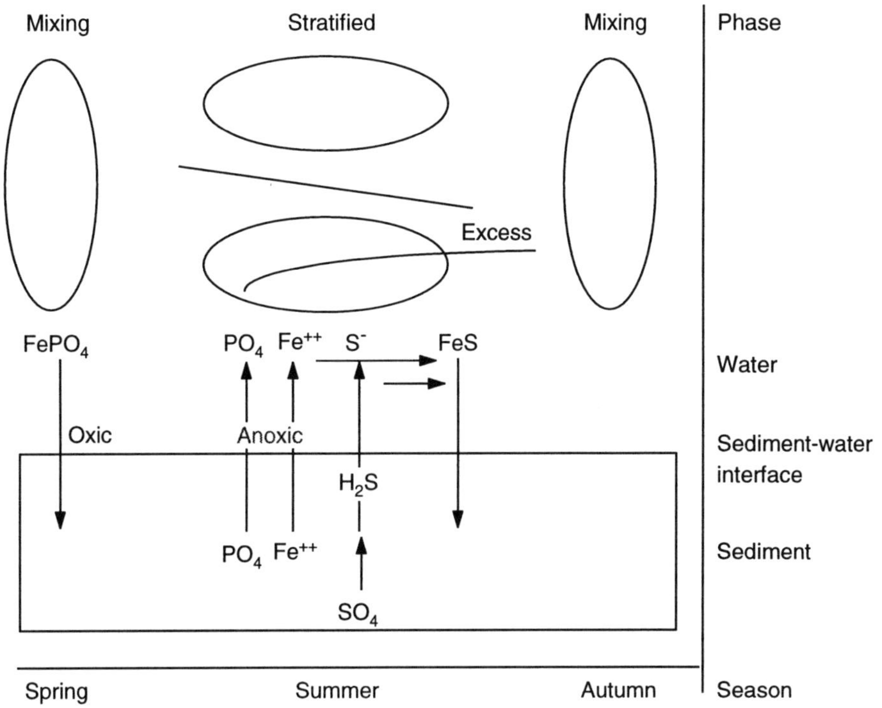

Figure 3.1. Interactions of phosphorus, sulfur, and iron in oxic and anoxic conditions in the sediment–water interface.

The dynamics of many other important elements in lake ecosystems – above all phosphorus, nitrogen, sulfur, silica, and iron – are dominated by microbial activity, which is controlled primarily by DO concentration (redox potential), temperature, pH, and various concentrations of these elements. In general, increasing temperature accelerates reaction rates. Decreased redox potential caused by low DO availability affects many equilibrium reactions (*Figure 3.1*). Phosphorus (P) cycle is sensitive to DO, particularly the near-bottom concentration. Even a short, extreme anoxic period near the bottom can induce P mobilization from the sediment to the water body, and change the lake permanently to eutrophic. It is very difficult and expensive to reverse this development. Typically, if surface erosion is increased from the catchment, P input is also increased. Nitrogen (N) cycle is connected to the atmosphere, both through N fixation by blue-green algae and bacteria (source), and through denitrification (sink). With increasing temperature, the solubility of gases decreases and processes such as denitrification and N fixation are accelerated.

Bicarbonates, carbonates, sulfates, and chlorides of alkalis and alkaline salts dominate the ionic composition of fresh water. The four major cations are Ca^{++}, Mg^{++}, Na^{+}, and K^{+}; the four major anions are HCO_3^-, $CO_3^=$, $SO_4^=$, and Cl^-. Also, silica is often a significant constituent of the total salinity in hardwater lakes, as is iron under some conditions. The dominating processes controlling the salinity of lakes are weathering processes in the catchment, atmospheric precipitation, and accumulation of salts due to dominance of evaporation over outflow. Salinization problems are notable in areas that are arid or that are at risk of becoming arid.

3.4.2 Ecology and biology

Whereas the potential changes in the physical and chemical domain of a lake are relatively easy to list, the changes in the flora and fauna, and in the whole ecosystem, are more difficult to foresee. The low level of understanding of these complex topics has lead to considerable uncertainty. There are a number of strongly interrelated ecological issues with complex feedback effects to be considered (see Walker, 1991). They are driven by the hydrologic, hydraulic, chemical, and thermal environment.

Let us first discuss the composition of flora and fauna. Phytoplankton composition and dominant species are widely used as water quality indicators in limnology. Certain species are an important source of nutrition for zooplankton, and certain zooplankton species regulate the growth of many fish species. Additionally, some phytoplankton species are especially detrimental to water supply and recreation. Those species are most often blue-green algae (cyanobacteria), many of which are able to form nuisance blooms and toxic strains. Growth of blue-green algae is favored under a number of conditions (see Varis, 1993), among which are those relevant to studies of climate change: elevated water temperature, nutrient enrichment, low N:P ratios, low light-energy availability, high pH or low inorganic carbon concentration, resistance to turbulence, and ability to rest in anoxic sediments. When discussing the impact of climate change on the competitive ability of blue-green algae in a lake, shifts in these issues should be considered.

A number of studies have discussed the other biotic communities and food webs in lakes. For instance, zooplankton in Esthwaite Water, Cumbria, UK, is discussed by George *et al.* (1990), and zooplankton in Pääjärvi, Finland, is discussed by Arvola *et al.* (1992). Bottom flora and fauna, macrovegetation, and the fish of the Laurentian Great Lakes are discussed by Hill and Magnuson (1990) and Magnuson *et al.* (1990); similar studies were done by Trippel *et al.* (1991) in Lake Constance, Germany, and by Chang *et al.* (1992) in Douglas Reservoir, Tennessee, USA. Minns and Moore (1992) have made a comprehensive study on fish yields of whitefish, northern pike, and walleye in 406 watersheds in eastern Canada. The results showed considerable redistribution of fishery capability; areas with high productivity will become marginal and areas currently marginal will have optimal conditions for the species considered. Large artificial efforts were suggested to redistribute preferred fish species.

Changes in the length of the growth season, in retention time, and in other control factors may have impacts on the pattern of ecological succession. This is particularly noteworthy in areas with strong seasonality, where seasons are a major reason for succession patterns.

3.5 Impacts on Water Quality Problems

In Section 3.2 (see *Tables 3.1* to *3.3*) we grouped lake water quality problems into eight classes: eutrophication, oxygen depletion, hygienic problems, salinization, acidification, toxic and cumulative substances, turbidity and suspended matter, and thermal pollution. The potential impacts of climate change on the most important external and internal features concerning lake water quality, as described in Sections 3.3 and 3.4, are summarized in *Table 3.4* (see also Chapter 4, particularly *Figure 4.8*). The table includes the directions of interrelationships between those features, indicated as same, inverse, none, or nonuniversal. Many impacts are very case specific, and are a sum of several phenomena; in some cases, essential information is lacking and direction of influence is not universally definable.

What then are the implications of the changes described above for the water quality problems in lakes? *Table 3.5* contains the directions of these impacts, defined again as same, inverse, none, or nonuniversal. The following analysis concentrates on those impacts with nonuniversally definable direction, because the other cases are evident on the basis of Sections 3.3 and 3.4. Many of those relations are due to several processes, and require lake-specific consideration.

3.5.1 Eutrophication

There are several attributes potentially affected by climate change that do not have a universally definable direction within the context of lake eutrophication (*Table 3.4*). The quantity of solar radiation penetrating to the lake influences eutrophication by changing the thermal balance and

Table 3.4. Attributes potentially altered due to climate change and their typical influences on one another.

Influence from	Influence to St†	If	Fl	Dr	El	Rt	Wl	Ox	Ca	Nu	Sa	pH	Gs	Fs†	Bg
Climatic input															
Temperature	○	+	○	○		○	○	○	○		+		+	○	+
Humidity	○		+	−	○	−	+				○			○	
Solar radiation	○	+	○	○				○	○			○	+	○	○
Precipitation	○	○	+	−	+	−	+	○	○	○	−	○		○	
Winds	○	+	○	○	+			○	○	○	○	○		○	○
Thermal and hydraulic															
Stratification† (St)	□	○				○	○	○	○	○		○	○	○	○
Ice-free period (If)	○	□	○		○			○	○	○		○	+	○	+
Hydrologic															
Floods (Fl)	○	+	□	−	+	−	+	○	○	○	−	○	○	○	○
Droughts (Dr)	○	○	−	□	○	+	−	○	○	○	+	○	○	○	○
Erosion, loading (El)			○	○	□			○	○	+	+	○		○	○
Retention time (Rt)	○	○	−		○	□	○	○	○	+	+		+	○	+
Water level (Wl)	○	○	○	−	○	○	□	○	○	−	−			○	-
Chemical															
Oxygen (Ox)					○			□	○	−		○		○	○
Carbon dioxide (Ca)					○			○	□		+	−		○	○
Nutrient enrichment (Nu)								○	○	□	+	○		○	+
Salinity (Sa)	○	+									□			○	○
pH					○			○	○	○	○	□		○	+
Ecological															
Growth-season length (Gs)								○	○	○		○	□	○	+
Foodchains, succession† (Fs)								○	○	○		○		□	○
Blue-green algae (Bg)								○	○	+		+		○	□

+ indicates effect in the same direction, and − indicates effect in inverse direction.
○ indicates impact in indefinite, unknown, or case-specific direction.
No mark indicates no influence.
Nonquantitative attributes are marked with † (cf. *Table 3.5*).

stratification. Such changes are likely to be greater the further away from the equator the lake is located (see Chapter 9). Moreover, the light energy for primary production and the competitive ability of nuisance algae are altered. These both have many indirect effects, and therefore the direction of change is lake specific.

Winds are also a major driving force in the formation of stratification. They mix the water body and increase the DO content of the water.

Table 3.5. Attributes potentially altered due to climate change and their typical impacts on water quality problems. Note that these directions, their impacts, and the priority of the problems are case specific.

	Problem impacted							
Cause	E	O	H	S	A	X	M	T
Climatic input								
Temperature	+	+	○	+				+
Humidity				−				+
Solar radiation	○	+		+			+	+
Precipitation				−	+	○		−
Winds	○	−		+	○	○	○	−
Thermal and hydraulic								
Stratification	○	○	○			○	○	○
Ice-free period	+	○	○				○	+
Hydrologic								
Floods	○	○	+	−	○	○	○	−
Droughts	○	○		+	○		○	+
Erosion, loading	+	+	+	+	○	+	+	
Retention time	+	+	○	+	○	+	○	+
Water level	−	−	○	−	○	−	○	−
Chemical								
Oxygen	−	−	−			−	−	
Carbon dioxide	○	○		+	+			
Nutrient enrichment	+	+	○				+	
Salinity				+	−	○		
pH	+	+	−		−	○	○	
Ecological								
Growth-season length	+	+	○				○	+
Foodchains, succession	○	○	○				○	
Blue-green algae	+	+	+			+	+	
Problems								
Eutrophication (E)	□	+			−	+	+	
Oxygen depletion (O)	+	□				+	+	
Hygiene (H)			□					
Salinization (S)				□	−	○		
Acidification (A)	−				□	+	−	
Toxicity (X)						□		
Turbidity (M)	○	+	+			+	□	
Thermal (T)	+	+	○	+				□

E = eutrophication; O = oxygen depletion; H = hygienic problems; S = salinization; A = acidification; X = toxic and cumulative substances; M = turbidity and suspended matter; and T = thermal problems.

+ indicates effect in the same direction, and − indicates effect in inverse direction.

○ indicates impact in indefinite, unknown, or case-specific direction.

No mark indicates no influence.

Particularly in shallow lakes, the amount of resuspension of sediment is largely controlled by wind. An increase in windiness enhances resuspension. Phytoplankton composition is also sensitive to windiness, because of unequal buoyancy properties of different algae.

Depending on the quality of the floodwater and other input during a drought, eutrophication may be increased or decreased. This change is also associated with stratification. Increased inorganic carbon input to water increases the availability of carbon for primary production, which enhances eutrophication and acidifies the water. Both the availability of carbon, and the acidification are competitive drawbacks to the growth of blue-green algae. The phenomenon is complex and more research is needed.

Increased turbidity and suspended matter act as catalysts in eutrophication if nutrients including carbon are added in forms available to primary producers. Another mechanism enhancing eutrophication is the consumption of oxygen if turbidity is composed of biologically degradable constituents.

The changes in ecological succession and in foodchains can be so manifold that no universal patterns can be detected. This is true of corresponding impacts due to all the other water quality problems, as well.

3.5.2 Oxygen depletion

The problem of oxygen depletion is very closely related to eutrophication, as can be seen from *Tables 3.1* to *3.4*. Regarding impacts with no universal direction in eutrophication, reference is given to Section 3.5.1 concerning stratification, floods, droughts, ecological succession, and foodchains.

The length of a lake's ice-free period defines the length of the growth season. It also greatly influences the stratification and thermocline depth in deep lakes. If the ice-free period is extended, primary production is likely to increase, causing more degradation and oxygen consumption in lower water layers. Concurrently, the volume of the lower water layers may decrease, which enhances DO problems. In contrast, the DO problems may decrease under the ice cover.

Inorganic carbon dissociated in water and DO are closely linked through primary production and degradation of organic matter. With reference to impacts on eutrophication, no universal direction can be posited.

3.5.3 Hygiene

Increased water temperature and nutrient enrichment may contribute to the spread of some water-borne and water-transmitted diseases. This is particularly the case if the lake no longer freezes and the growth season is not suspended by winter. Such conditions introduce the risk for the spread of diseases such as malaria and schistosomiasis. On the other hand, degradation of fecal pollution is accelerated in conditions with high biological activity, i.e., it generally degrades more rapidly in warm water than in cold water, and in nutrient-rich water than in nutrient-poor water. Hygienic pollution is influenced by stratification, retention time, and other corresponding conditions in the basin.

3.5.4 Salinization

Increased salinity in a lake may be due to increased concentration of salts in inflow, or to increased evaporation – and decreased outflow – in the lake itself, or to both. Such changes are most marked in areas or zones where the climate turns from humid to arid; i.e., evaporation becomes greater than precipitation. In arid areas, salinization is a typical quality for all waters.

3.5.5 Acidification

If directions and intensities of winds are changed, the sulfur and nitrogen concentrations of rainwater may be altered. Also, other undefined directions of influence associated with acidification are dominated by hydrology. Depending on the rock and soil in the catchment and on the quality of floodwater, they may contribute to acidification in either direction. In areas with sulfur-rich deposits, changes in ground water level may mobilize large amounts of acidifying sulfur to water.

3.5.6 Toxicity

This group of pollutants is highly diverse chemically, biologically, and ecologically. Therefore, a discussion of the undefined influence direction is more complicated than in previous cases. The composition, distribution, and pathways of these substances are important (Boyce *et al.*, 1991).

3.5.7 Turbidity

The composition of turbidity is also highly case specific. It may be composed of eroded substances that are either inorganically dominated, such as clay, or are organic, such as humic acids. Also, wind and ice erosion must be taken into consideration.

Another type of substance that causes turbidity is organic material produced in water. In many areas, floods dominate the mass flows of eroded material to the water. These substances behave very differently. Typically, however, adsorption–desorption reactions of P, for instance, are sensitive to water pH, so that in decreasing pH, more P is dissolved. On the other hand, increasing pH may increase biological production.

3.5.8 Thermal problems

The prospective directions of influences due to thermal pollution are rather clear. Only stratification constitutes a nonuniversal component. Changes in stratification – induced by other factors described previously – may change the thermal transport and distribution in the water so that the fate and impact of thermal pollution are altered.

3.6 Conclusions

Lakes and reservoirs are complex systems that can be affected by climate change in several ways. Many of the components and processes affected are interrelated and correlated. Impacts are largely, but not totally, lake specific.

Many water quality problems are at risk of worsening because of the climate change, particularly eutrophication, oxygen depletion, salinization, acidification, and thermal problems. Perhaps most typically, the aging processes – such as getting filled up with sediments or plant residues, and increased production – are likely to be accelerated. In comparison with many other geological formations on the earth's crust, lakes have a relatively short life span (Hutchinson, 1957). In this respect, the origin and the history of the lake are crucial. Generalizations and classifications concerning lake type, climatic zone, and altitude are needed, because different lake types have different characteristic problems. In addition to lake-specific studies, more studies are needed on regional and global scales to detect critical regions, critical lakes, or critical lake types concerning specific problematic changes. Such studies have been conducted already, for example, by Minns and Moore (1992) on the fish in eastern Canada; by Meyer *et al.* (Chapter 9) on thermal balances, including stratification, and on ice cover using hypothetical and real lakes located around the globe; and by Hondzo and Stefan (1993) on a number of lakes in Minnesota.

It is not always self-evident which changes in a lake will become most critical to water use. There is a connection between the uses of the lake and the supply–demand conditions of water in the region. Identifying the most relevant problems to be targeted requires thorough, holistic consideration. The checklists presented in *Tables 3.4* and *3.5* have been designed to assist in this procedure. Both the means and the variability of climatic and lacustrine variables must be considered: some impacts are due to shifts in the mean behavior of the hydroclimatic system, but many impacts are introduced by extreme events, or by increased or decreased frequency of such events (Katz and Brown, 1992). Apparently, studies of climate change should be more in line with conventional approaches to water resources management, where extreme design conditions are often the critical criteria. At the ecosystem level, the impacts also will depend on the rate of changes and on the ability of the ecosystems to adapt to these changes (Walker, 1991).

Climate change impact assessment is a task with many aspects. Generally, it is not independent of other impact assessment needs or of development scenarios due to other factors such as pollution trends and changes in land use. Impact assessment is subject to several methodological and philosophical problems (Laurmann, 1991), and to extreme uncertainties, which almost invariably have been discussed and emphasized in the case studies reviewed. Even the probable directions of climate changes, not to mention quantitative estimates of climate change, are often obscure at the level of any specific case. The spectrum of assessment methodology used should be wide, and the choice of the approach should be a conscious one (see Varis, 1994). In our opinion, relying only on simulations, statistics, risk analysis, knowledge-based techniques, or intuition is not the right approach. Clearly, computational techniques that are able to manage extreme uncertainty deserve more attention than they have been given in present studies (see Kundzewicz and Somlyódy, 1993). Methodological research and development should be focused in that direction.

Does current knowledge and understanding of climate change give reason to alter water resources management practices and policies? Many of the hydrological references (see Section 3.3) suggest this, at least for the issue of water quantity. In contrast, in the case studies reviewed in this chapter, water quality management options (see *Table 3.3*) have been discussed far less, especially as concrete recommendations for action. Whether climate change scenarios have been used, or will be used in the near future, for defining rational wastewater treatment level, land use, aeration, or any other water quality management option is currently unclear to us; however, we doubt that this will occur extensively. What is clear is that the currently very high level of uncertainty in water quality management now has an additional uncertainty – the potential impact of climate change. There is evidently a need to integrate, in an adaptive manner, the chance of climate change into water quality management. As suggested by Fiering and Rogers (1989), conventional stationary-climate policy statements should be periodically checked and updated by responsible experts as to the rationality of the assumption of an unchanging climate.

References

Arvola, L., Kankaala, P., Ojala, A., and Tulonen, T., 1992, Impact of climate change on carbon cycle in freshwater ecosystems, in M. Kanninen and P. Anttila, eds., *The Finnish Research Programme on Climate Change, Progress Report*, Valtion Painatuskeskus (VAPK), Helsinki, Finland.

Assel, R.A., 1991, Implications of CO_2 global warming on Great Lakes ice cover, *Climatic Change*, **18**(4):377–395.

Assel, R.A., Snider, C.R., and Lawrence, R., 1985, Comparison of 1982–1983 Great Lakes winter weather and ice conditions with previous years, *Monthly Weather Review*, **113**(3):291–303.

Balling, R.C., 1991, Comments on "The Potential Impacts of Climate Change on the Great Lakes," *Bulletin of the American Meteorological Society*, **72**(6):833–834.

Bárdossy, A., and Plate, E.J., 1992, Space-time model for daily rainfall using atmospheric circulation patterns, *Water Resources Research*, **28**(5):1247–1259.

Boyce, F.M., Schertzer, W.M., Hamblin, P.F., and Murthy, C.R., 1991, Physical behavior of Lake Ontario with reference to contaminant pathways and climate change, *Canadian Journal of Fisheries and Aquatic Sciences*, **48**(8):1517–1528.

Byron, E.R., and Goldman, C.R., 1991, The potential effects of global warming on the primary productivity of a sub-alpine lake, *Water Resources Bulletin*, **26**(6):983–989.

Chang, L.H., Railsback, S.F., and Brown, R.T., 1992, Use of a reservoir water quality model to simulate global climate change effects on fish habitat, *Climatic Change*, **20**(3):277–296.

Changnon, S.A., 1993, Changes in climate and levels of Lake Michigan: Shoreline impacts at Chicago, *Climatic Change*, **23**(3):213–230.

Cohen, S.J., 1991, Possible impacts of climatic warming scenarios on water resources in the Saskatchewan River sub-basin, Canada, *Climatic Change*, **19**(3):291–307.

Cooke, G.D., Welch, E.B., Peterson, S.A., and Newroth, P.R., 1986, *Lake and Reservoir Restoration*, Butterworths, Boston, MA, USA.

Croley, T.E., II, 1990, Laurentian Great Lakes double-CO_2 climate change: Hydrological impacts, *Climatic Change*, **17**(1):27–47.

Dooge, J.C.I., 1992, Hydrologic models and climate change, *Journal of Geophysical Research – Atmospheres*, **97**(ND3):2677–2686.

Fiering, M.B., and Matalas, N.C., 1990, Decision making under uncertainty, in P.E. Waggoner, ed., *Climate Change and U.S. Water Resources*, John Wiley & Sons, New York, NY, USA.

Fiering, M.B., and Rogers, P., 1989, Climate change and water resources planning under uncertainty, Report for US Army Institute for Water Resources, Fort Belvoir, VA, USA.

Frederick, K.D., 1993, Climate change impacts on water resources and possible responses in the MINK region, *Climatic Change*, **24**(1):83–115.

George, D.G., Hewitt, D.P., Lund, J.W.G., and Smyly, W.J.P., 1990, The relative effects of enrichment and climate change on the long-term dynamics of *Daphnia* in Esthwaite Water, Cumbria, *Freshwater Biology*, **23**(1):55–70.

Gulati, R.D., Lammens, E.H.R.R., Meijer, M.-L., and van Donk, E., eds., 1990, *Biomanipulation: Tool for Water Management*, Kluwer, Dordrecht, The Netherlands.

Hartmann, H.C., 1990, Climatic change impacts on Laurentian Great Lakes levels, *Climatic Change*, **17**(1):49–67.

Hill, D.K., and Magnuson, J.J., 1990, Potential effects of global climate warming on the growth and prey consumption of Great Lakes fish, *Transactions of the American Fisheries Society*, **119**(2):265–275.

Hondzo, M., and Stefan, H.G., 1991, Three case studies of lake temperature and stratification response to warmer climate, *Water Resources Research*, **27**(8):1837–1846.

Hondzo, M., and Stefan, H.G., 1993, Regional water temperature characteristics of lakes subjected to climate change, *Climatic Change*, **24**(2):187–211.

Hutchinson, G.E., 1957, *A Treatise in Limnology*, Vol. 1, John Wiley & Sons, New York, NY, USA.

Jacoby, H.D., 1990, Water quality, in P.E. Waggoner, ed., *Climate Change and U.S. Water Resources*, John Wiley & Sons, New York, NY, USA.

Kaczmarek, Z., and Krasuski, D., 1991, Sensitivity of Water Balance to Climate Change and Variability, WP-91-047, International Institute for Applied Systems Analysis, Laxenburg, Austria.

Katz, R.W., and Brown, B.G., 1992, Extreme events in a changing climate: Variability is more important than averages, *Climatic Change*, **21**(3):289–302.

Kauppi, L., Frisk, T., Forsius, M., Huttula, T., Posch, M., Bilaletdin, Ä., Kämäri, J., Kallio, K. Peltonen, A., and Saura, M., 1992, Effects of climatic change, air pollutants and land use on lake ecosystems, in M. Kanninen and P. Anttila, eds., *The Finnish Research Programme on Climate Change, Progress Report*, Valtion Painatuskeskus (VAPK), Helsinki, Finland.

King, D., 1970, The role of carbon in eutrophication, *Journal of the Water Pollution Control Federation*, **42**(12):2035–2051.

Kundzewicz, Z.W., and Somlyódy, L., 1993, Climatic Change Impact on Water Resources: A Systems View, WP-93-30, International Institute for Applied Systems Analysis, Laxenburg, Austria.

Kuusisto, E., 1987, An analysis of the longest ice observation made in Finnish lakes, *Aqua Fennica*, **17**(2):123–132.

Lashof, D.A., 1989, The dynamic greenhouse: Feedback processes that may influence future concentrations of atmospheric trace gases and climatic change, *Climatic Change*, **14**(3):213–242.

Laurmann, J.A., 1991, Global warming and the credibility of science, *Climatic Change*, **18**(1):107–109.

Lettenmaier, D.P., and Gan, T.Y., 1990, Hydrologic sensitivities of the Sacramento–San Joaquin River Basin, California, to global warming, *Water Resources Research*, **26**(1):69–86.

Lettenmaier, D.P., and Rind, D., 1992, Hydrological aspects of global climate change: Preface, *Journal of Geophysical Research – Atmospheres*, **97**(ND3):2675–2676.

Magnuson, J.J., Meisner, J.D., and Hill, D.K., 1990, Potential changes in the thermal habitat of Great Lakes fish after global climate warming, *Transactions of the American Fisheries Society*, **119**(2):254–264.

McCormick, M.J., 1990, Potential changes in thermal structure and cycle of Lake Michigan due to global warming, *Transactions of the American Fisheries Society*, **119**(2):183–194.

Miller, J.R., and Russell, G.L., 1992, The impact of global warming on river runoff, *Journal of Geophysical Research – Atmospheres*, **97**(ND3):2757–2764.

Minns, C.K., and Moore, J.E., 1992, Predicting the impact of climate change on the spatial pattern of freshwater fish yield capability in eastern Canada, *Climatic Change*, **22**(4):327–346.

Moore, L., and Thornton, K., eds., 1988, *Lake and Reservoir Restoration Guidance Manual*, US Environmental Protection Agency, 440/5-88-002, Corvallis, OH, USA.

Orlob, G.T., Meyer, G.K., De George, J., and Christensen, C.L., 1990, Impacts of global warming on water quality in river-reservoir systems, *Journal of Hydrologic Science and Technology*, **5**(1–4):10–19.

Quinn, F.H., and Croley, T.E., 1983, Climatic basin water balance models for Great Lakes forecasting and simulation, *5th Conference on Hydrometeorology*, Tulsa, Oklahoma, American Meteorological Society, Boston, MA, USA.

Rannie, W.F., 1983, Breakup and freeze-up of the Red River at Winnipeg, Manitoba, Canada, in the 19th Century and some climatic implications, *Climatic Change*, **5**(3):283–296.

Robertson, D.M., Ragotzkie, R.A., and Magnuson, J.J., 1992, Lake ice records used to detect historical and future climate changes, *Climatic Change*, **21**(4):407–427.

Ruttner, F., 1940, *Grundrisse der Limnologie*, Walter de Gruyer & Co., Berlin (English translation: Fundamentals of Limnology, 3rd ed. 1963, University of Toronto Press, Toronto and Buffalo).

Schindler, D.W., Beaty, K.G., Fee, E.J., Cruikshank, D.R., DeBruyn, E.R., Findley, D.L., Londsey, G.A., Sherer, J.A., Stainton, M., and Turner, M.A., 1990, Effects of climatic warming on lakes of the central boreal forest, *Nature*, **347**:967–970.

Shapiro, J., 1973, Blue-green algae: Why they become dominant, *Science*, **179**:382–384.

Shapiro, J., 1990, Current beliefs regarding dominance by blue-greens: The case of importance of CO_2 and pH, *Verhandlungen des internationalen Vereins für Limnologie*, **24**(1):38–54.

Smith, J.B., 1990, From global to regional climate change: Relative knowns and unknowns about global warming, *Fisheries*, **15**(6):2–6.

Smith, J.B., 1991, The potential impacts of climate change on the Great Lakes, *Bulletin of the American Meteorological Society*, **72**(1):21–28.

Stumm, W., and Morgan, J.J., 1981, *Aquatic Chemistry: An Introduction Emphasizing Chemical Equilibria in Natural Waters*, 2nd ed., John Wiley & Sons, New York, NY, USA.

Szilágyi, F., and Somlyódy, L., 1991, Potential impacts of climatic changes on water quality in lakes, in G. Kienitz, ed., *Hydrological Interactions Between Atmosphere, Soil and Vegetation: Proceedings of the International Union of Geodesy and Geophysics*, 11–24 August, at Vienna, Austria, IAHS Press, Institute of Hydrology, Wallingford, UK.

Thomas, G., 1990, The regional hydrologic impacts of global climate change: The role of climate models, *Global and Planetary Change*, **82**(3–4):343–368.

Trippel, E.A., Eckmann, R., and Hartmann, J., 1991, Potential effects of global warming on whitefish in Lake Constance, Germany, *Ambio*, **20**(6): 226–231.

Varis, O., 1988, Temporal sensitivity of *Aphanizomenon flos-aquae* dominance: A wholelake simulation study with input perturbations, *Ecological Modelling*, **43**(2):137–153.

Varis, O., 1989, Simulated impacts of flow regulation on blue-green algae in a short retention time lake, *Archiv für Hydrobiologie, Beihefte Ergebnisse der Limnologie*, **33**(1):181–189.

Varis, O., 1993, Cyanobacteria dynamics in a restored Finnish lake: A long-term simulation study, *Hydrobiologia*, **268**(1):129–145.

Varis, O., 1994, Water quality models: Tools for the analysis of data, knowledge, and decisions, *Water Science and Technology*, **30**(2):13–19.

Walker, B.H., 1991, Ecological consequences of atmospheric and climate change, *Climatic Change*, **18**(2–3):301–316.

Wetzel, R.G., and Grace, J.D., 1983, Aquatic plant communities, in E.R. Lemon, ed., *The Response of Plants to Rising Levels of Atmospheric Carbon Dioxide*, AAAS Selected Symposium Series 84, Westview Press, Boulder, CO, USA.

Whitehead, P., and Jenkins, A., 1989, *Climate Change and Water Quality*, Institute of Hydrology, Wallingford, UK [mimeo].

Williams, P., 1989, Adapting water resources management to global climate change, *Climatic Change*, **15**(1–2):83–93.

Williamson, P., and Gribbin, J., 1991, How plankton change the climate, *New Scientist*, **129**(1760):48–52.

Chapter 4

Impacts of Climate Change on Water Quality

Gerald T. Orlob, Gabriela K. Meyer, László Somlyódy, Danuta Jurak, and Karolý Szesztay

4.1 Introduction

Relatively little attention has been given to date to the implications of climatic changes on the quality of natural and artificial water bodies (e.g., streams, lakes, reservoirs, and estuarine systems, and their indigenous ecosystems), which are intimately linked to the circulation and distribution of water on earth and to atmospheric processes. The fates of aquatic ecosystems, which in many cases are already stressed near the threshold limits of temperature, dissolved oxygen, salinity, and toxicity (see the discussion in the previous chapter) have yet to be carefully examined in a quantitative framework to reveal the potential impacts of global climate change.

In this chapter we explore the climatic factors governing the quality of natural and artificial freshwater bodies and the viability of the life forms they support. Interactions between aquatic ecosystems and the landscape are also examined. We will attempt to identify relevant literature addressing the relationships between climate and water quality. We will present some fundamental relationships governing exchanges of energy and mass between the atmosphere and the hydrosphere, and we will illustrate them using simple examples. We will endeavor to set forth some preliminary methods and models for quantitative assessment of the uncertainties imposed on the thermal budget and water quality of impoundments by potential changes in

Z. Kaczmarek et al. (eds.),
Water Resources Management in the Face of Climatic/Hydrologic Uncertainties 70–105.

climate, whether occurring naturally or introduced anthropogenically. Finally, potential areas for future research on climate effects will be outlined.

4.2 Climate, Landscape, Water Quality, and Aquatic Ecosystems

Aquatic ecosystems are the joint products of climate and landscape. Climate changes drive global, regional, and local hydrological and biogeochemical cycles; landscape sets the boundary conditions for the formation and evolution of aquatic ecosystems. Understanding and quantification of the socially significant impacts of climate change require analysis of the interactions among five major groups of factors and processes:

- The macroscale and regional setting, which defines the distribution and quantity of water within drainage areas and the surrounding landscape.
- The flux of water, thermal energy, sediment, and the major biogeochemical elements into or out of aquatic systems.
- The distribution of water, thermal energy, sediment, and biogeochemical elements within the aquatic environment by physical, chemical, biochemical, and biological processes.
- Interactions among aquatic plant and animal species and the environmental factors that govern viability.
- Specific anthropogenic influences, either inadvertent or deliberate, that change the quality of the aquatic environment.

Long-range planning and policy formulation for integrated hydrologic basin development and management require identification of those components of aquatic environments that are of intrinsic social and economic value and that must be preserved, or that represent future potentials that might be mobilized. Quantification of climatic impacts on these components (e.g., indigenous fisheries, recreation, water supply, etc.) requires specific knowledge of interactions between climate-affected hydrologic and water quality processes and methods for assessment of the consequences of change and the associated uncertainties. With regard to methodology requirements, some of the important issues that must be addressed are posed in the following questions:

- How reliable are the tools (e.g., models, statistical methods, etc.) for prediction of future consequences?
- To what degree are the uncertainties inherent in these tools propagated or magnified through successive phases of the assessment process?

- How are these uncertainties related to other non-climate-induced risks or uncertainties?
- What are the critical levels of the environmental state variables that control the fate of aquatic ecosystems and that foretell catastrophic collapse of ecosystems?

With these goals in mind, let us turn now to the specific structure of the aquatic system and its interrelationship with climatic factors.

Coupling of atmospheric (climatic) water mass, thermal energy, and water quality states with those same states of the terrestrial hydrosphere is illustrated schematically in *Figure 4.1*. The representation builds on the conceptualization, presented by Eagleson (1982), of the dynamic coupling of land surface and atmospheric processes. In this case, however, the system is extended to two dimensions and includes the elements of water quality. The system shown identifies the principal reservoirs of water mass (liquid, solid, and gas), thermal energy, and water quality, and the major processes governing exchanges of these attributes within and between four interrelated aquatic environments: a snow shed, a vegetated soil horizon, a ground water system, and a surface water system. The schematic is considered two-dimensional in space and applicable to the nonsteady case. Air quality and water quality are shown as being linked to reservoirs of water mass. The influences of the various exchange processes (e.g., precipitation, evaporation, net radiation, sensible heat, conduction from the ground, etc.) are indicated as being either unidirectional, feedback, or reversible by the direction of the arrow head(s) on the flux pathways.

It is apparent from the schematic of *Figure 4.1* that perturbation of any of the principal atmospheric variables, whether due to natural causes or induced anthropogenically, has the potential to change water quality in certain parts of the system. For example, a change in atmospheric temperature, possibly due to an increase in the concentration of CO_2 in the atmosphere, might be expected to increase net long wave radiation to the earth's surface, to elevate the snow line in mountainous areas, to increase the melt rate, and to advance the onset of the melt. The shift in the quantity and timing of the flux of cold water coupled with a likely increase in surface water temperature in equilibrium with the atmosphere will change the spatial distribution and the total supply of thermal energy in a downstream reservoir. Thus, it may be expected that the hydromechanical and water quality behavior of the impoundment, both of which are dependent upon temperature, will also be altered.

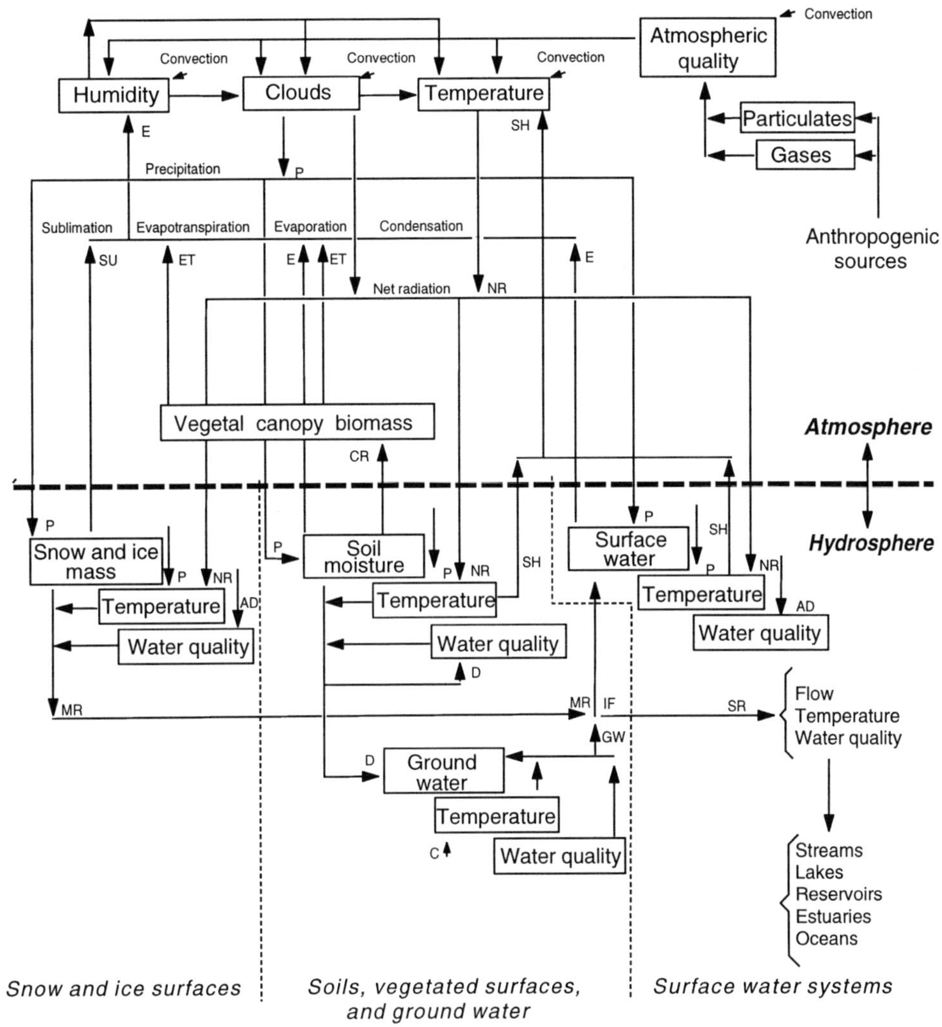

Legend: AD - atmospheric deposition
C - conduction
CR - capillary rise
CV - convection
D - drainage water
E - evaporation
ET - evapotranspiration
GW - ground water runoff
IF - interflow
MR - melt water return
NR - net radiation
P - precipitation
SH - sensible heat
SR - surface runoff
SU - sublimation

Figure 4.1. Atmosphere–hydrosphere interactions relative to effects of climate change on water quality.

The quality of water in natural systems is generally defined as a vector of state variables, including certain physical attributes (e.g., density, viscosity, surface tension, etc.) and concentrations of substances (conservative and nonconservative, dissolved and suspended, inorganic and organic, abiotic and biotic, and mineral and gaseous), that may be identified as intrinsically belonging to a sample of the water in one of its three physical states – liquid, solid, or gas. Changes in water quality are induced by exogenous forces, primarily exchanges in water mass and thermal energy associated with the hydrologic cycle, and by physical, chemical, and biological interactions among the various state variables. Depending on the number and type of state variables to be considered, water quality models range from being very simple to being exceedingly complex. Interactions between systems of differing physical environments (e.g., streams, lakes, ground water, soil moisture, etc.) add to the complexity.

As *Figure 4.1* implies, the interactions between atmospheric, hydrologic, and water quality variables are exceedingly complex. Processes are interrelated, subsystems are nonlinear, relationships between variables are often empirical, and parameters are inherently uncertain. Unfortunately, there currently are no models capable of simulating all of the processes of a system as intricate as the one conceptualized in *Figure 4.1*. However, useful models of some of the components of this system are available (WRE, 1967; Huber *et al.*, 1972; Chen and Orlob, 1975; Orlob, 1984, Jorgensen and Gromiec, 1989) and may be suitable for preliminary studies of climate change effects. Among the best documented of these are the stream model, QUAL 2E (Brown and Barnwell, 1987), and the water quality for river-reservoir systems model, WQRRS (HEC, 1978), which simulate some of the key interactions between the atmosphere and the aquatic system. The potential application of such models as tools for the assessment of climate change effects is explored in the following section.

4.3 Impact of Climate on Energy Budget and Water Temperature

4.3.1 Net heat exchange at the air–water interface

To illustrate the interactions between the atmosphere and the hydrosphere, let us examine the comparatively simple case of a stream flowing from its origin in a mountain catchment to an impoundment, as more or less suggested by the schematic in *Figure 4.1*. First, consider those processes by which the thermal energy content of the stream is modified along its route. According to the conservation of energy principle, the net change in thermal energy in a control volume of the flowing stream is represented as the sum of net heat exchange at the air–water interface, net heat conduction through the soil–water interface, heat energy advected to or from the control volume, and any explicit sources or sinks of heat.

The evolution of thermal energy in a one-dimensional flowing stream can be described in the following way:

$$c_p \rho h \left(\frac{\partial \theta}{\partial t} + \overline{u} \frac{\partial \theta}{\partial x} \right) = q_n + q_g \qquad (\mathrm{W\ m^{-2}})^* \ , \tag{4.1}$$

where h is mean depth of flow (m); $\overline{u}$ is mean velocity (m/min); θ is temperature (°C); c_p is specific heat (W·min/g·°C); ρ is density (g/cm^3); x is distance (m); t is time (min); q_n is net heat exchange at the air–water interface (W m^{-2}); and q_g is net heat exchange at the soil–water interface (W m^{-2}).

The term $\overline{u}\frac{\partial \theta}{\partial x}$ represents advected fluxes of heat energy along the x-axis. Longitudinal dispersion is omitted if it is assumed that gradients are very small in the direction of flow. The control volume for which heat energy balance is achieved is represented by the depth h, with a unit horizontal surface area. Sources or sinks are omitted from the formulation. The influence of climatic variables on the heat energy balance is revealed in the descriptions of the components of q_n and q_g, heat exchanges at the air–water and soil–water interfaces, respectively.

Consider first the exchange of thermal energy between the atmosphere and the hydrosphere. The net rate of thermal energy exchange through an air–water interface can be represented as the sum of surface heat fluxes, as given by

$$q_n = q_{sn} + q_{at} - q_{ws} - q_e - q_h \qquad (\mathrm{W\ m^{-2}}) \ , \tag{4.2}$$

where q_{sn} is net solar (short wave) radiation flux; q_{at} is net atmospheric (long wave) radiation flux; q_{ws} is water surface radiation flux; q_e is evaporative latent heat flux; and q_h is sensible (conductive) heat flux.

Each of the heat flux terms in equation (4.2) is affected in some way by some properties of the atmosphere. Several are dependent on properties of the hydrosphere (e.g., water temperature), and some are interrelated. Together they form a complex nonlinear model of the rate of thermal energy exchange through the air–water interface. To appreciate the linkages

*1 langley (ly) = 1 calorie per cm^2, another unit of thermal energy flux 1ly/min = 697.7 W m^{-2}.

between these terms and their dependence upon specific atmospheric observations, it is necessary to examine each in some detail.

Explicit quantitative expressions for the several flux terms in equation (4.2), can be found in several sources [see Brutsaert (1982), Sellers (1965) or Tennessee Valley Authority (TVA, 1972)]. Of these investigations, perhaps the one used most extensively for water quality modeling, and one in which interdependent processes can be easily identified, is that of the TVA, from which the following formulations had been derived.

Net Solar (Short Wave) Radiation (q_{sn})

Net solar radiation penetrating a water surface is the residual of radiation received at the top of the earth's atmosphere after this flux is reduced by scattering and absorption in the atmospheric column, by interception and absorption by clouds, and by reflection at the water surface, such that

$$q_{sn} = q_0 f(A)[1 - 0.65C^2][1 - R_{ws}] \quad (\mathrm{W\ m^{-2}}) \ , \tag{4.3}$$

where q_0 is radiation intensity of the solar beam at the top of the earth's atmosphere; $f(A)$ is the atmospheric attenuation factor, a function of optical properties of the air column, dust content, gaseous composition, moisture content, and terrestrial surface reflectance; C is cloudiness in decimal fractions of the sky covered, $0 < C < 1.0$ (the empirical factor 0.65 was derived by TVA as best characterizing the ratio of cloudy- to clear-sky radiation as a function of cloudiness); and R_{ws} is reflectivity of the water surface, a function of cloud cover and solar altitude, $0.02 < R_{ws} < 0.06$.

Solar radiation at the top of the earth's atmosphere for a given location and time is given by

$$q_0 = \frac{I_0}{r^2} \sin \alpha \quad (\mathrm{W\ m^{-2}}) \ , \tag{4.4}$$

where I_0 is a solar constant, equal to the incident solar radiation at the top of the earth's atmosphere for a mean distance between the earth and the sun, $I_0 \cong 1395.4$ W m^{-2} (2 ly/min); r is radius vector, the ratio of the actual earth–sun distance to the annual mean, a function of time of year varying from 0.9833 to 1.0167; and α is solar altitude in radians, a function of latitude ϕ, the sun's declination δ, and the hour angle of the sun (h), $0 < \alpha < \pi/2$.

Equation (4.4) gives the instantaneous value of q_0 corresponding to the altitude of the sun, defined by

$$\sin \alpha = \sin \phi \sin \delta + \cos \phi \cos \delta \cos h \ , \tag{4.5}$$

where ϕ is latitude of location, in radians; δ is declination of the sun, in radians, $23.45\,(\pi/180)\cos\{2\pi/^{365}(172-D)\}$; D is the Julian calendar day; and h is local hour angle of the sun in radians (measured westward from the upper celestial meridian).

Because $q_0 \geq 0$ only during daylight hours, the daily mean of extraterrestrial radiation is obtained by integrating equation (4.3) from sunrise to sunset, i.e., between hour angles of h_{sr} and h_{ss}. Generally, it is assumed that $\alpha = 0$ at sunrise and sunset (unless there are obstructions), so that $h_{ss} = h_{sr}$, and

$$h_{ss} = \arccos\left[\frac{-\sin\phi\sin\delta}{\cos\phi\cos\delta}\right] \quad . \tag{4.6}$$

By integrating equation (4.4) over the interval from sunrise to sunset, as indicated by equations (4.5) and (4.6), mean daily extraterrestrial solar radiation becomes

$$\bar{q}_0 = \left[\frac{1440}{\pi}\right]\left[\frac{I_0}{r^2}\right][h_{ss}\sin\phi\sin\delta + \cos\phi\cos\delta\cos h_{ss}] \quad (\mathrm{W\ m^{-2}}) \quad . \tag{4.7}$$

Atmospheric Radiation (q_{at})

Atmospheric radiation at the water surface is the net long wave flux from the overlying atmospheric air mass, including clouds, after reflection from the water surface. It is a function of absolute air temperature in accordance with the Stefan-Boltzman law, which gives the radiation of a "gray" body atmosphere by

$$q_r = \varepsilon\sigma T_a^4 \quad (\mathrm{W\ m^{-2}}) \ , \tag{4.8}$$

where ε is emissivity, the ratio of actual long wave radiation of the air mass to that of an ideal "black" body; σ is the Stefan-Boltzman constant $= 5.669 \times 10^{-8}$ W m^{-2} °K^{-4} (8.125×10^{-11} ly min^{-1} °K^{-4}); and T_a is the absolute temperature of the radiating air mass (°K).

Swinbank (1963) proposed representing emissivity as a function of air temperature alone. His empirical observations indicated that

$$\varepsilon = 6.538 \times 10^{-3} T_2{}^2 (0.937 \times 10^{-5} T_2{}^2) \quad (\mathrm{W\ m^{-2}}) \ , \tag{4.9}$$

where T_2 is absolute temperature ($T_2 = 273.2 + \theta_2$, with θ_2 in °C), usually at a height of two meters above the water surface (°K).

For $T_2 = 293$°K, ($\theta_2 = 20$°C), the air mass emissivity according to Swinbank is about 0.80. Among the formulations of 12 different researchers reviewed by TVA (1972), values of ε ranged between 0.74 and 0.81.

The presence of clouds provides an additional source of radiation from water and ice particles, the magnitude of which depends on the elevation, composition, and cover of cloud formations. Following the approach of Bolz (1949), the added effect of clouds is proportional to the square of cloudiness:

$$q_{at} = q_r(1 + kC^2) \qquad (\mathrm{W\ m^{-2}}) \ , \tag{4.10}$$

where k ranges between 0.04 and 0.24, depending on the type of cloud formation. The TVA (1972) recommends a value of 0.17 for general application.

Because a water surface is practically opaque to long wave radiation (virtually all of the radiation being absorbed in the first centimeter below the surface), the emissivity of the water surface is very high and reflectivity is very low; the reflectivity (R_a) is probably not more than 3%.

Incorporating the Swinbank relation, the cloudiness effect given by Bolz, and the reflectivity at the water surface, a general relation for net atmospheric radiation flux from a cloudy sky through the air–water interface becomes

$$q_{at} = 6.538 \times 10^{-3} \sigma {T_a}^6 (1 + 0.17C^2)(1 - R_a) \qquad (\mathrm{W\ m^{-2}}) \ . \tag{4.11}$$

Water Surface Radiation (q_{ws})

Water surface radiation "back" to the atmosphere follows the Stefan-Boltzman law:

$$q_{ws} = \varepsilon \sigma T_{ws}^4 \qquad (\mathrm{W\ m^{-2}}) \ , \tag{4.12}$$

where T_{ws} is temperature of water surface (°K) = 273.2 + θ_0; θ_0 is average water surface temperature (°C); σ is the Stefan-Boltzman constant [5.669 $\times 10^{-8}$ W m^{-2} °K^{-4} (8.125 $\times 10^{-11}$ ly/min °K^{-4})]; and ε is emissivity $\cong$ 0.97 for a water surface.

Evaporative Heat Flux (q_e)

Evaporative heat flux occurs as a result of the change of the state of water from a liquid to a gas (water vapor), sacrificing the latent heat of vaporization:

$$q_e = C \rho_w L_v E \qquad (\mathrm{W\ m^{-2}}) \ , \tag{4.13}$$

where ρ_w is the density of water (g/cm^3); L_v is latent heat of vaporization = 597.1–0.57 θ_0 (cal/g); E is the net evaporation (condensation) rate (cm^3/cm^2·min), and C = 697.7 W/m^2.

A key to the estimation of the evaporative heat flux is the determination of the net evaporation rate E. This quantity can be measured directly as a meteorological quantity using standardized evaporation pans, it can be estimated by water or heat energy budget, or it can be evaluated by empirical relationships, which include the fundamental variables governing mass transfer of water across the air–water interface. The empirical relationship most commonly used in temperature prediction models such as QUAL 2E (Brown and Barnwell, 1987) and WQRRS (HEC, 1978) is of the general formulation

$$E = (a + b\overline{u})(e_0 - e_a) \qquad (\mathrm{cm}^3/\mathrm{cm}^2 \cdot \mathrm{min}) \ , \tag{4.14}$$

where $\overline{u}$ is wind velocity (km/day); e_0 is saturation vapor pressure at water surface temperature (mb or cm Hg); e_a is actual water vapor pressure, usually measured at a height of two meters above water surface = saturation vapor pressure at dew point (mb or cm Hg); and a and b are site-specific empirical constants.

An alternative form of equation (4.14) that takes into account the stability of the overlying air mass (determining the rate of convection of saturated air away from the water surface) is

$$E = N\frac{P}{P_0}\overline{u}(c_0 - c_a)f(R_i) \qquad (\mathrm{cm}^3/\mathrm{cm}^2 \cdot \mathrm{min}) \ , \tag{4.15}$$

where

$$\frac{P}{P_0} = (1 - 0.0226Z)^{5.256} \qquad (\mathrm{cm}^3/\mathrm{cm}^2 \cdot \mathrm{min}) \ \text{ and} \tag{4.16}$$

P is atmospheric pressure (mb); P_0 is standard atmospheric pressure at sea level = 1013 mb; $\overline{u}$ is mean wind velocity (km/day); Z is altitude in meters above sea level; c_0 and c_a are water vapor concentrations (kg vapor/kg air); $f(R_i)$ is function of Richardson's number, a ratio of potential energy to kinetic energy in the air column, usually represented by empirical functions derived for specific geographic locations; and N is an empirical constant.

Jurak (1985) used the heat energy budget approach to estimate E for shallow ponds in Poland where the exchange of energy through the pond bottom had already been estimated. Estimation of heat fluxes through the ground–water interface is essentially based on empirical evidence from field observations.

Sensible Heat Flux (q_h)

Sensible heat is transferred between air and water by conduction and is transported away from the interface by convection in the air mass. It is a function of the difference in temperatures of water and air:

$$q_h = \frac{k_h}{k_w} f(R_i) \rho_w c_p N \frac{P}{P_0} \overline{u} (\theta_0 - \theta_a) \qquad (\mathrm{W\ m^{-2}}) \ , \tag{4.17}$$

where c_p is specific heat of air at constant pressure (W·min/g°C); θ_0 is water surface temperature (°C); θ_a is dry bulb air temperature (°C), usually measured two meters above the water surface; and k_h/k_w is the ratio of turbulent heat and mass diffusivities, usually assumed = 1.0. Other terms are as previously defined.

An alternative to equation (4.17) that is commonly employed in hydrologic heat budget analysis is to relate sensible heat flux to evaporative flux. The proportionality is expressed as

$$q_h = B q_e \qquad (\mathrm{W\ m^{-2}}) \ , \tag{4.18}$$

where B is Bowen's ratio $= 0.66 \left[\frac{\theta_0 - \theta_a}{e_0 - e_a}\right] \frac{P}{P_0}$, and all terms are as previously defined.

Net Heat Exchange at Water Surface (q_n)

Combining the flux terms results in the following expression for net heat exchange at the air–water interface:

$$\begin{aligned} q_n \ = \ & \left[q_0 f(A)[1 - 0.65C^2](1 - R_{ws})\right] + \\ & \text{short wave radiation} \\ & \left[6.538 \times 10^{-3} \sigma T_a{}^6 (1 + 0.17C^2)(1 - R_a)\right] + \\ & \text{atmospheric radiation} \\ & [\varepsilon \sigma T_0{}^4] + \\ & \text{water surface radiation} \\ & [\rho L_v N \overline{u_a} (e_0 - e_a)] + \\ & \text{evaporative heat flux} \\ & 0.66 L_v N \overline{u_a} (\theta_0 - \theta_a) P/P_0 \qquad (\mathrm{W\ m^{-2}}) \ . \\ & \text{sensible heat flux} \end{aligned} \tag{4.19}$$

For a specific geographic location, altitude, and time, net heat exchange is seen to be a complex function of atmospheric properties, e.g., air temperature, vapor pressure, wind velocity, cloudiness, air pressure, and temperature of the water surface. Given observations of atmospheric quantities and values of empirical and universal constants, the value of q_n can be estimated for a specified water temperature, θ_0. Also, equation (4.19) provides a direct means of assessing the potential change in heat exchange and the consequences to water temperature of changing climatic conditions. Before demonstrating the application of equation (4.19), however, let us consider an additional important heat flux associated with the soil–water interface.

4.3.2 Net heat exchange at the soil–water interface

As can be inferred from the schematic of *Figure 4.1*, heat can be transferred to or from a surface water body through the boundary with the ground. The rate of exchange is proportional to the gradient of temperature in the water-saturated soil or the solid medium near the interface, thus the ground–water heat flux can be represented as

$$q_c = k_g \frac{(\theta_0 - \theta_g)}{\Delta z} \qquad (\mathrm{W\ m^{-2}}) \ , \tag{4.20}$$

where θ_0 is the water temperature at the interface (°C); θ_g is the water-saturated soil temperature at depth Δz (m) below the interface (°C); and k_g is the conductivity coefficient (W m^{-1} °C^{-1}).

A more convenient practical formulation is to relate the heat flux to only the temperature differential between the water column above the interface and the saturated soil below the interface:

$$q_c = K_g(\theta_0 - \theta_g) \qquad (\mathrm{W\ m^{-2}}) \ , \tag{4.21}$$

where $K_g = k_g/\Delta z$ (W m^{-2} °C^{-1}).

The value of K_g can be estimated from observations of ground water and surface water temperatures using the equilibrium temperature concept (as described in the following section). The ground heat flux q_g can be adjusted for the differences in water surface and wetted perimeter, if necessary, to allow it to be added to the net water surface flux, q_n. Ground water temperature is perhaps the most difficult variable in the heat budget equation to measure directly. Reasonable estimates can be obtained from observations of well waters adjacent to the stream and supplied by the common gravity aquifer. The conductivity coefficient K_g can be derived from the properties of the saturated porous media following the procedure outlined by Farouki (1982).

4.3.3 Equilibrium energy and water temperature models

Although each of the processes contributing to the thermal energy balance of a surface water body is dynamic at the usual scales of interest, the cumulative effect of all the fluxes is to drive the thermal energy balance toward equilibrium, i.e., a state where the net exchange is zero. At a particular time when the exogenous fluxes q_{sn} and q_{at} are known, there is a unique water temperature that determines the values q_{ws}, q_e, q_h, and q_g necessary to balance the energy budget equations, so that

$$q_{ng}(\theta_e) = q_n(\theta_e) + q_g(\theta_e) = 0 \quad . \tag{4.22}$$

This temperature is the so-called equilibrium temperature, θ_e.

Consider the case of a stream at steady state, i.e., at constant discharge. The temporal rate of change of temperature θ of a unit mass of water moving downstream is

$$c\rho h \frac{\partial \theta}{\partial t} = \frac{\partial q_{ng}}{\partial \theta}(\theta - \theta_e) = k(\theta - \theta_e) \qquad (\mathrm{W\ m^{-2}}) \ , \tag{4.23}$$

where k characterizes the slope of the relationship between q_{ng} and θ. In the region close to $q_{ng} = 0$, where $\theta \cong \theta_e$, the equilibrium temperature, k may be considered nearly constant. A solution of equation (4.23) for a constant k is

$$\theta_t = \theta_e + (\theta_1 - \theta_e) \exp\left[\frac{-kt}{c\rho h}\right] \qquad (^\circ\mathrm{C}) \ , \tag{4.24}$$

where the variables and properties are as previously defined.

Therefore, under the assumption of steady state hydraulic conditions, the water temperature of the stream tends toward the equilibrium temperature with time (or distance) from the source.

For a given site, date, and time of day, and for specific atmospheric conditions given by $C, T_a, \theta_a, \overline{u}$, and e_a, as defined in equations (4.7) to (4.14), the net heat exchange at the air–water interface is a complex function of water surface temperature, i.e., $q_n = f(\theta_o)$. Also, the heat conduction flux through the soil–water interface q_g depends on water temperature, but it also requires predetermination of θ_g, the ground water temperature (assumed to be identical to the temperature of the soil water matrix), and the conductivity coefficient K_g.

Given estimates of q_n and q_g, it is possible to generate the relationship $q_{ng} = f(\theta_o)$ and determine q_e for a given location and time.

A similar approach to that described above was developed by Jurak (1978) for the energy budget of shallow lakes with negligible inflow or outflow. Equating the rate of heat energy change in a lake of depth $\overline{h}$ (m) to the sum of heat exchange fluxes (using the notation and definitions presented earlier), Jurak's equation for the energy budget is

$$c\rho\bar{h}\,\frac{\partial\overline{\theta}}{\partial t} = q_{sn} + q_{at} - q_{ws} - q_e - q_h + q_g \qquad (\mathrm{W\ m^{-2}}) \ . \tag{4.25}$$

The right-hand side of equation (4.25) is a function of the mean water temperature $\overline{\theta}$ in the lake, assuming complete mixing. To determine the net rate of heat exchange, it is convenient to represent the heat flux–temperature relationship as quadratic:

$$q_{ng} = c\rho\bar{h}\,\frac{\partial\overline{\theta}}{\partial t} = \alpha + \beta\overline{\theta} + \gamma\overline{\theta}^2 \qquad (\mathrm{W\ m^{-2}}) \ . \tag{4.26}$$

The coefficients α, β, and γ can be obtained by a quadratic least squares fit or can be evaluated by the method of Jurak (1985). In general, the solution of equation (4.26) provides a dynamic description of temperature change as the water body interacts with the atmosphere and the ground. When there is no net change in the temperature, the equilibrium temperature θ_e is found by setting $q_{ng} = 0$ and solving the quadratic for the positive root,

$$\theta_e = -\frac{\beta}{2\gamma} + \left[\left(\frac{\beta}{2\gamma}\right)^2 - \frac{\beta}{\gamma}\right]^{1/2} \qquad (^\circ\mathrm{C}) \ , \tag{4.27}$$

for $\theta_e > 0$.

An example of the equilibrium temperature model concept is presented to demonstrate the method and the utility of the simple model in the investigation of the effects of climate change on water quality.

The case of interest is a hypothetical situation for a northern temperate zone in which it is necessary to estimate the influence of climate change on the temperature of a stream system that constitutes the principal inflow to an impoundment. In this case, it is assumed that the time of flow, rate of discharge, and atmospheric conditions are such that, over periods longer than a few days, the mean daily temperature of the water reaches equilibrium by the time it reaches the reservoir. The goal of this exercise is to assess the change in equilibrium temperature due solely to an increase in mean atmospheric temperature resulting from climate change.

The stream system considered in this example is located in a mountainous area of northern California. It is assumed that the temperature of the principal inflow to a major reservoir has reached its equilibrium value during its transit from its sources to the reservoir in mid-summer, say on

July 1. The questions to be resolved are: What is the equilibrium temperature under present atmospheric conditions? And what change in equilibrium temperature can be expected as a result of a 4°C increase in atmospheric temperature predicted by a GCM?

The conditions prevailing at the site are described as follows:

Geographic Location

Location:	Northern California
	longitude = 120°W; latitude = 42°N; $\phi = 42^{\circ}$.
Altitude:	$z = 500$ meters above sea level.
	Atmospheric pressure $p = 958$ mb.
Time:	July 1.
	Declination of the sun on July 1:
	$\delta = 23^{\circ}\ 07'$.

Atmospheric Conditions, Base Case

Atmospheric short wave attenuation	$f(A) = 0.8$.
Cloudiness	$C = 0.3$.
Mean wind velocity	$\overline{u} = 300$ km/day.
Mean daily air temperature	$\theta_a = 25$°C.
Lake Hefner constant	$N = 0.0003$.
Relative humidity	= 50%.

Atmospheric Conditions, Climate Change Case

Mean daily air temperature	$\theta_a = 25 + 4 = 29$°C.

Additionally, the ground temperature was observed to be 10°C for the base case, but was assumed to increase to 11°C for the climate change case, allowing for feedback between atmosphere and hydrosphere. (Note: In this case, a more rigorous approach would require an integrated model of surface water and ground water systems.) The result of applying the equilibrium model [equation (4.22)] under these assumptions is illustrated in *Figure 4.2.* Net heat exchange as a function of water temperature over a range from 0°C to 40°C was calculated for both the base case and the climate change case.

Typical patterns of $q_{ng} = f(\theta_0)$ are noted in both the base case and climate change case. The indicated equilibrium temperatures for both were

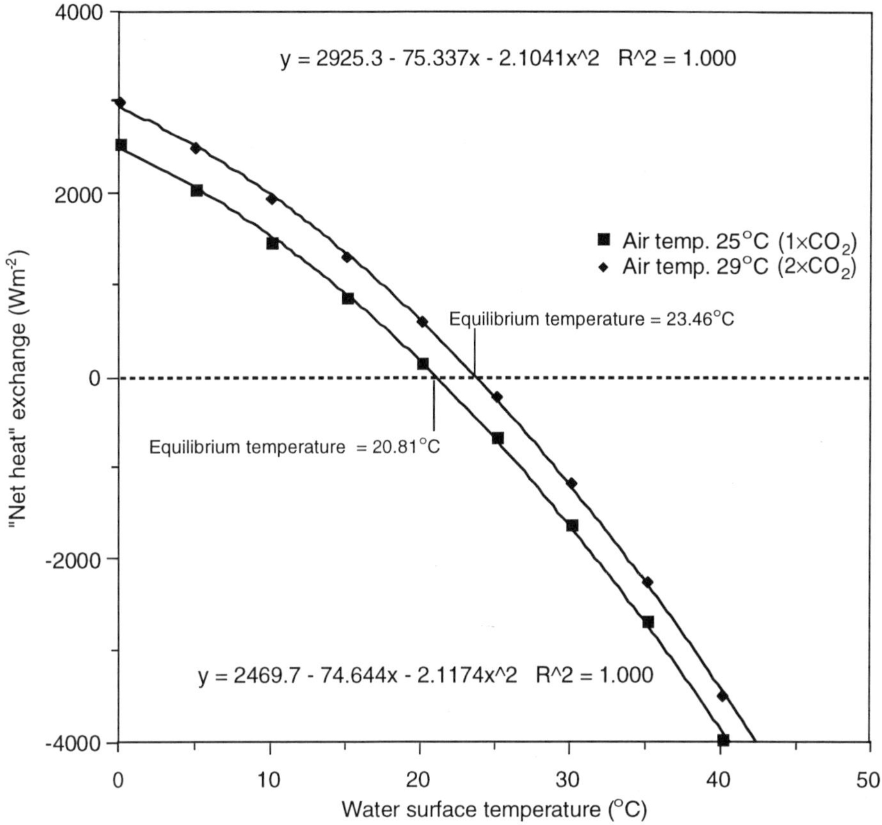

Figure 4.2. Climate change test case.

estimated by solving the best-fit quadratic of calculated values of q_{ng} [equation (4.27)] at intervals of 5°C. For the base case, the equilibrium temperature $\theta_e = 20.81$°C; for the climate change case (4°C increase in θ_a) $\theta_e =$ 23.46°C. The indicated difference of 2.65°C between the two equilibrium temperatures represents the effect of climate change due solely to a 4°C change in atmospheric temperature. It should be noted, however, that this change in θ_e is the cumulative result of changes in all terms dependent on atmospheric temperature, i.e., q_{at}, q_e, q_h, and q_g. The result is a combined effect of increased warming by atmospheric long wave radiation and sensible heat conduction and decreased cooling by evaporation and conduction between the water column and the ground beneath the stream. The net change in equilibrium temperature is the result of an increase of about 215 W m^{-2} in the heat budget.

4.3.4 Dynamic water temperature models

The general heat budget equation (4.19) can be used to describe the dynamic changes of temperature in a water body if each of the key variables can be directly measured or estimated at the necessary time scales. For a given time of year, geographic location, altitude, and set of atmospheric conditions, values of net short wave and atmospheric long wave radiation can usually be accurately estimated if not directly measured. The remaining terms in equation (4.2) are functions of water temperature itself, and for non-equilibrium conditions (e.g., $q_n \neq 0$) must be simulated dynamically given times series of key parameters, i.e., air temperature, wind velocity, humidity, cloudiness, etc. The most troublesome variable, as noted above, is the evaporation rate, E.

Jurak (1985) has applied the dynamic energy budget approach to obtain the temperature histories for a number of hypothetical shallow lakes in Poland. In this study, it was assumed that each lake had a surface area of 1 km^2 and a mean depth of 4 m, values representative of a number of shallow water bodies at various locations in the country. At each location studied monthly mean values of air temperature, wind speed, vapor pressure, sunshine duration, and cloudiness were determined from meteorological records over a 40-year period from 1951 to 1990. Incoming short wave radiation (q_{sn}) was estimated for Poland. Albedos for lake surfaces were considered to be the same for all years on record, and were distributed seasonally according to experimental data collected at several lakes in Poland. Conductive fluxes from the ground into (or out of) the bottom of the water body were estimated by Rossinski's (1975) method.

It is noted that values of ground heat conduction were generally small and were negative in the late winter and early spring. The values became increasingly negative in the late spring; that is, heat moved from the water to the ground, with a maximum negative flux of about –16 W m^{-2} in May or June. A maximum positive flux of about 13 W m^{-2} (ground to water body) was reached in October or November.

Some preliminary results obtained by the model in a study of the effects of climate change on shallow lake water temperatures are illustrated in *Figure 4.3.* In this study, the model was applied first to simulate a base condition under climate conditions observed at Suwalki Station in northern Poland (longitude 22.9°E, latitude 54.1°N). The resulting cycle of mean temperature in a shallow lake at this location is shown as the lower curve in *Figure 4.3.*

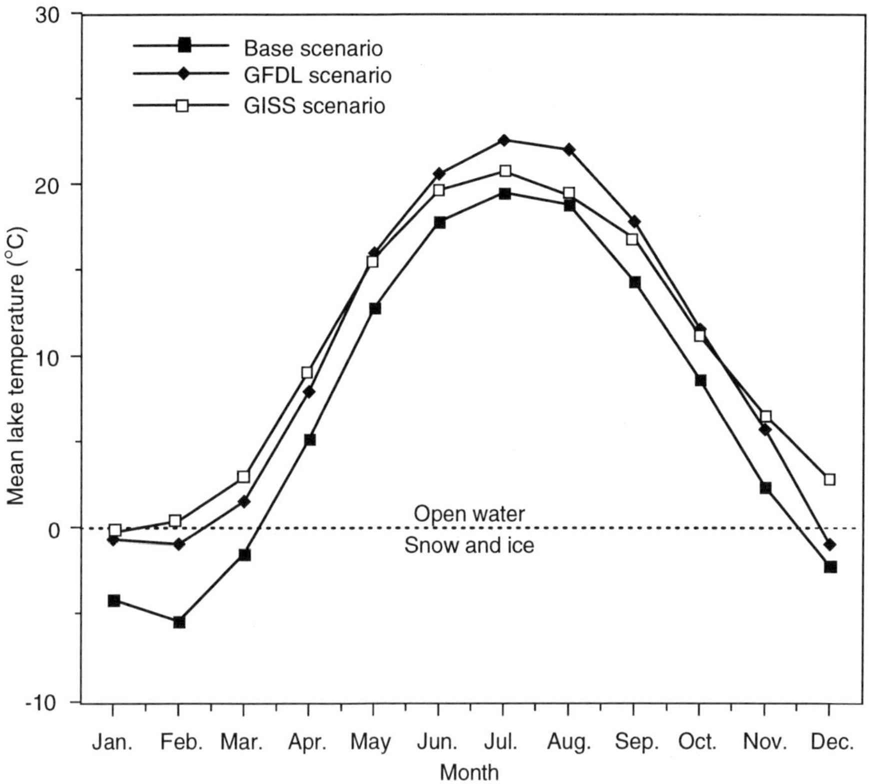

Figure 4.3. Predicted water temperature changes in a shallow lake due to climate changes estimated by two global climate models; Suwalki Station, northern Poland.

Climate change scenarios were created from the results of GCM simulations for $2{\times}CO_2$ concentrations in the atmosphere, adding the estimated difference in air temperature between the $1{\times}CO_2$ and the $2{\times}CO_2$ cases to the observed air temperature at Suwalki Station. Two additional conditions were imposed on the simulations: the short wave energy flux was reduced by 2.5 W m^{-2}, as indicated by Bultot *et al.* (1988) for the case of CO_2 doubling; also, it was assumed that the relative humidity was unchanged between the two climate change scenarios.

Results of these simulations suggest increases in the mean temperatures of shallow lake water at Suwalki Station in the range of 2°C to 5°C over the annual cycle. Also, there is an indication with both GCM cases that there is likely to be greater warming during the late winter (January to

March) when normal conditions exhibit lowest temperatures, usually below freezing. This suggests a significant reduction in the period of ice cover over the lake. Summertime temperatures could be expected to increase by 2°C to 3°C, a result that is in accord with the previously cited case for northern California.

4.4 Water Quality Changes in Impoundments

Climate changes can be expected to affect water quality in lakes and reservoirs – as discussed in Chapter 3 – in a number of ways:

- By modifying the hydrologic regime of tributary basins, thus changing the quantity of surface runoff and its temporal distribution.
- By changing the temperature of inflow and its temporal distribution.
- By altering the hydrodynamic behavior of impounded water as a result of changes in the physical properties of water that are dependent on temperature, e.g., density, viscosity, etc.
- By changing the rates of thermal energy and mass exchange at the water surface.
- By changing the rates of chemical and biological reactions and interactions among various water quality constituents, especially those that are dependent on water temperature.

As we shall see, these effects on the quality of impounded water are generally driven by the dynamics of mass and energy transfer between the atmosphere and the hydrosphere. Brief descriptions of some of the key processes will serve to illustrate this point.

4.4.1 Changes due to hydrologic regime

Critical examination of the structure and output of GCMs reveals the inadequacy of these models in the simulation of hydrologic processes. There still exists an enormous gap between the description of the driving atmospheric processes on a global scale and the important physics-based processes on a basin scale. The lack of mesoscale atmospheric models capable of translating GCM output to the spatial and temporal scales relevant to hydrologic analysis also limits quantitative assessment of the climate change impacts on water quality that are intimately linked to basin hydrology. Despite such deficiencies, some GCM outputs are instructive in the examination of possible effects. For example, estimates of the seasonal redistribution of precipitation may provide guidance in the estimation of probable changes in the timing of surface runoff. Prediction of changes in air temperature can assist in the evaluation of the contribution of snowmelt to the total yield of the basin.

The quantity of surface runoff estimated by GCMs seems generally unreliable. The so-called bucket method, which seems to be the norm for most GCMs, predicts runoff merely as a difference between precipitation and evaporation, without regard to factors such as antecedent conditions in the hydrologic basin, e.g., soil moisture, vegetation state, and snow and ice cover.

Somewhat more confidence may be placed in estimates of the temporal distribution of precipitation. The combined effects of precipitation and air-temperature changes, coupled with increases in atmospheric temperature that would reduce snowfall and snow accumulation, decrease the period of snowmelt runoff, and raise the temperatures of surface flows, may significantly modify impoundment responses. Although one may assume that there will be no significant change in total precipitation due to a doubling of CO_2, the indication by GCMs of increased rates of precipitation in normally wet periods and reduced rates in dry periods will profoundly affect the runoff cycle. Studies by Gleick (1988) of potential changes in runoff of northern California streams due to temperature changes alone indicated that even with no change in total annual precipitation, three GCMs predicted higher rates of runoff in winter and lower rates in summer. Roos (1990) concluded that a 3°C increase in atmospheric temperature would raise the snow line in California's Sierra Nevada range by about 460 meters and would hasten the onset of snowmelt. Under such conditions snow storage in northern California basins could be reduced by as much as 75% on the average. More detailed investigations of runoff changes related to precipitation and temperature changes estimated by GCMs can be found in Wood (1990), Waggoner (1990), Arnell *et al.* (1990), Bultot, *et al.* (1988), Cohen (1986), McGabe and Ayers (1989), Mimikou *et al.* (1991), Schaake (1990), Lettenmaier and Gan (1990), and Gleick (1986). All of these studies refer to specific conditions in a given water basin. Moreover, sensitivity analyses of impacts of global climate change on hydrologic processes (Nemec and Schaake, 1982; Kaczmarek and Krasuski, 1991) show that the changes in runoff due to global climate changes may range between −75% and 180%, depending on combined effects of meteorologic variables, such as precipitation and temperature.

A review of the literature suggests that for regional projections GCMs are best suited to the estimation of percentage changes of hydrologic records (Gleick, 1986; McCabe and Ayers, 1989). This argument is supported by the contention of McCabe and Ayers that "[GCMs'] internal consistency and strong physical basis may provide more accurate estimates of relative changes in the variables they predict."

Figure 4.4 illustrates a possible result of climate change on the seasonal distribution of runoff. In this illustration, the runoff cycle of a large mountainous basin in northern California with significant snow storage potential is depicted. Actual observed runoff and runoff simulated for a climate change study are compared. In the climate change case, it was assumed that runoff rates could be made proportional to the seasonal patterns of precipitation predicted by the Geophysical Fluid Dynamics Laboratory (GFDL) climate model. The modified runoff pattern predicted for the Shasta Lake basin is compared to the observed inflow to the lake for the period from 1964 to 1977. A notable feature of the changed hydrology is the greater runoff occurring in the winter period. This response of the basin to climate change is balanced in this instance by a redistribution of flows between wet and dry periods, significantly reducing the indicated runoff in the normally drier months; this result suggests an increased stress on water quality variables during such periods.

4.4.2 Temperature of inflow

It is expected that the thermal energy input associated with inflow to an impoundment would also be significantly altered by climate change, both in magnitude and temporal distribution. Advected heat energy proportional to the product of runoff rate and water temperature would change seasonally as the seasonal cycles of these variables change. Generally, the thermal energy influx to the impoundment would be expected to be highest in the winter, due to the proportionately greater runoff in this period, and lowest in the summer. Although temperatures of surface runoff would generally increase in all months of the year as a result of climate change, the thermal flux to the impoundment could be lessened in the summer period due to reduced runoff.

The Shasta Lake case cited above also serves as an example of the effect of climate change on advected thermal energy entering the impoundment. *Figure 4.5* shows the seasonal distribution of thermal energy fluxes into the lake under conditions of climate change predicted using the GFDL model with the assumptions previously described in the estimation of equilibrium temperatures for surface water flows. In these cases, inflows were taken as shown in *Figure 4.4*. Base case temperatures were assumed to correspond to the means of the historic observations. Climate change case temperatures

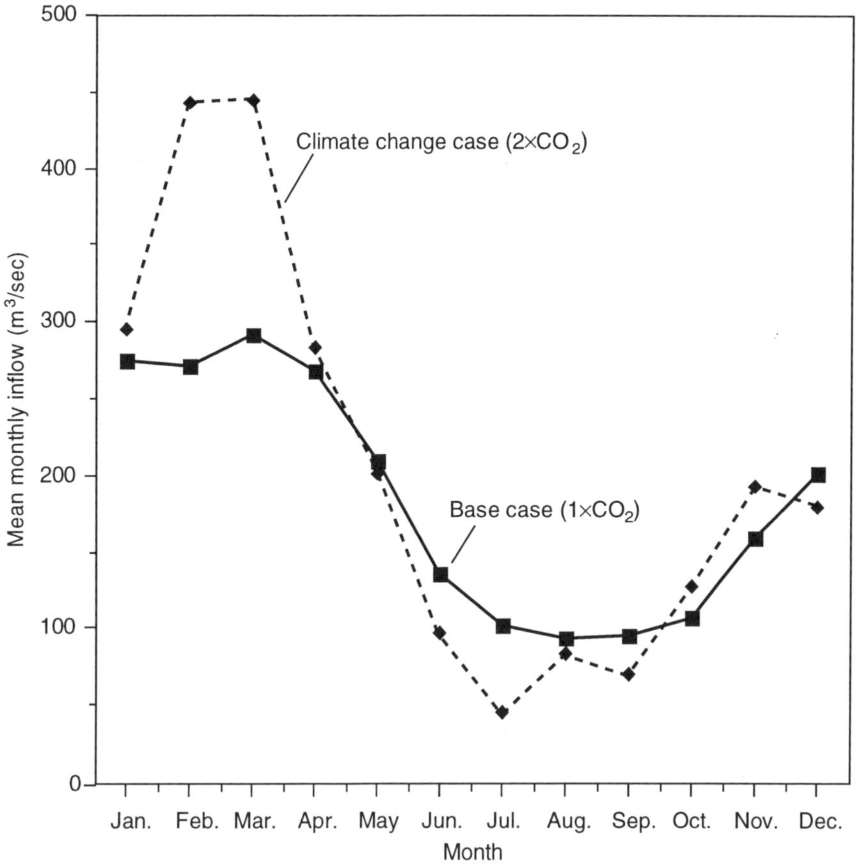

Figure 4.4. Runoff into Shasta Lake, climate change study.

were adjusted by adding a constant 2.65°C, as estimated according to *Figure 4.2* for a 4°C increase in atmospheric temperature, approximately the mean annual change predicted by the GFDL model for the Shasta Lake location. *Figure 4.5* shows a greatly increased thermal energy flux to the lake during the winter period, nearly doubled in February and March, and a moderate reduction in energy input during the summer months. In this example, the total annual influx of thermal energy to the lake was estimated to increase by about 32% as a consequence of climate change.

In this case, it is perhaps most important to recognize that the increase in energy advected to the lake means less cold water is available to regulate temperatures in the river downstream. Given a temperature threshold for stream regulation of 13.3°C, as suggested for maintenance of a salmon

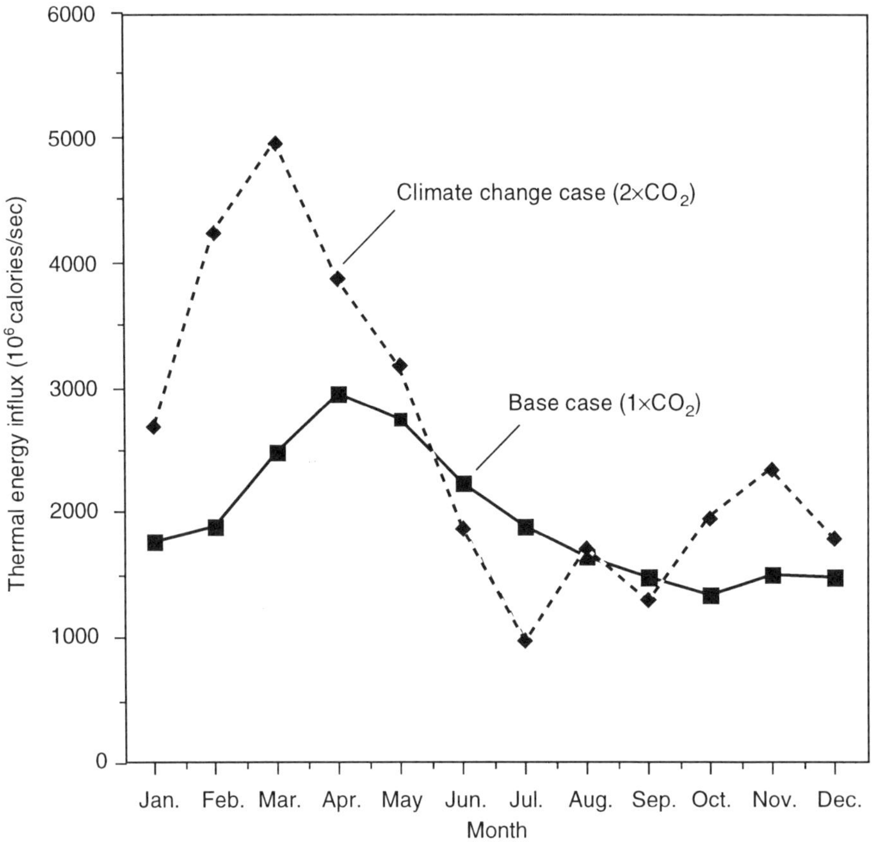

Figure 4.5. Thermal energy influx to Shasta Lake, climate change study.

fishery in the Sacramento River (SCDFG, 1968), it is estimated in this preliminary study that the annual "cold" water storage would be reduced by about 37%. This reduction is significant because it shortens the period during which the salmon habitat could be maintained at temperatures favorable to reproduction. The consequences of such alterations of thermal energy balance due to climate changes are explored further in the case study presented in Chapter 10.

4.4.3 Hydrodynamic effects

The distribution of water entering the impoundment is governed primarily by the difference between its density and that of the water in storage. Because the density of the inflow is determined mainly by temperature (although salinity and suspended solids also affect density), it will be distributed within strata of comparable temperatures. It may be mixed vertically and laterally depending on local density gradients. In deep, thermally stratified impoundments, vertical mixing may be relatively intense in the upper strata because of wind and wave action, and because of diurnal cooling, which sets up convective circulation. In the lower strata of such impoundments, mixing induced initially by kinetic energy input at the surface is gradually overcome by the potential energy represented by differences in density between successively deeper layers of colder water. At some depth below the surface, roughly at the level where short wave radiation is fully dissipated, a comparatively strong gradient, a thermocline, may develop. This thermocline may virtually divide the impoundment physically into two distinctive zones, an epilimnion above and a hypolimnion below. This division takes on special significance from the water quality viewpoint, because these zones are often biologically and chemically unique.

Changes in the vertical distribution of temperature in an impoundment are also dependent on the manner of outflow. In a lake, the warmest water in storage is released at the surface of the water body and the colder water that entered in the prior winter season is retained. The temperature of the released water is likely to be in equilibrium with the atmosphere due to its prior exposure. In a reservoir, on the other hand, withdrawal may occur at a fixed elevation below the surface such that the temperature of outflow has been predetermined by the prior history of basin runoff, thermal energy exchanges in both stream and reservoir, and mixing processes within the reservoir.

4.4.4 Exchanges of thermal energy within a stratified impoundment

Simulation of the complex histories of thermal energy balances within stratified water bodies has been the subject of much investigation and development since the mid-1960s, resulting in a number of useful tools for the investigation of water quality changes in stratified lakes and reservoirs (Orlob, 1984). Generally, these models have been constructed on the conservation of energy principle, as given previously by equation (4.1), using formulations of thermal energy exchange similar to those outlined in a previous section of this chapter [equations (4.2) to (4.21)]. Extending these concepts to the case of the stratified water body requires a few adjustments in the energy budget approach.

For present purposes, a deep, stratified impoundment is conceptualized as a one-dimensional continuum of isothermal elements that are linked through application of the equations of mass and energy conservation as follows.

Mass Conservation

For a control volume of discrete thickness ΔZ the mass conservation equation is

$$\partial V_j/\partial t = Q_j - Q_{j+1} + Q_{ij} - Q_{oj} \quad , \tag{4.28}$$

where V_j is volume of jth control element (m^3), $= a_z \Delta Z$; Q_j and Q_{j+1} are vertical advection (m/min); Q_{ij} is inflow; Q_{oj} is outflow; a_z is mean horizontal area (m); Z is depth below surface (m); and ΔZ is thickness of control element (m). Other terms are as previously defined and all flows are functions of time.

Energy Conservation

The general equation for thermal energy conservation is

$$\begin{aligned} \frac{\partial V_j \theta}{\partial t} &= \underbrace{[Q_i\theta_i - Q_0\theta_0]_j}_{\text{local advection}} + \underbrace{\left[\frac{1}{c\rho}\int_z^{z+\Delta z} q_{sz} a_z d_z\right]_j}_{\text{short wave radiation}} + \\ &\quad \underbrace{[Q_j\theta_j - Q_{j+1}\theta_{j+1}]}_{\text{vertical advection}} + \underbrace{\left[Ea\frac{\partial\theta}{\partial t}\right]_j - \left[Ea\frac{\partial\theta}{\partial t}\right]_{j+1}}_{\text{vertical ``effective'' diffusion}} \quad , \end{aligned} \tag{4.29}$$

where θ_j is temperature (°C); q_{sz} is $q_{sn}(1-\beta)\exp(-\eta z)$, in which β is the short wave radiation fraction absorbed in surface layer; η is bulk extinction coefficient (m^{-1}); E is vertical "effective" diffusion coefficient (cm^2/min); and all other variables are as previously defined.

Equations (4.28) and (4.29) form the mathematical representation of the temporal and spatial distributions of thermal energy in a stratified impoundment. Application of this model requires information on both hydrologic and atmospheric boundary conditions (e.g., air temperature, wind, cloudiness, flows, and temperatures of inflows, etc.), as well as properties of the water body (e.g., bulk extinction coefficient, effective diffusion, etc.). Implicit in this representation is the assumption that the distribution of flows entering the impoundment is ordered in elevation according to the density

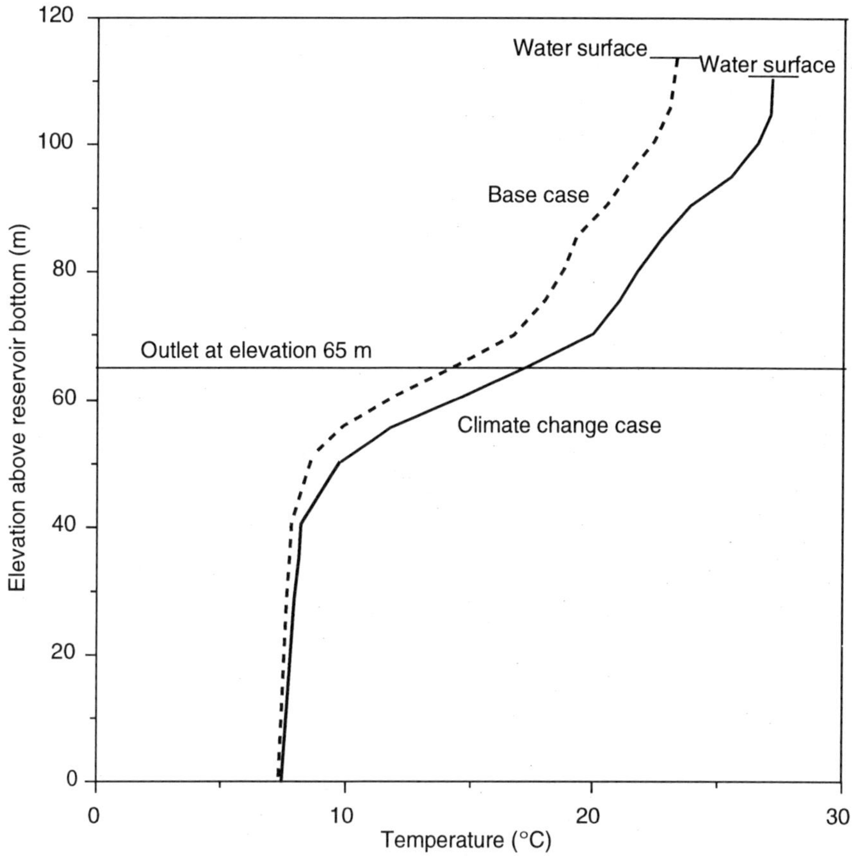

Figure 4.6. Effect of climate change on temperature profile in a stratified impoundment; Shasta Lake, September conditions.

(temperature) structure of the impoundment. Thus, the model forms the foundation for characterizing other quality attributes within the water body, which in turn determine the quality of water that is released downstream. It also serves as a means for quantitative assessment of the effects of changes in boundary conditions, for example those induced by climate change.

Typical distributions of thermal energy within a stratified impoundment are shown in *Figure 4.6.* The figure depicts the temperature profiles simulated for the month of September in Shasta Lake under two conditions: a recent historic scenario (1987), and a hypothetical climate change scenario in which the mean monthly historical atmospheric temperature was increased 4°C in each month of the year (Orlob *et al.*, 1990). In accordance with the hydrologic changes indicated by Gleick (1988), inflows for the climate change case were increased by about one-third during the winter months. Inflow temperatures were calculated for both cases using the equilibrium temperature concept described previously [equation (4.22)].

It is important to recognize that in this illustration the September profiles shown represent the cumulative energy balance experienced to this point in the annual cycle, combining the antecedent effects of hydrologic, hydrodynamic, and heat energy exchange processes. From the standpoint of water quality/ecologic changes induced by climate change, this simple example suggests a number of important consequences.

- The temperature at the level of the outlet (about 65 m above the reservoir bottom) is increased, in this case by about 2.7°C. This occurs at a time that is considered critical for the reproduction of salmon species in the river downstream.
- The volume of cold water ($\theta < 13.3$°C) retained in storage at this time would be reduced by climate change, suggesting future difficulties in meeting and maintaining downstream temperature control targets.
- The heat accumulated in the impoundment, both above and below the level of the outlet, is increased substantially by the change in climate, suggesting increased rates of water quality processes that are temperature dependent.

A case study investigation of climate change effects on the thermal energy balance over the full annual cycle in a stratified impoundment is presented in Chapter 10.

4.4.5 Water quality/ecologic processes

The dominant water quality/ecologic relationships in a stratified impoundment are illustrated schematically in *Figure 4.7* (see also Chapter 3). The sketch depicts the flow of energy (or mass) in various forms, both abiotic and biotic, through a succession of trophic levels in an aquatic environment. This environment is characterized by various water quality state variables, such as temperature, dissolved oxygen, dissolved minerals, primary nutrients, suspended and deposited sediments, and aquatic biota. Energy (or mass) transformations are seen as being regulated by physical, chemical, biochemical, and biological processes within the impoundment; by environmental quality parameters, such as temperature, light, and primary nutrients; and by external sources and sinks of energy. The water quality/ecologic

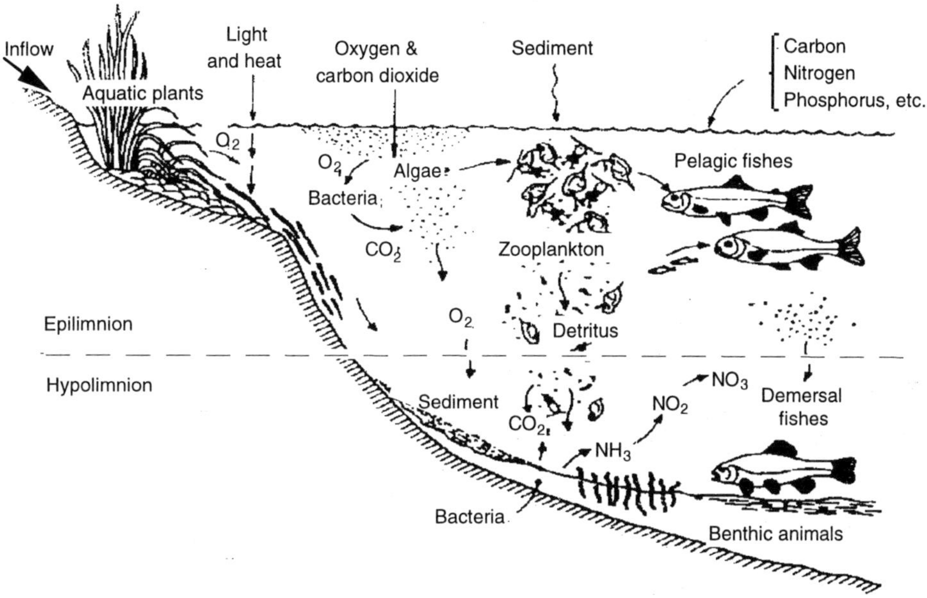

Figure 4.7. Water quality/ecologic processes in a stratified impoundment.

system is highly coupled; that is, balances of energy or mass between individual compartments are invariably linked to several others. These linkages are illustrated by the matrix of variables and coefficients shown in *Figure 4.8*, which are required in the formulation of the energy and mass balance equations implied in *Figure 4.7*.

The ability to simulate the thermal energy distribution in a stratified water body provides the foundation for description of the transformation processes outlined in *Figure 4.7*. This has been accomplished in the formulation of several useful models, which are currently available for assessment of climate change effects on water quality. Among these, the WQRRS model (HEC, 1978) exemplifies the simulation approach. The case study presented in Chapter 10 uses this model for the assessment of effects of climate change on a reservoir system in northern California. Two typical balance equations drawn from WQRRS will serve to illustrate the general approach to formulation.

Mass Conservation

The general mass conservation equation for an element j of a stratified impoundment is

$$\begin{aligned}\frac{\partial VC}{\partial t} = \; & \underbrace{Q_j C_j - Q_{j+1} C_{j+1}}_{\text{vertical advection}} + \underbrace{\left[Ea\frac{\partial C}{\partial z}\right]_j - \left[Ea\frac{\partial C}{\partial z}\right]_{j+1}}_{\text{vertical ``advective'' diffusion}} + \\ & \underbrace{C_j \frac{\partial V_j}{\partial z}}_{\text{volume change}} \pm \underbrace{V_j \frac{dC}{dt}}_{\text{total derivative}} \quad , \qquad (4.30)\end{aligned}$$

where C_j is the concentration of any quality constituent that is transported with the fluid mass, and the total derivative dC/dt represents all processes other than advection and dispersion that act to change the concentration of the state variable C_j in the control volume V_j. Equations of this form are written for each of the water quality state variables, either independently or coupled through definition of the local derivative. To illustrate this further, consider the formulation of a mass balance of dissolved oxygen in a stratified impoundment.

As an intrinsic property of water in a natural environment, dissolved oxygen moves with the fluid mass, i.e., it is advected and dispersed. It is supplied to the impoundment by the inflow, by mass transfer across the air–water interface, and by photosynthetic oxygenation. It is consumed by the processes of chemical and biochemical oxidation, and by respiration of biota.

Considering each of the important processes, the oxygen mass balance equation for a control volume V_j becomes

$$\begin{aligned}\frac{\partial VO}{\partial t} = & - \underbrace{A}_{\text{advection}} + \underbrace{D}_{\text{diffusion}} - \underbrace{O\frac{\partial V}{\partial t}}_{\text{volume change}} \\ & + \underbrace{a_s k_2 (O^* - O)}_{\text{reaeration}} - \underbrace{k_1 L V}_{\text{biochemical oxygen demand}} - \underbrace{k_s a_b (OS)}_{\text{benthic demand}} \\ & - \underbrace{\beta_1 \alpha_1 (\mathrm{NH_3}) V}_{\text{ammonia oxidation}} - \underbrace{\beta_2 \alpha_2 (\mathrm{NO_2}) V}_{\text{nitrite oxidation}} - \underbrace{\beta_3 \alpha_3 (DET) V}_{\text{detritus oxidation}} \\ & - \underbrace{k_b (r_p P) V}_{\text{biota respiration, (e.g., algae)}} + \underbrace{k_b (\mu_p P) \gamma V}_{\text{photosynthesis}} \pm \underbrace{S_0}_{\text{external sources and sinks}} \quad , \qquad (4.31)\end{aligned}$$

where O is dissolved oxygen concentration (mg/m^3); O^* is dissolved oxygen concentration at saturation (mg/m^3); V is volume (m^3); A is advection (mg/s); D is diffusion (mg/s); a is horizontal area at elevation element (m^2); a_s is surface area (m^2); k_2 is the reaeration coefficient (m/s); k_1 is the oxygen demand coefficient (s^{-1}); L is carbonaceous biochemical oxygen

Dependent variable \ Independent variable	Temperature	Cons. subs.	BOD	Detritus	Org. sed.	Carbon	NH_3	NO_2	NO_3	PO_4	Coliforms	Algae	Zooplankton	Benthic animals	Nekton	DO	Light penet.	Growth	Respiration	Mortality	Sediment	Ammonia rel.	Reaeration	Decomposition	Oxidation	Coli die away	Dispersion	Advection
Temperature												•					•										•	•
Cons. subs.	•																										•	•
BOD	•																							•			•	•
Detritus	•				•								•		•									•			•	•
Org. sed.	•			•																	•			•				
Carbon	•		•	•								•	•	•	•				•				•				•	•
NH_3	•			•	•																				•		•	•
NO_2	•						•																		•		•	•
NO_3	•							•				•															•	•
PO_4	•			•	•							•															•	•
Coliforms	•																										•	•
Algae	•					•			•	•			•				•	•	•		•						•	•
Zooplankton	•			•								•			•			•	•	•								
Benthic animals	•														•			•	•	•								
Nekton	•													•		•		•	•	•								
DO	•		•	•			•	•				•							•				•				•	•
Light penet.				•								•																
Growth	•								•	•							•											
Respiration	•																											
Mortality	•																											
Sediment	•																											
Ammonia rel.	•																											
Reaeration	•															•												
Decomposition	•																											
Oxidation	•																											
Coli die away	•																											
Dispersion	•																											
Advection	•																•											

Figure 4.8. Connectivity matrix for formulation of water quality/ecologic system. Cons. subs. = conservative substances; BOD = biochemical oxygen demand; Org. sed. = organic sediment; DO = dissolved oxygen; Light penet. = light penetration; Ammonia rel. = ammonia release.

demand (mg/m^3); k_4 is benthic sediment oxygen demand coefficient (s^{-1}); a_b is bottom area associated with element $j = a_j - a_{j-1}$ (m^2); $(OS)^*$ is organic sediment accumulation (per unit of bottom area (mg/m^2); NH_3 is ammonia nitrogen concentration (mg/m^3); NO_2 is nitrite nitrogen concentration (mg/m^3); DET is detritus concentration (mg/m^3); $\alpha_1, \alpha_2, \alpha_3$ are the stoichiometric equivalents with oxygen for ammonia, nitrate, and detritus, respectively (mg/mg); $\beta_1, \beta_2, \beta_3$ are the decay coefficients for ammonia, nitrite, and detritus, respectively (s^{-1}); P is phytoplankton concentration

(mg/m^3); r_p is a respiration coefficient (s^{-1}); μ_p is a growth coefficient (s^{-1}); k_b is biota activity, dimensionless; γ is a stoichiometric oxygenation factor for algae growth (mg/mg); and S_0 are external sources and sinks.

Equation (4.31) characterizes major elements of water quality/ecologic models, ranging from simple mass transport to complex interactions among abiotic substances and biota. The first three terms of the equation represent the primary physical processes of oxygen transport by fluid motion, i.e., advection, dispersion, and volume change. Implicit in the advection term are both vertical and local advection rates. The next three terms – mass transfer, biochemical oxygen demand, and benthic demand – correspond to the classic Streeter-Phelps formulation for oxygen balance in one-dimensional stream systems. Oxygen demand for biochemical oxidation of nitrogen species are included next. The demand for organic detritus is tacit recognition of the natural process of conversion of soluble organic matter to the particulate form and the contribution of biota through the processes of growth, excretion, and death to the overall balance of oxygen. Respiration of living forms, typified in this example by phytoplankton, is a constant sink for oxygen at the level in the impoundment where viable algae exist. Oxygen may be supplied to the system by photosynthesis, either by free-floating algae or by attached algae and green plants along the margins of the impoundment within the photic zone. Finally, a source-sink term is included to account for local advection of oxygen to or from an element.

To complete the formulation of the comprehensive model, similar balance equations must be written for each of some 16 independent state variables. Detailed descriptions of the mass balance equations can be found in Orlob (1984).

It is especially important to recognize the strong dependence on water temperature of most of the primary water quality/ecologic state variables, as well as of those parameters that govern the rates of chemical, biochemical, and biological reactions. This dependence is illustrated by a matrix of key variables and parameters presented in *Figure 4.8*. With the exception of conservative substances and light, all quantities listed as dependent are directly affected by water temperature change, hence also by climate change.

The matrix representation of the model equations also illustrates the interdependence of many of the key state variables. Notable examples are nutrients and algae. Algal growth rates are regulated by the availability of nutrients such as carbon, nitrogen, and phosphorus, the concentrations of which are reduced by algal uptake. Similar interdependencies exist between the other successive trophic levels – zooplankton:algae and nekton:zooplankton. It will be recognized that not all of these linkages feed forward in the system's response to external forcing by climate change. There are both positive and negative responses, often compensating or offsetting one another. An example is the self-shading effect of algal growth, where the increase in algal cell concentration reduces light availability for photosynthesis of algal biomass. Also, an increased rate of growth of algae due to elevated temperatures may be offset by the more rapid uptake of a limiting nutrient, thus imposing a restraint on growth. While it is clear that a modified climate will change the internal dynamics of the stratified impoundment, it is also evident that the water quality boundary conditions will also be modified, perhaps in ways that will diminish the net effect on a particular state variable.

On the other hand, there exist significant prospects for multiplicative adverse effects or combined stresses on ecosystems due to changes in the water quality/ecologic dynamics of aquatic systems. An obvious example is the combined effect on fish species of elevated temperatures and reduced dissolved oxygen concentrations, both of which could occur in stratified impoundments as results of increased atmospheric temperatures. Oxygen depletion in the hypolimnion due to increased rates of decomposition and lower oxygen saturation levels in inflowing water, both of which are due to higher water temperatures, could also enhance the potential for anoxic conditions and solubility of toxic metal ions from the sediment.

Some water quality processes in impoundments may be modified by the gradual changes in air quality overlying the water surface, e.g., increases in CO_2 associated with climate change. Szilagyi and Somlyódy (1991) have shown that a twofold increase in the partial pressure of CO_2 could induce a 30% increase in HCO_3^- concentration in lake waters, thereby increasing water hardness. Also, an increase in carbon available as a primary nutrient could influence eutrophication rates, although the effect on primary production in reservoirs is not yet known. Furthermore, such changes in CO_2 could lower pH by about two-tenths, which would be superimposed on the fluctuations associated with primary production and respiration. In hardwater lakes such changes in pH may be more important than changes associated with acid deposition.

4.5 Concluding Comments

The aquatic environment and its indigenous ecosystems are inherently sensitive to climate change. The dominant effects are those that modify the hydrologic regime of natural aquatic systems and the thermal energy balances achieved between the atmosphere and hydrosphere. Increases in atmospheric temperature and changes in the amount and distribution of precipitation and runoff can greatly modify the thermal equilibrium of streams, lakes, reservoirs, and even estuarine systems. The prospect of elevated water temperatures in northern temperate climatic zones suggests generally enhanced biological activity in aquatic environments, increased rates of decomposition of organic substances, reduced saturation levels of dissolved gases in equilibrium with the atmosphere, greater potential for release of nutrients and toxic metals from the benthos, and greater stresses imposed on sensitive aquatic species, particularly those at the margin of survival.

In climate zones where the annual climate cycle normally includes a period of freezing, which temporarily isolates the water body from the atmosphere, the effect of climate change may prove to be beneficial in reducing the period of ice cover and generally enhancing biological activity. In these areas, generally warmer water and longer ice-free periods could induce shifts in the character of the aquatic ecosystems that could be maintained; it also would induce an elevation of the overall level of sustainable biomass.

While the impact on terrestrial systems is not the focus of this chapter, it should be appreciated that the boundary conditions that drive water quality/ecological activity include accretions of nutrients, sediment, toxins, and other substances from tributary watersheds, which are likely to be changed in amount and character by climate change. For example, a shift in a basin's hydrologic response that results in increased rates of surface runoff in the early winter period after a prolonged, climate-induced dry summer may greatly increase erosion rates and the quantity of suspended sediment carried by streams and deposited in reservoirs. Likewise, increased production of vegetal matter in the watershed due to more favorable growing conditions could result in enhanced potential for eutrophication of waters receiving the runoff from such areas. Extending the growing season may also imply extended application of fertilizers and pesticides, both significant from the water quality point of view.

Comprehensive water resources planning and management in the face of climate change will require special consideration of the potential effects of that climate change on water quality and aquatic ecosystems. In areas where water is already a limited resource, climate changes that would put aquatic species at greater risk will force re-examination of priorities for water allocation and delivery schedules. Trade-offs will undoubtedly have to be made on socioeconomic bases, considering gains and losses attributable to climate changes.

The methodology for assessment of the water quality impacts of climate change is still in the development phase and needs improvement. Some well-documented simulation models are reasonably capable of deterministic simulation of water quality responses of water bodies to changes in boundary conditions. What appears to be missing are reliable means for fixing the boundary conditions themselves. In particular, the hydrologic conditions forced by climate changes are poorly simulated, especially regarding the interaction between surface water and ground water and the contribution of snowmelt runoff. Also, heat exchanges between surface waters and soil are ill-defined; they are only empirically represented in existing models. Finally, there is need for much improvement in the estimation of reliability in existing water quality models, especially considering the large number of state variables and uncertain parameters that these models include.

References

Arnell, N.W., Brown, R.P., and Reynard, N.S., 1990, Impact of Climate Variability and Change on River Flow Regimes in the UK, Report No. 107, Institute of Hydrology, Wallingford, UK.

Bolz, H.M., 1949, Die Abhängigkeit der infraroten Gegenstrahlung von der Bevölkerung, *Zeitschrift für Meteorologie*, **3**:5–6.

Brown, L., and Barnwell, T.O., 1987, QUAL 2E-UNCAS, Enhanced Stream Water Quality Model, Program Documentation, EPA 600/3-87/007, Environmental Research Laboratory, US Environmental Protection Agency, Athens, GA, USA.

Brutsaert, W., 1982, *Evaporation Into the Atmosphere: Theory, History, and Applications*, vol. of *Environmental Fluid Mechanics*, Kluwer, Boston, MA, USA.

Bultot, F., Coppens, A., Dupriez, G.L., Gellens, D., and Meulenberghs, F., 1988, Repercussions of a CO_2 doubling on the water cycle and on water balance: A case study for Belgium, *Journal of Hydrology*, **99**:319–347.

Chen, C.W., and Orlob, G.T., 1975, Ecological simulation of aquatic environments, in B. Patten, ed., *Systems Analysis and Simulation in Ecology III*, Academic Press, New York, NY, USA.

Cohen, S., 1986, Impacts of CO_2-induced climate change on water resources in the Great Lakes Basin, *Climatic Change*, **8**(2):135–153.

Eagleson, P.S., 1982, Hydrology and climate, in Geophysics Study Committee, *Studies in Geophysics: Scientific Basis of Water Resource Management*, National Academy Press, Washington, DC, USA.

Farouki, O.T., 1982, Evaluation of Methods for Calculating Soil Thermal Conductivity, United States Army Corps of Engineers, Cold Regions Research and Engineering Laboratory (CRREL report 82-8), Hanover, NH, USA.

Gleick, P.H., 1986, Methods for evaluating the regional hydrologic effects of global climate changes, *Journal of Hydrology*, **88**:97–116.

Gleick, P.H., 1988, *The Implications of Climate Change for California*, Pacific Institute for Studies in Development, Environment, and Security, Berkeley, CA, USA.

HEC (Hydrologic Engineering Center), 1978, WQRRS: Water Quality for River-Reservoir Systems, Computer Program Description, US Army Corps of Engineers, HEC, Davis, CA, USA.

Huber, W.C., Harleman, D.R.F., and Ryan, P.J., 1972, Temperature prediction in stratified reservoirs, *Journal Hydraulics Division*, American Society of Civil Engineers, Paper No. 8839:645–666.

Jorgensen, S.E., and Gromiec, M.J., 1989, *Mathematical Submodels in Water Quality Systems*, Elsevier Science Publications, Amsterdam, The Netherlands.

Jurak, D., 1978, Heat and mass exchange coefficients for open water surfaces, IAHS Publ. No. 125, International Association of Hydrological Sciences, Oxfordshire, UK.

Jurak, D., 1985, Model prognozowania temperatury wody w jeziorach niestratyfikowanych naturalnych i obciazonych zrzutami wod podgrzanych, *Wiadomosci IMGW*, **8(29)**(1–2):23–32 [in Polish].

Kaczmarek, Z., and Krasuski, D., 1991, Sensitivity of Water Balance to Climate Change and Variability, WP-91-047, International Institute for Applied Systems Analysis, Laxenburg, Austria.

Lettenmaier, D.P., and Gan, T.Y., 1990, Hydrologic sensitivities of the Sacramento–San Joaquin River Basin, California, to global warming, *Water Resources Research*, **26**(1):69–86.

McCabe, G.J., Jr., and Ayers, M.A., 1989, Hydrologic effects of climatic change on the Delaware River Basin, *Water Resources Bulletin (AWRA)*, **25**(6):1231–1242.

Mimikou, M., Kouvopoulos, Y., Cavias, G., and Vayainos, N., 1991, Regional hydrological effects of climate change, *Journal of Hydrology*, **123**:119–146.

Nemec, J., and Schaake, J.C., 1982, Sensitivity of water resources systems to climate variations, *Hydrological Sciences Journal*, **27**(3):327–343.

Orlob, G.T., 1984, Water quality modeling, streams, lakes, and reservoirs, in G.T. Orlob, ed., *Mathematical Modeling of Water Quality,* Vol. 12 of *IIASA State of the Art Series*, Wiley Interscience, London, UK.

Orlob, G.T., Meyer, G.K., DeGeorge, J., and Christensen, C.L., 1990, Impacts of global warming on water quality in river-reservoir systems, *Journal of Hydrologic Science and Technology*, **5**(1–4):10–19.

Roos, M., 1990, Possible climate change and its impact on snow melt and water supply in California, Presentation at Western Snow Conference, Sacramento, CA, April 1990, in *Global Climate Change: Potential Impacts and Policy Recommendation*, Vol. II, Report of California Energy Commission, Sacramento, CA, USA.

Rossinski, A.H, 1975, Termičeskij režim vodochranilišč, lzd. Nauka, Moscow, Russia [in Russian].

Schaake, J.C., 1990, From climate to flow, in P. E. Waggoner, ed., *Climate Change and U.S. Water Resources*, John Wiley & Sons, New York, NY, USA.

SCDFG (State of California Department of Fish and Game), 1968, Recommendations on Thermal Objectives for Water Quality Control, Policies on Interstate Waters of California, Water Project No. 7, SCDFG, Sacramento, CA, USA.

Sellers, W.D., 1965, *Physical Climatology*, University of Chicago Press, Chicago, IL, USA.

Swinbank, W.C., 1963, Long-wave radiation from clear skies, *Quarterly Journal of the Royal Meteorological Society of London*, **89**(July).

Szilagyi, F., and Somlyódy, L., 1991, Potential Impacts of Climatic Changes on Water Quality in Lakes, IUGG General Assembly, Vienna, Austria.

TVA (Tennessee Valley Authority), 1972, Heat and Mass Transfer Between a Water Surface and the Atmosphere, Engineering Laboratory Report No. 14, Norris, TN, USA.

Waggoner, P.E., ed., 1990, *Climate Change and U.S. Water Resources*, John Wiley & Sons, New York, NY, USA.

Wood, E.F., 1990, *Land Surface–Atmosphere Interactions for Climate Modeling: Observations, Models, and Analysis*, Kluwer Academic Publishers, Dordrecht, The Netherlands.

WRE (Water Resources Engineers, Inc.), 1967, Prediction of Thermal Energy Distribution in Streams and Reservoirs, Report to California Department of Fish and Game, Walnut Creek, CA, June, revised August 1968.

Chapter 5

Implications of Climate Change for Future Water Demand

Surendra N. Kulshreshtha, Zdenek Kos, and Valentina Priazhinskaya

5.1 Introduction

5.1.1 Background

There is enough water to meet all human needs: the world water supply is estimated to be 40,700 km^3, which is theoretically adequate to sustain about 20 billion people (Woodwell, 1990). In practice, however, availability of water does not coincide with requirements: the supply of water is either in the wrong place, or is available at the wrong time or in the wrong form and quality. The demand for water has grown and the supply has diminished due to human activities. The number of water-deficient regions in the world will rise unless decisive action is taken. This crisis may be accentuated by the climate changes that are expected in the next century. The scientists of the Intergovernmental Panel on Climate Change (IPCC) have shown that, in the absence of significant efforts to stem greenhouse gas emissions, the earth is leaving a period of substantially stable climates and entering a period of rapidly and continuously changing climates marked by global warming of extraordinary speed. Scientists worldwide agree that, if left uncontrolled, growth in the emission of greenhouse gases will commit human societies to more radical and disruptive environmental changes than anything experienced during recorded human history (Mintzer, 1989). According to Kerr (1990), there is a virtual unanimity among greenhouse gas

Z. Kaczmarek et al. (eds.),
Water Resources Management in the Face of Climatic/Hydrologic Uncertainties 106–167.

experts that climate warming is on the way and that the consequences will be serious. Scientists are showing how the activities of modern societies and the resulting anthropogenic changes, driven by the burgeoning third world population, can alter the fragile natural balance, which may not necessarily be self-correcting. Benedick (1989) has suggested that greenhouse warming is a complex issue, involving many contributing factors, wide-ranging and uncertain consequences, and difficult economic choices.

Global warming is likely to have socially significant effects; climate fluctuations affect human behavior, which in turn may alter the water supply–demand balance in different regions of the globe. A short-run weather deviation resulting in a meteorological drought, for example, may become an agricultural drought, and subsequently, perhaps, a perturbation in the wider economy. Even in the absence of climate changes, pressures on existing water resources are growing due to increases in population, industrial activity, and other developments. Where water resources are shared, as is the case with international river basins or lakes, the possibility of friction and conflicts exists. According to a report by the United Nations (1978), freshwater resources in the world are widely shared. For example, nearly 50 countries on four continents have more than three-quarters of their total land area falling within international river basins. Globally, 47% of all land area falls within international river basins. Over 200 river basins are multinational, including 57 in Africa and 48 in Europe. With the recent events leading to the breakup of the former Soviet Union and the former Yugoslavia, this number is bound to be even higher during the next decade. Such events may occur if there is a shortfall in supply, and may even be accelerated by increases in water demand. Although the impact of climate change on the supply (or availability) of water is generally accepted, its effect on water demand is a relatively unstudied area.

5.1.2 Concept of water use/demand

In the water resources literature, one commonly encounters the use of terms such as *water demand*, *water use*, *water requirements* or *water needs*, and *water consumption*. Of all these terms, "water consumption" is distinct from the others because it implies water lost (or consumed) by a given type of use, either through direct use in the production process or through other losses. The other terms are somewhat closer to each other in meaning. The term "water demand" is an economic concept and refers to the relationship between the quantity of water purchased by a given user and a specific price of water. "Water requirements" or "needs" are based on a technical relationship between the final product (use) and the quantity of water necessary for its satisfaction. These levels are generally technology dependent, and are completely independent of the price of water. The term "water use" is a more generic term and refers to the quantity of water withdrawn to sustain the level of a given type of use. However, some water uses are of the non-withdrawal type, where water is not removed from its source but is still an input in producing user satisfaction. In this chapter, water use and water demand are used interchangeably.

5.1.3 Objectives

An analysis of the impacts of climate change on water demand is a fundamental component of water resources management. Although the climate changes cannot be predicted precisely, the sensitivity and vulnerability of ecological, economic, and social systems to these changes can be evaluated with the help of climate scenarios. In this chapter, the relationships between change in the climate and the manner in which water demand levels or patterns may be expected to change are outlined. The discussion is based on the available literature in this area.

The following are the objectives of this chapter:

- To provide a conceptual basis for the relationship between climate change and water demand.
- To show current methods for predicting climate-induced water demand.
- To identify major methodological and empirical problems in estimating the climate-induced water demand.
- To find major implications of climate-induced impacts on water demand for water resources management and other related areas.

In addition to climate change, the existing trends in population growth and economic development in various regions may also have an impact on water demand, and thereby on regional vulnerability to climate change. In some regions, these changes may be accompanied by environmental and socioeconomic changes leading to dramatic changes in demographic structure and population migration (e.g., desertification of some regions may lead to such changes at the local level). In this chapter, identification of vulnerable regions in the world is carried out under conditions of known population changes.

5.2 Global Environmental Changes and their Effect on Water Demand

5.2.1 The need for water demand–supply integration

With the increasing demand for water resources, the first task of water resources managers is to solve water problems in an integrated demand–supply framework. At the same time, the environmental and climatological aspects must also be taken into account.[1] Because the task of supply enhancement is time consuming, planning for future events is not only prudent, it is also necessary in order to minimize major human suffering in various countries.

Water resources projects are generally constructed to serve multiple objectives through meeting various types of water demands and services (see Dee *et al.*, 1973). For instance, a large reservoir created by a high dam may supply water for irrigation or for municipal and industrial water demand. It may also improve navigation and flood control, generate electric power, provide fishing and recreation facilities, maintain suitable water quality, provide storage for droughts and ground water recharge, improve related land use, and through these factors enhance regional development (see Hufschmidt, 1965). All these goals and water demands may, in some way, be influenced by climate change. Some influences may be beneficial, while other impacts may cause problems for water resources management.

The need for integrated management of water resources can be viewed at three levels:

- Integration of various aspects of water resources – supply, demand, hydrology, economics, and so on.
- Integration of water within other systems, such as land use systems.
- Integration of the water system with other social, economic, and political systems.

In the context of climate change, integrated water resources management must be viewed in all of these contexts.

Predictions of likely water demands under climate change should be based on the application of a systems approach. This will ensure that the important interactions among the various control structures and users take place, and that the interrelationships among various users of a water resources system are explicitly considered. This approach permits the evaluation of the physical and economic consequences of the climate changes on water demands in a water resources system (Loucks *et al.*, 1982).

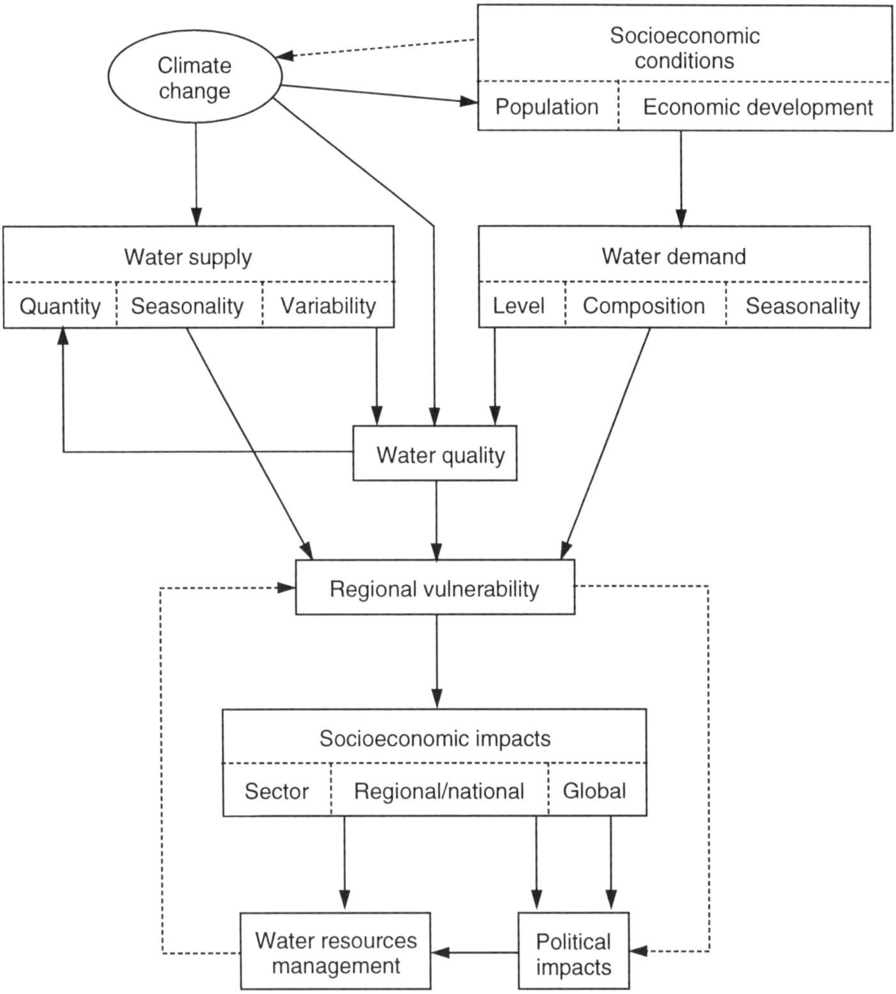

Figure 5.1. Water demand and climate change in an integrated framework.

In a broader context, climate change can be viewed both as an instigator of change and as a factor affected by human activity. Climate change can affect three major aspects of water – its availability, its demand, and its quality – as shown in *Figure 5.1.* All of these changes working together may have an impact on human activity, as well as on the ecosystem. The climate change that influences water quantity will also have an impact on water quality, and vice versa. The climate-induced changes may influence both the natural processes that form the inputs of pollutant constituents, and the wastewater discharges from municipal, industrial, or agricultural activities.

A significant outcome of these interrelated changes in supply, quality, and demand would be in terms of the vulnerability of some countries or regions to the availability of water resources. Many of these vulnerable countries may no longer be able to pursue their social and economic policies. This, in turn, may lead to major population migrations and conflicts among various adjoining countries or regions. Recent experiences in the former Soviet Union, such as the depletion of the Aral Sea and the ecological disaster in the adjacent area, and the fluctuations in the level of the Caspian Sea and the related policy issue of interbasin water transfers to the Volga River, demonstrate the need for the integration of water demand, supply, and quality.

5.2.2 Conceptual impacts of climate change on water demand

Conceptual Linkages Between Climate Change and Water Demand Level

Linkages between climate change and water demand level can be divided into two broad classes: direct linkages and indirect linkages. In the former case, the climate change affects water demand through adjustments in the requirements for various economic processes or social needs. Examples of such uses include increased use of water for lawn watering, increased use of water for personal hygiene purposes, and increased use of water for supplementary irrigation, in addition to the water requirements due to high evapotranspiration.

In the second type of linkage, the change in water demand is induced, not directly by the weather variable, but indirectly through a change in the activities that use water. For example, a change in climate may alter the power requirements in a country through the need for more air conditioning. This change would induce the generation of more electricity. Depending on the type of power generation, hydroelectric or thermal electric, there would be some impact on the amount of water used in that country.

In reality, the direct and indirect changes in water demand would operate together. Because they are triggered by a host of other factors as well, the effect of climate change on water use should be studied by initially keeping these linkages separate.

Climate Change Attributes and Water Demand

There are four attributes of climate change that may have a profound effect on water demand:

1. General global warming.
2. A change in the intra-year variability in temperature and precipitation.
3. An increase in inter-year variability in precipitation and temperature, leading to a higher probability of extreme events, such as droughts and floods.
4. An increase in the sea level.

An increase in the global mean temperature, along with an increase or decrease in the precipitation, could have a number of significant effects: water requirements for humans, livestock, and vegetation may increase; the relationship between water and yields of various crops may be altered; there may be an increase in the water requirements of various industries accompanied by an increase in power requirements; and the water quality may deteriorate, which would affect humans, livestock, and ecosystems. These changes may have an effect on the food-production potential of the region, and therefore on the food self-sufficiency goals of many countries.

The second attribute of climate change would alter the seasonal pattern of water demand, and thus may also affect the annual level of usage. Seasonality of indirect water demand, such as electric power generation, would also be altered, because power requirements in different seasons may also change under the climate change scenario.

The third attribute of climate change – the occurrence of extreme events – would also have a profound impact on humans, livestock, and ecosystems in any region. Occurrence of a drought can have severe negative impacts on the regional economy. Continued occurrence may change the community and social systems in the region. Past major droughts have been linked to major migration patterns in some countries. Other extreme events, such as floods, have similar consequences. These extreme events affect property value and increase the risks of production of agricultural goods, and thus affect human welfare.

The fourth attribute of climate change, the rise in sea level, may affect various countries in several ways. The loss of fresh water through seawater intrusion, the inundation of coastal areas through flooding, and the resulting effects on the ecosystem are three major consequences of a sea-level rise. The impacts will be felt mainly in the coastal areas and by the agricultural sector located therein. According to Hekstra (1990), the coastline will move several hundred meters in many places, depending on the characteristics of the shoreline. In addition, salt water will move upstream via rivers into lowland freshwater pockets and ground water aquifers. Countries most prone to inundation are Bangladesh, Egypt, Thailand, China, the western parts of Denmark, the USA (Louisiana), and Indonesia. The impact of this type of change would be the loss of potential agricultural lands, which would have implications for the countries' food security and may lead to some pressures to improve yields through irrigation.

5.2.3 Impact of climate change through water demand levels

In any discussion of global climate change, one must keep in mind that not all regions of the world would be affected the same way. Most of the predicted climate changes are given in terms of global averages. However, the temperature is expected to increase much more in middle and higher latitudes and over land surfaces. Thus, each region of the world must be examined on its own merit before any specific conclusions regarding the impacts of climate change can be drawn.

Given the nature of the linkages between water demand and climate change, we can formalize different types of impacts of climate change on society through water resources. These can be categorized as first-order effects, second-order effects, and third-order effects. Each of these impacts is discussed below.

First-Order Effects: Climate Change and Water Demand Patterns

The first-order effects include the direct effects of climate change on water demand levels, and through these on various water users. First and foremost, the impacts of climate change will be realized by the direct water users. To domestic water users, this would translate into a higher water demand, and thus, if there is a charge for water, a higher outlay for obtaining water. To the irrigators, this may translate into an altered level of profitability for irrigating various crops and a change in the decision about whether or not to irrigate.

In developed countries where there is more leisure time, climate changes may increase the demand for water-related recreation. However, there are negative health aspects connected with this. The concentration of people in resort areas, often with inadequate housing and sanitary facilities, increases the danger of water-related diseases caused by the pollution of drinking water or exposure of the skin to polluted water.

Second-Order Effects: Impact on Water-using Sectors

These impacts would follow the first-order effects. For example, the first-order impacts from adjustments in irrigation would lead to altered income levels for agricultural producers. This may be a result of higher outlay for water unless new water use technology can be adopted. Some of these water-saving technologies are capital intensive, and adoption of these measures concurrently with climate change would have economic implications for farming in the region. Livestock production may also require a change in inputs under the changed climate, which would translate into a change in the relative competitiveness of livestock production, particularly in colder and harsher climates.

Water resources are involved, directly or indirectly, in 80% of all diseases, according to the World Health Organization (WHO, 1978). Both surface waters and ground waters receive large amounts of industrial waste, which can cause health hazards, namely cancer or cardiovascular diseases. On the other hand, in many respects water contributes to our good health due to its recreational and environmental value. Mineral waters are used to treat rheumatic and digestive problems by bathing in them or drinking them. All of these health aspects may be influenced by future climate changes.

The indirect effects of climate change may be associated with the following types of changes. First, there may be a change in the consumer demand pattern. The new climate would result in some replacement of products by those that become more urgently needed under the global climate change. Food consumption, as well as demand for other consumer goods, would undergo some changes due to the changed climate. Second, there may be a change in energy demand patterns. In developed countries, climate change may mean higher power needs, which could increase the water requirements in the region. In developing countries, it may result in the use of other energy sources, such as cow dung. As more cow dung is used for domestic energy-related needs, less would be left for use as manure in crop production, which may result in a subsequent decrease in food production.

Third-Order Effects: National and Regional Level

These impacts would be a culmination of the previous two types of impacts, and would be felt more at aggregate levels – regional or national levels, and in some instances the international level.

The impact of climate change will be different in developed countries, in countries in economic transition, and in developing countries. In developed countries, major problems will be connected with the increase of water pollution due to reduction of flows. In the transition economies, the general tendency to cope with the climate changes will be the same; however, the process may be hindered by the lack of capital investment, and the negative impacts of the climate changes may occur more quickly than would the economic counterbalance actions. The developing countries may be catastrophically affected by the climate change. For the many regions not served by public water supply, diseases caused by polluted drinking water are the most important aspect of water-related disease. Climate changes can make this situation even worse by increasing pollution, reducing the water resources available for drinking purposes, and reducing the scarce capital that can be allocated to the development of water resources systems with the objective of supplying drinking water. In 1980, around 25,000 people died every day because of a lack of clean drinking water (WHO, 1980). The warmer temperature caused by global warming will increase drinking-water requirements and, unless drinking-water quantity is increased and drinking-water quality is improved, water-related diseases will increase in frequency and severity.

The climate change may necessitate additional irrigated agricultural areas. In developing countries, irrigated agricultural land may be the source of increased incidence of schistosomiasis. The irrigation ditches provide an ideal habitat for the snail hosts, and the irrigation workers become the host for the schistosome worm.

Adjustments in food supply could trigger a variety of socioeconomic and political problems. The trade patterns among countries or regions may be altered, which may ultimately affect the inflation rate and balance-of-payment situation. Increased use of water will result in increased competition for water, which may lead to more conflicts among water users, regionally and internationally.

5.3 Prediction of Climate-induced Water Demand

Forecasts are needed to develop any plan.[2] Water demand forecasting is important in the context of climate change because society must prepare to adapt to the inevitable changes. Prediction of climate-induced water demand change must encompass the relationship between water demand and characteristics that can alter the demand level or its pattern of use. It also must consider ways the climate change would alter the relationship between water demand and those characteristics. Because water demand is diverse, with a different set of characteristics affecting each kind of water use, it is desirable to analyze total water demand by categorizing it into major types. There is a different impact of climate change on demand for water for human settlements, industry, electric power generation, agricultural use (combined with changes of water quality, especially salinity problems), navigation, waterways, and water-related recreation. Therefore, the vulnerability of society to climate change reflected in these different categories of water demand should be analyzed separately. It should also be recognized that demand forecasts developed at different levels may not be consistent with each other. Thus, according to Turnovsky (1969), a need exists for a critical evaluation on different levels.

Four major water uses are discussed in this section: domestic, industrial (including power generation), agricultural, and *in-situ* use, such as recreation.

5.3.1 Domestic water demand levels

Domestic Water Demand Patterns

Domestic water demand generally refers to the water used to satisfy human needs in rural and urban communities. In some instances this is combined with municipal water use. Although domestic water use is the dominant category within municipal water use, the two are not synonymous: the distinction between the two uses is that municipal water use includes domestic, industrial, and commercial water use, and public water use, such as for fire fighting, street cleaning, and watering public lawns and gardens. In some small rural settlements, potable water delivered to farmers is also included, some of which may be for use by livestock.

The average daily per capita domestic water use is 15 to 20 liters in rural areas and 100 to 350 liters in urban areas. However, there are great differences between countries, especially those located in different climate zones – arid, semiarid, or humid. For example, the average water use in the USA is approximately 295 liters per capita per day (l/c/d), compared with only 104 l/c/d for the Netherlands and 215 l/c/d for Sweden (Hanke and de Maré, 1984). In Saskatchewan, domestic water use in small communities (fewer than 1,000 inhabitants) is around 150 to 250 l/c/d, whereas in larger urban centers (with a population of 175,000 or more) it is around 350 l/c/d

(Brockman *et al.*, 1987). Public water use may be about 10% of the average domestic water use. The conveyance losses are typically 15% of the domestic use, although they may be as high as 30%, particularly when the water supply network is obsolete.

Factors Affecting Domestic Water Demand

Significant variations in residential water demand occur according to the time of day and the season.[3] Residential water use also depends on climatic conditions, cultural practices, the economic characteristics of the water users, and the cost of water to the user (Dworkin, 1975). The change of climate will be reflected directly in the change in the domestic water requirement and indirectly in socioeconomic changes induced by climate change.

The general form of a water demand relationship for domestic use can be specified as follows:

$$Q_{\mathrm{DOM}} = \beta_0 + (\beta_1/P) + (\beta_2/N) + (\beta_3/SZ) + (\beta_4/CL) + \epsilon \quad , \tag{5.1}$$

where Q_{DOM} is the quantity of water used by a resident (in a single billing period); P is the price per m^3 of water per billing period; N is the income of the household per annum; SZ is the size of the family; CL are the climatic attributes of the region where the household is located; the β terms are parameters to be estimated; and ϵ is a stochastic error term. The size of the family can be measured in several different ways: some studies combine the number of adults and children in the same measure, whereas others separate them into two variables. Conceptually, it is possible to make a combined measure of family size by using some water use conversion coefficients based on relative water use, as well as by taking into account working versus non-working adults.

The effect of climate on this equation can be direct, or it can be indirect through the relationship between the characteristics of the households and their respective water use. An example of an indirect effect of climate on water demand is an increase in requirements for cooling. The direct relationship between climate and water demand is represented by the parameter β_4. The indirect impact of climate change can be specified as follows:

$$\beta_k = f(CL, CUL, INST) \quad \text{for} \quad k = 1, 2, 3 \quad , \tag{5.2}$$

where CL are the climatic attributes of the region; CUL is the cultural reference for the region; and $INST$ is the institutional framework relevant to the households. To put this another way, the relationship between price of water, household income, and size of household would vary from one region to another in response to climatic variability, cultural differences, and the institutional framework. The latter is reflected in a variety of measures related to the pricing of water. Let us consider the following examples:

- Domestic consumption is higher when no measurement of the amount of water used is carried out.
- The level of water use varies when households face different types of pricing – minimum charge or flat charge versus a block-pricing system. In general, flat-charge pricing provides the smallest incentives for conservation.
- Under the block-pricing system, domestic customers conserve more water under increasing block pricing than under either decreasing block-pricing or constant block-pricing structures. Increasing block pricing refers to a system of pricing where successively higher quantities of water cost more. In contrast, under the constant block-pricing system, each block is priced the same; and under the decreasing block-pricing system, each successive block of water used costs less to the customer.

The culture-specific relationship is more difficult to decipher. In some cases, the cultural patterns of water use and climate-induced patterns evolve together, which makes it difficult to isolate one from the other. A sociological study of households in Texas found that in areas with a higher proportion of the population of Mexican origin, there was a significant negative effect on the residential water consumption level (Murdock *et al.*, 1986).

The climatic influence on water use is reflected in seasonal variations in the level of water use, as well as in terms of the reaction of customers to longer-term changes in weather and the factors affecting water use. Griffen and Chang (1990) have shown that in Texas communities, residential water users were twice as responsive to a price change during the summer period than to a price change in the winter period: the price elasticity for the summer months was –0.37 versus only –0.19 for the winter period.

In a cross-sectional study of water demand patterns, for example, regional climate was an important determinant of water demand levels. Brockman *et al.* (1987), in a study of Saskatchewan communities, have shown that a 1% increase in the net evapotranspiration in the sample increased water use by 0.406%. The net evapotranspiration was defined as the evapotranspiration for the region less precipitation, and was measured in millimeters. Communities for which this variable was a significant determinant of water use levels were those where customers were billed on a quarterly basis. This finding suggests that in a warmer climate water use for residential purposes would increase significantly.

Effect of Climatic Variability on Domestic Water Demand

The effect of short-term variability in climate has been studied by Kulshreshtha and Spriggs (1982) and by Cohen (1987). The study by Kulshreshtha and Spriggs was carried out to estimate the effects of drought on municipal water use, and within that on residential water use. Two alternative forms of relationships were specified. In the first specification, the climate variable, measured as net evaporation, appeared in a linear fashion, as shown in equation (5.3):

$$WU_t = f(NEP_t, Y_t, P_t, D_1) \quad , \tag{5.3}$$

where WU_t is water use per capita per annum in an urban center during the year t; NEP_t is the net evaporation from May to August of year t (mm); P_t is the population of the urban center in year t; Y_t is taxable income per capita in constant dollars, expressed in 1971 Canadian dollars; and D_1 is a binary variable for an urban center.

In the second formulation, change in the climate variable was hypothesized to have a nonlinear and discontinuous effect on water use level. Binary variables were introduced in the following manner:

$$WU_t = f(NEP_{1t}, NEP_{2t}, NEP_{3t}, NEP_{4t}, NEP_{5t}, Y_t, P_t, D_1) \quad , \tag{5.4}$$

where NEP_{1t} is a binary variable that takes a value of 1 if the net evaporation during the year $t \leq 45$ mm, and takes zero otherwise; NEP_{2t} is a binary variable that takes a value of 1 if during the year 45.1 mm $\leq NEP \leq$ 75 mm, and takes zero otherwise; NEP_{3t} is a binary variable that takes a value of 1 if during the year 75.1 mm $\leq NEP \leq$ 105 mm, and takes zero otherwise; NEP_{4t} is a binary variable that takes a value 1 if during the year 105.1 mm $\leq NEP \leq$ 135 mm, and takes zero otherwise; NEP_{5t} is a binary variable that takes a value of 1 if $NEP \geq$ 135.1 mm during the year, and takes zero otherwise. Other variables are as defined for equation (5.3). Equations (5.3) and (5.4) were developed using data for two urban centers in Saskatchewan for the period from 1952 to 1980. The hypothesis that the weather affects residential water demand in a nonlinear fashion was not accepted on the basis of sample evidence. The elasticity coefficient for the net evaporation variable was estimated to be 0.43, which suggests that for every 1% increase in net evaporation the residential water use increases by about 0.5%.

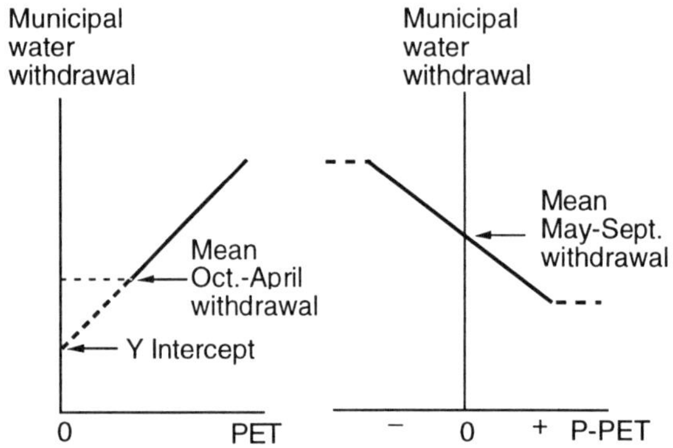

Figure 5.2. A general model of municipal water withdrawal and climate change.

The above results were used to predict changes in water use level in a drought scenario. A drought was defined as an increase in the net evaporation two standard deviations above the mean. The average net evaporation over the period from May to August in the sample region was 90 mm. In the drought scenario this was estimated to increase to 150 mm. Estimates of water use changes under this scenario suggest an increase by as much as 30% on an annual basis over and above the normal water use level. One should keep in mind that these projections are for an extreme climate occurrence. For more probable climate changes, say a 10% increase in the net evaporation, the increase in annual water withdrawal would be around 4%.

Cohen (1985) has shown that there is a significant correlation between potential evapotranspiration (PET) and municipal water withdrawals during the summer months. Cohen (1987) used this basis to estimate municipal water demand under climate change in various municipalities in Ontario, Canada, and in Illinois, Michigan, Pennsylvania, and Ohio in the USA. His general model is shown in *Figure 5.2.* Correlation coefficients between water withdrawals and PET for different municipalities varied between 0.49 and 0.94. In addition, a model using net evapotranspiration, which was estimated as precipitation (P) less PET, was used; this produced slightly lower correlation coefficients between climate and water withdrawals. Two climate change scenarios – using the GISS (Goddard Institute for Space Studies) and GFDL (Geophysical Fluid Dynamics Laboratory) general circulation models (GCMs) – were used in the study to project municipal water demand in the Great Lakes region for the months of May to September. The results suggested that there would be an increase in the water demand by 5.6% and

Table 5.1. The 10 largest industrial water users in Saskatchewan, Canada.

Industry	Quantity of water (m^3) to produce one million dollars' worth of products in 1987 Canadian dollars: Withdrawal	Consumption
Primary metal industry	5,546	70
Pulp and paper industry	4,516	120
Slaughtering and meat processing	2,761	79
Metal-fabrication industry	2,566	1
Chemical and chem. prod. ind.	2,448	294
Distilleries and breweries	1,780	174
Fish products industry	1,475	28
Leather industry	1,353	0
Textile industry	1,097	57
Vegetable-oil mills	1,011	47

Source: Based on information in Kulshreshtha and Spriggs (1982). Output figures were updated to reflect 1990 values, using data from the Industrial Selling Price Index.

5.2% for the two scenarios, respectively. This translated into a 2.6% and a 2.4% increase on an annual basis, respectively, under the assumption that the winter water demand would remain unaffected by climate change.

There is no available evidence to determine the effect of climate changes on annual water demand in humid climates; but given that precipitation in these regions is expected to increase under future climate change, domestic water demand may not be affected in any appreciable manner.

5.3.2 Industrial and power generation water demand

Industrial Water Demand Patterns

Industrial water demand varies considerably with the type of industry. The 10 industries using the most water are listed in *Table 5.1*, based on the study in Saskatchewan, Canada. In comparison with other industries, primary metal, pulp and paper, and agricultural processing industries (such as slaughtering and meat-processing operations) require more water to produce a similar value of final product. These figures should be used with caution, because even for the same type of industry the water demand levels will vary from region to region depending on the level of technology being employed. Therefore, the effect of climate change on industrial water demand would be industry specific, as well as region specific.

Table 5.2. Manufacturing water intake by use in Canada, 1981.

	Type of water use (%)			
Industry	Processing	Cooling	Sanitary and other	Total
Food and beverages	30	60	10	100
Rubber and plastics	35	59	6	100
Textiles	37	61	2	100
Wood products	27	66	7	100
Pulp and paper	75	23	2	100
Primary metals	26	71	3	100
Metal fabrication	53	40	7	100
Transportation equipment	44	48	8	100
Nonmetallic minerals	23	71	6	100
Petroleum and coal, chemicals and chemical products	7	92	1	100

Source: Tate and Scharf (1985a).

The purpose of water use varies by industry. For example, in Canada more than two-thirds of the water used in industries like petrochemicals, chemical products, metal fabrication, wood products, and nonmetallic mineral products is used for cooling purposes, as shown in *Table 5.2.* A similar situation exists in England and Wales, where the industrial water use is mainly for cooling purposes.

A great amount of water is used in the generation of electric power. Here the water is used primarily as a coolant in the condensers. However, compared with other industries, a larger proportion of the water intake is recirculated and, therefore, water consumption is less than 2% of the intake.[4]

Climate Change and Industrial Water Use

Industrial water use may be affected by climate change in several ways. Major lines of influence are shown in *Figure 5.3.* The climate change would increase the need for cooling water in manufacturing, as well as in thermal power generation plants. This change would alter the average water requirement for these industries. However, two types of indirect effect of climate change may also be present. First, climate change may alter the demand for some products. This type of change would alter the product mix produced in the region. To the extent that products of industries that use more water are demanded, total industrial water use in a region would be higher. The second indirect influence may be through a change in power

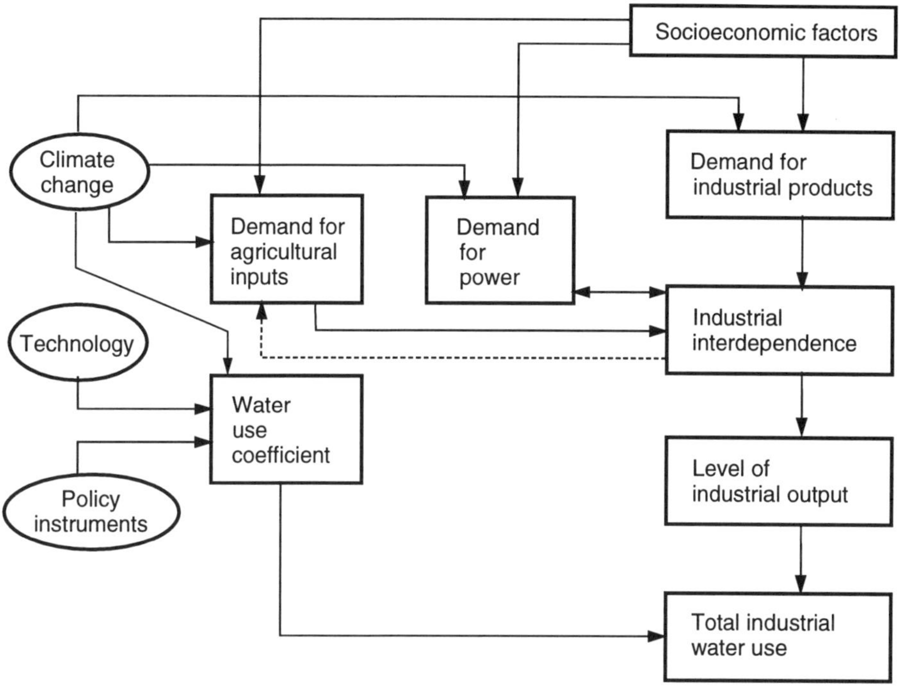

Figure 5.3. Direct and indirect water use in manufacturing and power generation.

demand. As the demand for certain goods changes due to climate change, the power requirements for the region may be altered, which in turn would affect the water demand levels. In addition, climate change may have a direct effect on power consumption in the region. The combination of these two effects would affect the water demand level for industrial uses.

Estimation of the Climate-induced Effect on Industrial Water Demand

To fully appreciate the manner in which climate change may affect the water demand level in industry, water demand in thermal power production is examined in detail. In thermal power generation, water is involved in the cooling processes in three principal ways: consumptive loss, condenser flow, and net withdrawal.

The determination of the condenser flow can be done by calculating the heat flow. The quantity of heat given up by the condenser for a given temperature change can be determined from the following equation (Dossat, 1978):

$$H = mc(T_2 - T_1) \quad , \tag{5.5}$$

where H is the heat energy transferred to the cooling water (or heat energy flow rate), in joules (J) or J/hr; m is the condenser mass or condenser mass flow rate (abbreviated condenser flow), in kilograms (measured in liters per hour, one kilogram of water being approximately equal to one liter); and c is the specific heat of water, in joules per kilogram per degree Kelvin (J kg^{-1} K^{-1}). For water, c = 4190 J kg^{-1} K^{-1} = 4.19 kJ kg^{-1} K^{-1}. The term T_1 is the initial temperature in K; T_2 is the final temperature in K; and $(T_2 - T_1)$ is the rise across the condenser in K.

From this equation, the necessary condenser flow m for the waste heat H is computed as follows:

$$m = \frac{H}{c(T_2 - T_1)} = \frac{H}{4.19(T_2 - T_1)} \quad . \tag{5.6}$$

The total quantity of heat energy flow rate input to the power plant for one kilowatt-hour of electricity (net heat rate) is related to the waste heat H and to the heat input lost to the atmosphere. Thermal power plants have average net heat rates of about 10,000 kJ kWh^{-1} (fossil-fuel plants) and 10,200 kJ kWh^{-1} (nuclear plants), of which 4,600 kJ and 6,700 to 7,100 kJ are dissipated by cooling water, respectively. When these figures are used in equation (5.6), with a condenser rise of 8 K, the water flow for cooling will be approximately 140 l kWh^{-1} for fossil-fuel plants and 200 to 210 l kWh^{-1} for nuclear plants. With a greater condenser rise, higher dissipation of heat into the air, and utilization of waste heat for heating, these figures can be substantially reduced.

The flow required for net withdrawals depends on the type of cooling system used, e.g., one-trough system, natural draft tower, or pond. In the one-trough system, the consumptive loss is equal to the increase of evaporation from the water body due to its increased temperature. For a typical fossil-fuel plant, it equals approximately 1.8 to 2.0 l kWh^{-1}. For the cooling pond, this loss is approximately doubled. For the natural draft tower, the consumptive loss is between these two values (approximately 3.0 l kWh^{-1}). Losses in nuclear power plants are greater by a factor of approximately 1.25. In water-scarce areas, air cooling can be used, with no water consumption; however, such systems are more capital intensive.

The impact of climate change on industrial water use will not be directly reflected in condenser flow at power plants, because the temperature rise across the condenser is a value related to the technology used and not to the environment. However, it is possible that this rise will increase from 8 to 10 K, the present level, to 13 or 14 K. This increase would reduce the condenser flow rate, more heat would be dissipated into the air, and the flow of waste heat into watercourses would be reduced.

The climate change, however, may increase the consumptive loss. This loss in streams and ponds can be computed in a way similar to the calculation of potential evaporation for irrigation demand. It is estimated that the consumptive loss increase due to an average temperature rise of 4 K would be approximately 7%. If the temperature rise is accompanied by a decrease in relative humidity (as indicated by regression analysis for semi-humid climates), the consumptive loss increase may even double, to approximately 15%. For the thermal power plants with cooling towers, where approximately 75% of the total heat is transferred to water evaporation, the consumptive loss will be reduced accordingly (5% and 11%, respectively).

A less reliable factor in the estimation of water requirements for power production under climate change is the total power requirement. In cooler climates, global warming may translate into a requirement for less power, whereas in the warmer arid and semiarid climates increased power demands may arise from the cooling needs of households and businesses. With increasing economic prosperity, the demand for power in many of these countries would increase over time. In addition, hydroelectric power may become more attractive as a means of both abating the greenhouse effect and adapting to increased power demands that might accompany it (Scott *et al.*, 1990).

5.3.3 Agricultural water use

Among the various uses of water, agriculture requires the largest amounts. It is estimated that agricultural water use accounts for almost 70% of the total water withdrawal in the world, and almost 90% of the total world water consumption (Shiklomanov, 1990). In fact, during the period from 1950 to 1980, 60% of the net increase in total water use was accounted for by agricultural water use.

The majority of world agricultural water use is for irrigation purposes. Although precise estimates are not available, in some countries this use accounts for more than three-quarters of the total agricultural water use. It is also the most sensitive use, because it is interrelated with food production, and for many nations self-sufficiency in food production is of paramount importance (Heady *et al.*, 1973).

Conceptual Linkages between Climate Change and Agricultural Water Use

Estimation of agricultural water use under climate change is a complex subject. The complexities can be viewed in terms of three types of impacts, some of which may occur simultaneously:

- The direct effects of the changed climate on water requirements.
- The indirect effects of climate change on water requirements.
- The policy-induced effects of climate change on water requirements.

Each of these impacts is conceptualized in *Figure 5.4*.

The direct effect of climate change on agricultural water use can be seen in terms of two major interrelationships, each of which could have a major impact on the agricultural water use in a region. These are:

- The change in the water production function.
- The change in the water requirements for stock watering.

The first effect would basically be biological. It is possible that under a warmer climate with a higher occurrence of droughts in a region more water would be required to produce the same quantity of cereals. Such a relationship is shown in *Figure 5.5*. Similarly, under global warming livestock may also require more water to sustain themselves. Both of these could increase the water requirements.

The indirect effects of climate change would be realized through a number of the direct effects of the climate change. These include the following:

- The effect on the efficiency of the water delivery system.
- The effect on rain-fed (dryland) food production.
- The effect on the level of demand for food and its composition.
- The change in the livestock production function, particularly with respect to cereal consumption.

Under a warming-climate scenario, more irrigation water may be required due to decreased dryland production, increased requirements for forage and cereals for livestock, and an increased demand for food in general. All of these effects would lead to a larger irrigated area. The climate change would have another indirect effect on irrigation water requirements: there would be a decrease in the efficiency of the water delivery system, unless investments are made in a more efficient water delivery infrastructure.

The third effect of climate change – the policy-induced impact – is even more indirect. Here, the motivation for increased irrigation may be caused

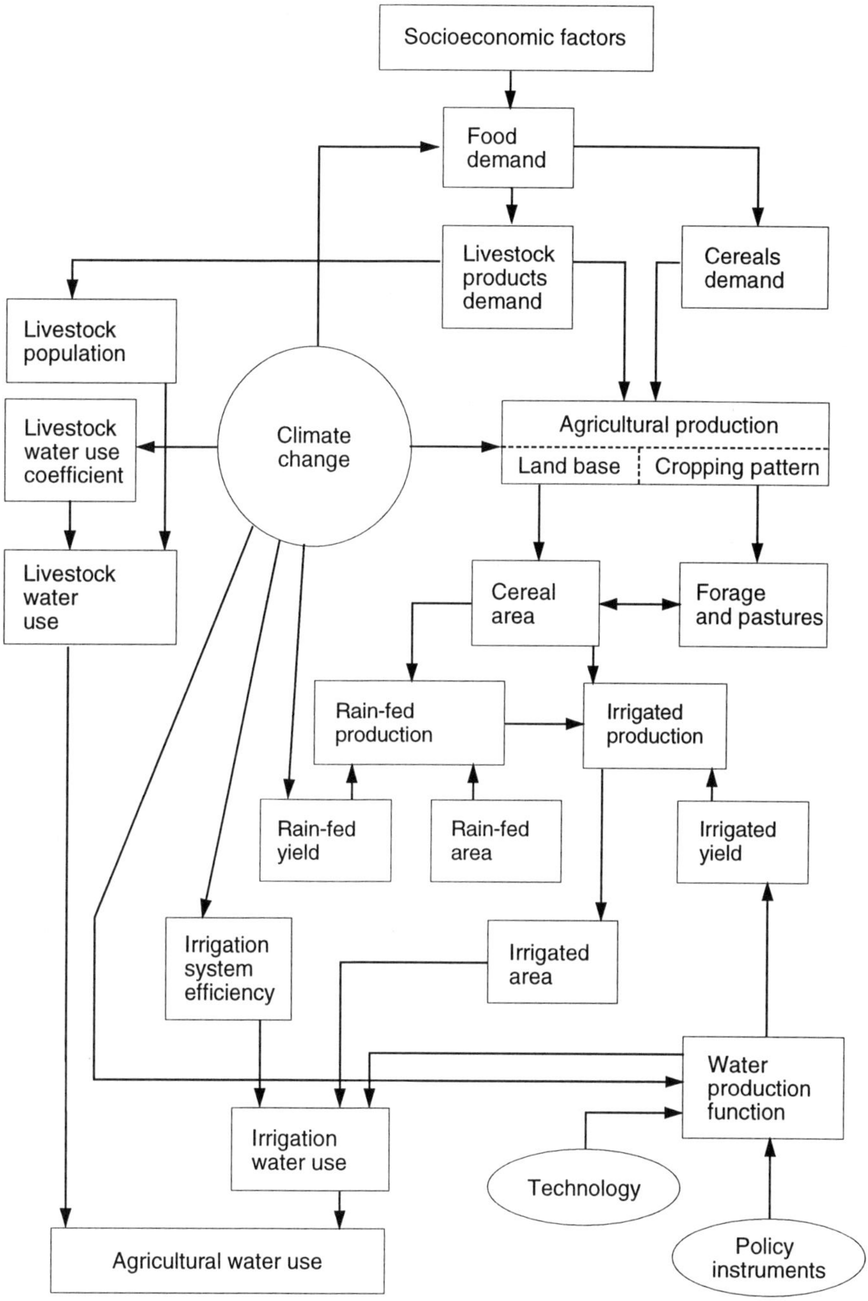

Figure 5.4. Interrelationship between climate change and agricultural water use.

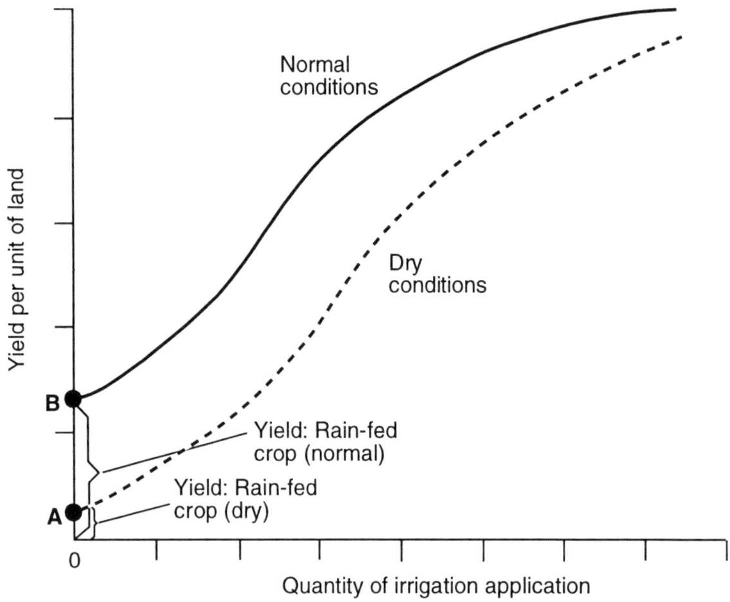

Figure 5.5. Production function for irrigation under climate change.

by a country's desire to become self-sufficient in food production. If rain-fed agriculture is affected to such an extent that irrigated production is more economically efficient, there may be more incentives for the farmers to increase the amount of land being irrigated. The final product of this type of change would be increased agricultural water use in the region (Skogerboe *et al.*, 1982). In some countries, the decision may be to meet the food shortfall through imports, which in the short term will reduce water use but in the long run will have profound socioeconomic impacts on the country.

The expected changes in climate and their consequent impacts will profoundly affect agricultural water use as a result of these three types of changes. However, precise information on many of these changes is not available, because most research efforts have focused on biological aspects of climate change. There is an urgent need to set these studies within a framework of water resources management, agricultural policy management, and economic analysis.

Model of Irrigation Water Requirements under Climate Change

Under a changing climate, one of the most obvious changes in agricultural water demand will result from increased crop requirements. However, this would not be the only reason for a change in the level of agricultural water use; there may be other contributing factors. These factors are discussed in this section.

Relationship of irrigation water demand to meteorological data. The impact of climate change on water requirements is an important issue in water- and irrigation-systems planning. Many analytical tools can be used for the analysis of the impacts of climate change. For example, mathematical models can be used to trace the structural relationships that might occur in water resources management and agricultural systems as a result of climate change.

The change in crop yields due to climate change depends on factors such as the crop species and farm technology, but its main dependence is on soil moisture. For optimal yield, soil moisture should not decrease below a critical value. This condition can be expressed as the water balance in the soil (Seuna, 1977) and can be estimated using equation (5.7):

$$S = EP - AE \quad , \tag{5.7}$$

where S is the change of soil water storage in the effective root zone; EP is the effective precipitation; and AE is the actual evapotranspiration. This equation can be written in the following fashion:

$$S = PRC - RO - SPG - AE \quad . \tag{5.8}$$

In other words, S is determined by the difference between actual precipitation (PRC) and a sum of subsurface runoff (RO), seepage into lower soil layers (SPG), which is expressed as unused part of precipitation, and the actual evapotranspiration (AE).

Effective precipitation is derived from values of precipitation that are measured directly (Fleming, 1975). Actual evapotranspiration is often based on different formulae, because the actual measurement of evapotranspiration in field conditions is not technically or economically acceptable: using Penman's formula (1954) for PET, or using the Morton model (1983), and then reducing it to the actual evapotranspiration are common procedures.[5] The formula for actual evapotranspiration often relates PET and soil moisture (Rodda *et al.*, 1976).[6] The PET can be expressed in relation to the potential evaporation (E) through a conversion factor f, as shown in equation (5.9):

$$PET = fE \quad . \tag{5.9}$$

According to Penman (1954), the potential evaporation is defined as:

$$E = \frac{DR + G(EV)}{D + G} \quad , \tag{5.10}$$

where G is the psychrometric constant; D is the slope of the dependence of the saturated water vapor pressure on air temperature; R is the energy budget (net radiation); and EV is the flow of water vapor.

The factors R and EV are related to the climatic variables in the following way:

$$R = r(\textit{Temperature}, \textit{Sunshine}, \textit{Albedo}, \textit{Dewpoint}) \quad , \tag{5.11}$$

$$EV = e(\textit{Temperature}, \textit{Dewpoint}, \textit{Wind speed}) \quad , \tag{5.12}$$

where r and e are functional operators. Because many of these variables would be altered under the climate change scenario, the model has an obvious application in the study of climate change.

Model of irrigation water requirements under climate change. The model of irrigation water requirements is based on equations (5.7) to (5.10) (i.e., on the water balance in the soil of the root zone), on Penman's formula (FAO, 1979) of PET, and on the assumption that actual evapotranspiration is a linear function of PET.[7] Using a time interval of one month, the model of irrigation water requirement can be expressed as follows:

$$W_t = k_0 + k_1 f_t \frac{R_t(D_t)}{(D_t + G)} + k_2 f_t \frac{EV_t(G_t)}{(D_t + G)} + k_3 P_t + k_4 W_{t-1} \quad , \tag{5.13}$$

where W_t and W_{t-1} are monthly irrigation water requirements during periods t and $t-1$, respectively; $k_0, \ldots, k_4$, are regression coefficients; f is a conversion factor as described in equation (5.9); R, D, G, EV are as described in equation (5.10); and P is precipitation. The model was calibrated on the basis of observed monthly values of irrigation water requirements from the routine irrigation of large-scale irrigation projects by Kos (1982b). The calibration produced the model that adequately describes the present practices of large-scale irrigation.

The model, as presented in equation (5.13), is capable of incorporating the effects of phenomena such as changes due to better practices or due to climate change. The model of climate-induced irrigation water requirements can be created by assuming that the present relation of temperature to other meteorological factors will be maintained under the changed climate and that a relationship between these variables and temperature, based on a linear regression, is an adequate approximation for the future. Thus, the transformation of meteorological factors under the climate change regime can be accomplished in the following manner:

$$Z_{it} = Y_{it} + \delta_i T_t \quad , \tag{5.14}$$

where Z_{it} is the transformed value of the meteorological factor i in time period t; Y_{it} is the measured value of the meteorological factor i in time t; δ_i is the regression coefficient of the relation of Y_{it} to temperature; and T_t is the expected change in mean temperature (°C).

Morton (1983) has indicated that the conventional techniques used in estimating areal evapotranspiration (ET) yield results completely divorced from reality. In his analysis, he introduced the concept of wet-environment areal evapotranspiration (ETW). In the hydrologic analysis of watershed systems or in the derivation of irrigation water requirements of large irrigation areas (e.g., 1,000 ha or more), a more precise estimate is based on the complementary responses of PET to changes in water available for ET.

The relationship between these values was estimated by Morton (1983) and can be expressed as in equation (5.15):

$$ETW = \frac{(ET + PET)}{2} \quad . \tag{5.15}$$

In other words, ETW is an average of ET and PET, estimated using Penman's formula.

Impact of precipitation change. The estimation of irrigation water requirements so far has been based on the assumption that a statistically significant relationship between change in temperature and other meteorological variables does exist. However, in some localities and for certain time periods such an assumption may not be valid.

Some climate change scenarios indicate that the annual total precipitation will not change significantly, but its distribution during the year may change, thereby affecting the precipitation available during the vegetation period. Some estimates have put this decrease at 20%. Furthermore, some hydrologists predict that the pattern of precipitation could also change and further decrease the effective precipitation. For instance, the changes in the synoptic situation may be more abrupt, resulting in lower effective precipitation (e.g., after a long period of drought, a severe storm with high runoff might follow and a new period of drought might appear).

Table 5.3. Irrigation water requirements in southern Alberta under alternative climate change scenarios.

Study scenario	Total regional irrigation water requirements m^3/year	Percentage change from base scenario
Base scenario (no change in climate)	12,247,550	–
Temperature increase of 2°C	18,433,450	50.5
Temperature increase of 2°C and a 10% decline in precipitation	20,239,030	65.3
Temperature increase of 4°C	23,980,640	95.8
Temperature increase of 4°C and a 10% decline in precipitation	26,456,000	116.0

Source: Estimated using information by Byrne *et al.* (1989).

Estimated Climate-induced Direct Water Use Level

This type of formulation has been used by Kos (1970) for monthly time series in 10 meteorological stations in Czechoslovakia. A change in temperature (T) of 1.5°C resulted in an average increase of irrigation water requirements of 20% to 35%, whereas a change of 4.5°C caused the level of water requirements under the present conditions to double.

Byrne *et al.* (1989) have assessed the impact of climate change on irrigation water requirements in southern Alberta. Their results are shown in *Table 5.3.* Under any climate change scenario one conclusion is inescapable: the water requirements for irrigation would rise significantly.

Priazhinskaya (1989) has estimated seasonal water requirements for an irrigation system in Russia. During the dry years, not only is the quantity different, but its monthly distribution is also different due to the change in irrigated crop patterns (see *Figure 5.6*).

Evidence of Other Factors Affecting Agricultural Water Demand

In addition to changing the irrigation water requirements, climate change may affect agricultural water demand through the following changes:

- The effects on agriculturally suitable areas.
- The indirect effects of a sea-level rise.
- The effects on the rain-fed (dryland) crop yields.
- The effects on agricultural pests and diseases.
- The impact of drought and other extreme events.

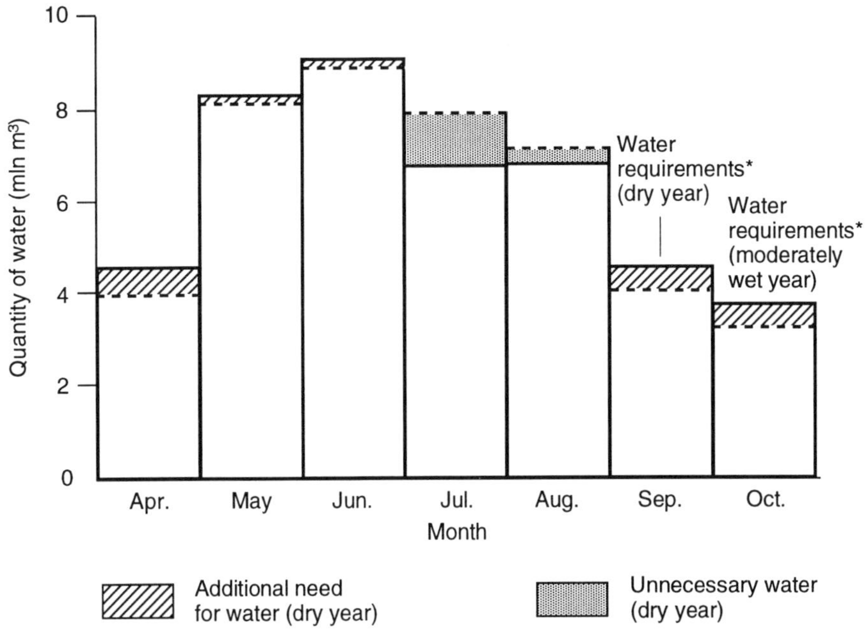

Figure 5.6. Seasonal irrigation water requirements under different climatic patterns. *Indicates difference due to change in crop requirements and change in pattern of irrigated crops.

Effect on area suitable for agriculture. According to the Second World Climate Conference (see Parry and Jiachen, 1991), an increase in temperature can be expected to lengthen the growing season in areas where agricultural potential is currently limited by insufficient warmth, resulting in a poleward shift of the thermal limits of agriculture. This effect is expected to be more pronounced in the Northern Hemisphere, although there may be some regional variation in the ability to exploit this shift. For example, Parry and Carter (1989) and Parry *et al.* (1989) have suggested that a 1°C increase in the global mean temperature would advance the thermal limits of cereal cropping by about 150 to 200 km. A major implication of this change would be an increased ability to produce cereal in the region, which may indirectly change the need for irrigation to meet the gap between demand and supply.

Effect of a sea-level rise. Global warming induced by greenhouse gases is expected to lead to a rise in the sea level, although some experts say that the magnitude is not clear (Swaminathan, 1991). If such changes result in a loss of fertile agricultural lands, the effect on water use for irrigation may be positive or negative. If the area lost is irrigated, and if it cannot be replaced by a similar area elsewhere, the result would be a decrease in the level of irrigated water use. Conversely, if the lost area is a rain-fed (dryland) area, this will reduce the food supply in the region and may lead to an increase in the area of land that is irrigated.

Effect on crop yields. Increased CO_2 in the atmosphere would lead to somewhat higher yields for certain crops. According to Cure and Acock (1986), this would lead to an increase in the rate of plant growth. However, crops such as wheat, rice, and soybeans would fare much better than would maize, sorghum, millet, and sugarcane. The implications of such changes would be in terms of both the global food security of many nations and the land use in many countries. For example, according to Swaminathan (1991), rice may become more important than it currently is in India's national food-security system because it can give high yields under a variety of growing conditions.

The effect on rain-fed agriculture should be viewed together with other changes not related to climate change. For instance, as the supply of firewood in some countries dwindles due to heavy deforestation, cow dung and crop residues will become the main fuel sources. In certain parts of India, this change has already taken place (Brown, 1989). If these trends continue land productivity will decline, and with increasing population shortfall in food supply will increase. This will have a significant impact on irrigation water use, because in many countries irrigation is being used to increase land productivity.

Effect on agricultural pests and diseases. Some studies have suggested that a temperature increase may be associated with extending the range of some pests that are currently being limited by temperature. In addition, there may be a greater occurrence of livestock diseases (Environmental Protection Agency, 1989). This may translate into an increase in the overwintering range and population density of some agricultural pests under a warmer climate at mid-latitudes; in cool, temperate regions there may be more insect pests and diseases than are currently present. These events would lead to decreased crop yields in some regions and a lowering of profitability of livestock operations.

Effect of extreme events. With changing climate, there is an increase in the probability of extreme events such as droughts, storms, heat stress, and severe frosts. Although crop yields exhibit a nonlinear response to heat or cold stress, as suggested by Mearns *et al.* (1984), very little is known about changes in the probabilities of such extreme events. A similar concern has been raised by Anderson (1982), and Rind *et al.* (1989). According to the Second World Climate Conference, there is a distinct possibility that as a result of high rates of evapotranspiration some regions in the Tropics and subtropics could be characterized by a higher frequency of drought, or by droughts of a similar frequency to those of the present but more severe in nature.

Summary. These impacts of climatic variability would have serious implications for current food production, and through that for future food security, particularly with increasing population. In some countries this may lead to reducing the food surplus to meet the growing needs; in others it may lead to a widening of the current gap between food supply and demand. However, the biggest uncertainty is technological change, which may override any negative effects of climate change. All of these factors would have significant socioeconomic implications for many countries.

5.3.4 Recreation, navigation, and ecosystem water use

Water-related recreation is important in modern society; its importance will grow as leisure time increases because of the need to satisfy basic functions of human life (i.e., rest, entertainment, personal development, etc.). Water quantity and quality requirements for water-related recreation vary according to the type of recreation, for example, whether there is direct or indirect contact with the human body (Baumann, 1969). Direct water contact activities include bathing and swimming in rivers, lakes, or bathing pools, boating, sailing, windsurfing, and waterskiing. Indirect water contact activities include sunbathing with short contact with water, sport fishing, camping on the shores of rivers and lakes, individual and collective water transport, and houseboating. During the winter, under suitable weather conditions, some winter sports can be performed on frozen rivers and lakes, including ice skating, ice hockey, ice sailing, and ice bowling. In addition to natural water courses, recreation activities also take place in swimming pools.

Various types of water-related recreation and sports will be influenced by the climate change in some way. The increasing temperature will increase the number of days suitable for activities such as bathing, swimming, boating, and sailing. This could result in the overcrowding of existing facilities, as well as in higher public demand for the planning of new recreational facilities. In some cases, it may lead to a need to change minimum flow requirements for rivers. In many cities, particularly those without any adjacent natural-water recreation areas, the need for large recreational facilities may be overwhelming, which would lead to higher water demand at the municipal level.

The evaluation of the recreation value of a river reach or a lake must take into account the fact that environmental quality and recreational use are closely connected (Stevens, 1967). Both of these objectives can be expressed in terms of their alternative use in the multipurpose water resources system and can be reflected in their shadow prices (Cicchetti *et al.*, 1972; David, 1968). The changes in climate conditions can influence these shadow prices: the shadow prices of other uses may increase under the climate change, which may have an adverse impact on the recreation and ecosystem water use (including other in-stream water uses).

The economic evaluation of the effect of climate changes on recreational activities and ecosystems is relatively difficult, but possible. Using the contingency valuation methods, the willingness to pay for recreational services can be estimated.[8] These values can be used for planning new recreational sites or for carrying out qualitative changes that are aesthetically pleasing to site users. One must be cognizant of the fact that recreational use is a function of many socioeconomic variables, such as personal income, change in social behavior, availability of leisure time, and preference for water-related sports. Many of these may also be related to the climatic variability.

Water use for navigation is generally expressed in terms of the requirements for navigation depth. These requirements can be translated either in terms of discharge or minimum flow. This water demand would depend on the cost of alternative transportation available in the region. Under future climate changes, the cost of maintaining the same level of discharge may increase, and this type of water use may only be feasible if the relative merits of inland water navigation are sufficiently high.

In some river reaches, in some seasons, navigation may be complementary to other in-stream uses; in others it may be competitive, depending upon the relative requirements and configuration of the water resources system. If the two uses are complementary, the climate change may have no effect on the navigational water use. However, under the competitive regime, allocation of water for navigational uses would have to be based on the opportunity cost of water.

Table 5.4. Selected indicators of effects of climate change on stream water quality.

Pollutant	Change in low flow	Increased temperature	Increased storm flow
Conventional			
Dissolved oxygen	X	X	X
Suspended sediments	X		X
Fecal coliform	X		X
Acidity	X		
Nonconventional			
Nitrogen	X		X
Phosphorus	X		X
Salinity	X		
Toxic chemicals	X		

Source: Jacoby (1990).

5.4 Climate Change, Water Quality, and Water Demand

Change in the quality of water, although not the focus of this chapter, is an important ingredient in the determination of water demand. Obviously there is an interdependency between water quality and its demand level (Middleton, 1981). If the quality of water goes too low and the cost of upgrading this quality increases, the demand for this water approaches zero.

The water quality impacts of climate change are dependent on society's response to water quality management through adoption of corrective measures. Without such measures a number of impacts can be foreseen. For rivers and streams, the quality of water is generally defined in terms of concentration of pollutants at the time of critical low flow. With a climate change, there may be a change in the stream flow and, more importantly, in the critical low flow. This would result in a decrease in the river and stream water quality. If the climate change brings about a different seasonal distribution of precipitation, as suggested by Gleick (1987), this may also result in the deterioration of river water quality during the drier periods. Selected indicators of water quality as affected by climate change are shown in *Table 5.4*. All measures of water quality in rivers and streams would be affected under the low-flow situation simply because of dilution.

Climate warming would also create water quality problems in estuaries and bays similar to those in rivers and streams. In addition, penetration of saline water from the oceans may also increase, thereby decreasing the quality of water. According to Jacoby (1990), the greenhouse gas-induced climate change would affect the quality of water in lakes through three mechanisms: changes in throughput and volume, higher water temperature, and reduction in ice cover. The change in temperature will affect the rate of bacteriological activity and the amount of oxygen to support it. Water quality problems would also emerge in coastal areas through saline-water intrusion. Some of the ground water may be particularly vulnerable.

Changes in water quality have significant implications for all types of water use. Of major importance will be water use for human consumption, which has health implications. A change in the water temperature caused by a change in the atmospheric temperature may further enhance the occurrence of water-borne diseases (Nemec, 1988).

Tourism and recreation are particularly vulnerable to changes in water quality. In some countries these activities have surpassed agriculture in job creation. However, the climate change may have a negative impact. If the temperature of a stream increases, the dissolved oxygen content in the water may drop below the level necessary for some kinds of fish to survive (Edinger *et al.*, 1968; O'Connor, 1967). This may have a negative impact on sport fishing and may further affect tourism and related economic activities.

5.5 Issues in Conceptualizing the Effect of Climate Change on Water Demand

The relationship between water demand and climatic variability, as demonstrated in this chapter, is a relatively unstudied subject. At this time, information is lacking on an appropriate methodology for assessing the impact of climate change on water demand levels and patterns. Some of this is already noted in the previous sections. At the same time, a number of methodological problems and empirical issues exist in the estimation of such impacts. These and related topics are discussed in this section. The discussion is divided into two broad categories: methodological issues and empirical issues.

5.5.1 Methodological issues

Major methodological issues relate to those areas where relatively less reliable information is available for use in the estimation of climate-induced impacts on the level and pattern of water demand. Major deficiencies in our understanding of the phenomena remain in some critical areas, as listed in the following sections.

Uncertainty in the Nature of Future Climate Change

In spite of the fact that a number of studies have addressed the issues of climate change, existing literature does not provide a unanimous account of the forthcoming changes. Budyko (1989) maintains that the mean global temperature can be estimated more accurately than can the forthcoming regional temperature changes. Predictions of future weather patterns that are based on different GCMs tend to be completely different. Part of the explanation for these discrepancies is the degree of resolution used in calibrating the model. According to Giorgi and Mearns (1991), GCMs are the main tools currently available for climate simulation, but they are run (and for the next several years will continue to be run) at resolutions that are too coarse to adequately describe mesoscale forcings and yield accurate climate details. Major refinements have to be made in the GCMs for the Tropics, where they are at present even less reliable (Chen and Parry, 1987).

Understanding the future climate (i.e., climate change over time) is important; but there is another equally important aspect, that of climate change over space, that may also be important for assessing regional impacts on water demand levels and patterns. According to Parry (1986), "climate as a resource varies greatly from one part of the globe to another and this climate change over space may be of greater consequence for human development than climate change over time." For example, research results by Manabe and Wetherald (1987) on soil moisture have suggested that soil wetness is controlled by the latitudinal profile of warming.

In the context of the impact of climate change on agriculture, one needs an even smaller level of resolution, that of the crop-yield models. Links need to be developed between the large-scale GCMs and the smaller-scale input required for predicting crop yields under alternate climates. A similar requirement exists for river-basin planning models.

Understanding the Socioeconomic Impacts of Extreme Climatic Events

Extreme climatic events such as droughts and floods are predicted to occur more frequently under a changing climate. At the same time, as noted earlier, very little is known about how such phenomena would affect water use and, furthermore, what socioeconomic implications these phenomena would have for various regions. Unfortunately, as suggested by Riebsame and Magalhaes (1991), the processes of regional development and social change are no better known than are those of climate. Nevertheless, these impacts are expected to be significant, as suggested by Warrick and Bowden (1981) who have shown hypothetical pathways of drought impacts on society. These pathways link local, individual-level impacts along a path that goes all the way to the international level. These impacts may be economic, resulting from changes in the pattern of food production, or they may be social problems resulting from stress caused by financial vulnerabilities faced by firms. Such drought-induced changes would only be compounded under the climate change, because occurrence of such extreme events would be coupled with the impact of climate change on water availability and on water demand in other years.

Some of the impacts of drought are more indirect in nature. For example, frequent droughts may result in decreased production of electric power. In the short run this may lead to the rationing of power in some regions or for some users. There may be a threshold level of rationing that would result in a switch to alternate sources of power, thereby permanently reducing the demand for electric power generation.

Adjustments in Domestic Water Use Under Climate Change

A number of unresolved issues remain in the prediction of domestic water demand in the context of climate change. Does the demand for water depend in part on its availability, and as the availability of water changes under climate change, would its use for domestic purposes be adjusted accordingly? What types of additional demands for water are created under climate change? Would all these changes be positive, or would some of the increases be counterbalanced by negative changes? Would municipal uses of water related to recreational and leisure activities increase under a warmer climate? These and related questions need to be answered to satisfactorily estimate the nature of the change in domestic water demand under conditions of climate change.

Issues Related to Agricultural Water Demand

Issues related to agricultural water demand are numerous and complex, because the impacts of climate change are both direct and indirect. These issues are therefore divided into various categories, according to the nature of the relationship between causation and climate change.

Direct water requirements for irrigation. The direct water requirements for irrigating various crops are determined by the underlying water production function for each crop. How would this production function change under a climate change? Would the marginal difference in the physical productivity of irrigated crops and rain-fed (dryland) crops remain the same as now? If not, how would these changes lead to different land use patterns, cropping mix, and cropping intensity in various regions of the world? In addition, the water requirements for crops would also be reflected in a change in regional carrying capacity of pasture and rangeland, which will also be affected under the climate change.

Direct water requirements for livestock. Similar to the impact on water requirements for crops, information on changes in livestock water requirements under climate change is also lacking. How would the relationship between sustenance of physiological functions for various types of animals and their needs for water (both for drinking and for cooling) change under the warmer climate?

Related to water requirements is the question of feed requirements, and the subsequent effect on the competitiveness of raising certain types of livestock. For example, under cooler climates some animals need higher levels of feed intake to maintain their body heat; if some regions were to get warmer, how would that affect feed and forage requirements? A drastic change in these levels would alter the relative competitiveness of raising livestock in various regions and may lead to further socioeconomic impacts at the regional level.

Demand for irrigation water. The relative change in the irrigated and rain-fed (dryland) crop production functions and the relative competitiveness of raising livestock would lead to a change in demand for irrigation water. However, such a demand would be site specific and would need to be estimated by taking into account various socioeconomic factors affecting the need for, and adoption of, irrigation on farms.

Associated with the demand for irrigation water is the question of irrigation technology. The total water requirement for an irrigated area depends significantly on the nature of the technology used. How would the choice of technology change in response to climate changes?

Effect of climate change on the pattern of food demand. A warmer climate may affect agricultural water demand through adjustments in the nature of demands for food and clothing, and their relative composition. The manner in which such changes would take place, or if they would take place, is a relatively unstudied subject. More research is needed to identify the relationship between the level of demand for agricultural products, such as food and fibers and change in climate.

Climate Change and Industrial Water Demand

As is the case with food and fibers, industrial water usage cannot be estimated accurately at this time because of insufficient knowledge. There are major issues that need to be investigated.

Direct water requirements for industrial production. A changing climate would alter the water needs of various industries for cooling purposes. Would all industries be affected in the same way? How would the change in climate affect an industry's potential for recycling water? An industry may find that it is to its advantage to change its cooling technology under a climate change.

Demand for electric power generation. A changing climate would alter the effective demand for electrical power. This change would arise due to the need to cool residential, office, and industrial buildings. The change in availability of water in the region may also determine the relative economics of producing electric power in alternate ways. For example, frequent droughts in the region may necessitate a lower reliance on hydroelectric power, and may increase the production of thermal and nuclear technologies. In some cases a switch to more conventional sources of power may also take place. How would a changing climate in different regions stimulate a change in the technology of power production on the one hand, and power production demand on the other?

Demand for industrial goods under climate change. Consumers' tastes and preferences for various industrial goods may change along with the changing climate. The manner in which such changes would take place is not certain. Prediction of these changes is necessary for better estimates of climate-induced industrial water demand.

Other Water Uses Under Climate Change

In addition to the three major water uses – domestic, industrial, and agricultural – there are a number of other water uses of interest. These include freshwater fisheries, waste disposal, navigation, recreation, and tourism. At this time, very little is known about the impact of climate change on these uses, and therefore more research activity should be directed in these areas.

5.5.2 Empirical issues

In addition to methodological issues, the estimation of climate-induced change in water demand is complicated by a number of issues that are encountered in the empirical estimation process. These issues are discussed in this section.

Choice of the Study Region

In any empirical application the choice of the region is a crucial decision. From a hydrologic point of view the logical unit of study should be a watershed – a river basin or a lake. From the point of view of data availability, however, an administrative unit would be preferred. Thus, a researcher often encounters one of two types of problems. The first is that a watershed may be shared jointly by a number of countries; its boundaries will rarely coincide with political boundaries or those of different administrative regions. This often means that the researcher has to deal with different sources of data, and frequently with different approaches to water demand problems. The nations often pursue different, barely comparable, multi-objective policies. In addition, their economic potential may be different. Water under these conditions becomes a strategic, scarce resource and the development decision often involves conflicting interests.

The second problem is that a political region may contain more than one watershed. Here the problem is availability of data of good quality. Disaggregation of national data into regional aggregates is fraught with problems of approximation and guesstimation. Nonetheless, the river basin is the preferred unit of accounting for the study of climate change. As noted earlier, until the GCMs are modified to a lower level of resolution, the researcher also faces the problem of obtaining accurate information regarding climate change at a river basin.

Identification of the Model

Over a period of time when the climate is changing and both water supply and water use levels are being affected, a problem of model identification is encountered. This is because at any point in time if the market is functioning, the supply and demand will come to equilibrium at some point through a change in the price level. Thus, one has a situation such as is shown in equation (5.16) for demand and in equation (5.17) for supply:

$$P = f(Q_d, C) \ , \tag{5.16}$$

and

$$Q_s = f(P, C) \ , \tag{5.17}$$

where Q_d is the quantity of water demanded; Q_s is the quantity of water supplied; C is the climate variable determining quantity; and P is the price of water per unit.

The interaction of these variables over time is such that supply and demand functions are not statistically distinguishable. A common practice is to estimate the relationship between price and quantity using available data. This may result in severe bias in the estimated parameters. For example, as shown in *Figure 5.7* the observed data may portray neither a demand function nor a supply function; in fact, they may be tracing a mongrel relationship. Therefore, it is important that in any study of climate-induced change an attempt is made to properly identify the two relationships prior to estimation.

Choice of Functional Form

In an estimation of climate-induced impact, the magnitude may be determined by the choice of functional form for the estimated function. The relationship between climate change and the level of water demand may be linear, interactive with other explanatory variables, or nonlinear in nature. Every attempt must be made to empirically test the nature of the functional form if the existing theory does not help guide the researcher.

Mathematical Modeling of Water Demand

In any empirical study of water demand forecasting, there is a variety of models to choose from. The simplest model of water use forecasting is a trend model, where the classic assumption that the future will be a continuation of the past is made. Such models are not appropriate for the

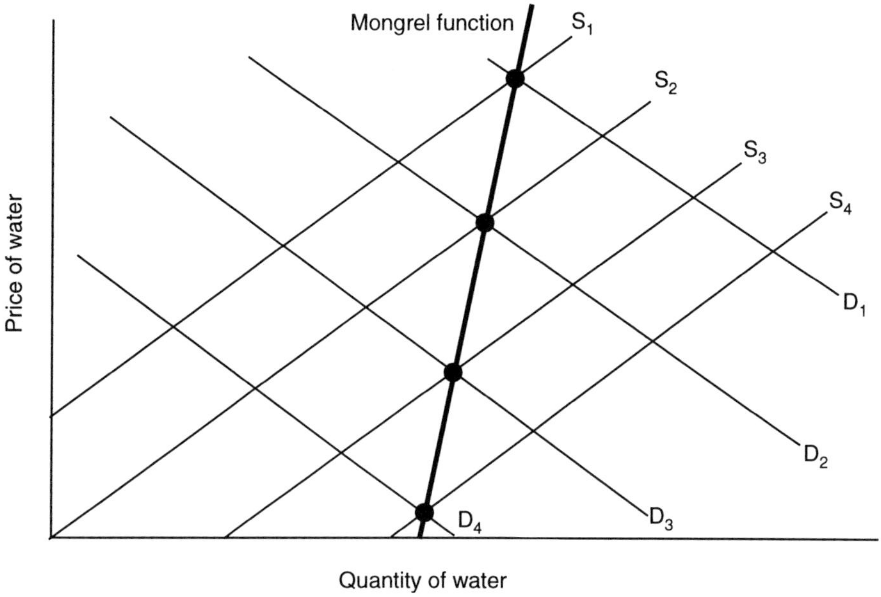

Figure 5.7. Shifting of demand and supply functions with climate change.

estimation of climate-induced changes because past trends are rather poor predictors of the future, particularly under climate change conditions.

Statistical modeling of water use functions is a method commonly used to estimate the relationship between climate variables and other structural and institutional variables affecting water demand levels. By explicitly including climate- and non-climate-related variables, the risk of overestimating the impact of climate change is minimized. However, in these situations, the selection of the functional relationship between climate variables and water demand is a critical decision. Furthermore, these models are more susceptible to specification errors, particularly those resulting from an exclusion of relevant variables.

Optimization models allow detailed consideration of the behavior of economic agents under a changing environment. As economic and non-economic factors affecting water demand change, statistical models may not be able to portray the desired behavior, particularly if the change does not form a major part of the historical evidence used for estimation purposes. Such models can be developed at an enterprise level, at a regional level, or at a national level. The most common techniques are linear programming, quadratic programming, and simulation.[9]

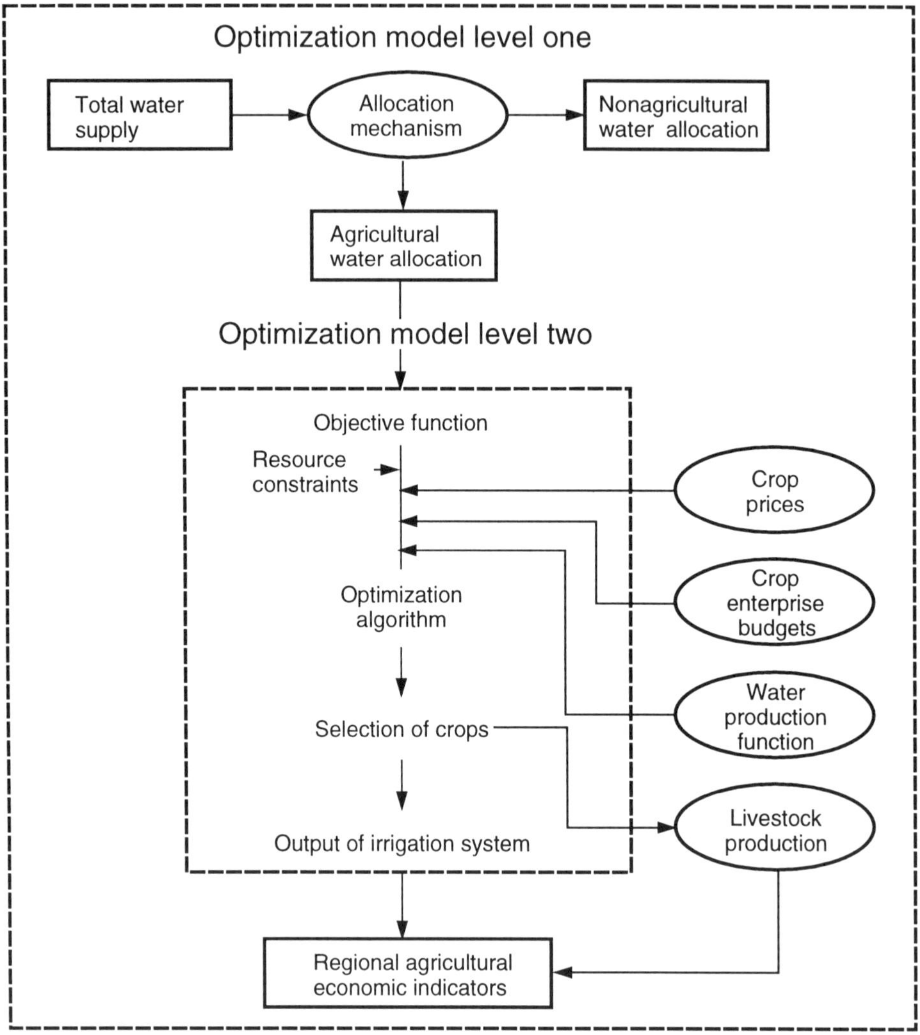

Figure 5.8. Scheme of modeling irrigation water use based on an optimization model.

An example of an optimization model for agricultural water use is shown in *Figure 5.8.* The sequence of calculations starts with the determination of water availability in the region, then water for nonagricultural uses is deducted. The remaining water is used for agricultural production. The activities in which this water is used are selected using the optimization routine given economic factors and the relationship between water and production. Water availability can now be varied and a new set of computations can be made. The difference between the results of the model for two water supply levels is the impact of change on the supply. If this supply change is related to climatic variability, then this result provides us with the impact of climate change on agricultural water use.

5.6 Implications of Climate-induced Water Use

Future climate change will affect the socioeconomic conditions facing many nations through its effect on water use levels and patterns. In some regions the effects may be catastrophic; some countries may benefit from the change. Although each situation warrants an examination on its own merit, some generic implications of climate-induced water demand can be drawn. Some of these implications are relatively obvious; others are more implicit. In this section, major implications of such a change are outlined.

5.6.1 Increased competition for water

Climate change will increase the competition for water in many countries and in regions within a country. Competition for fresh water between cities and rural areas is intensifying in many countries. Evidence of such competition can be found in the southwestern part of the USA, where increasing urban domestic water demand has forced some municipalities to bid for water rights from agricultural water users. The end result of these changes will be more disputes among water users within the same region. In some regions, a need for interregional transfers of water may arise; this would require sensitive handling in the wake of uncertainty created by climate change.

5.6.2 Water demand assessment

The collection of adequate and comprehensive information is essential for sound decision making. To make such decisions in anticipation of climate change, we first need to better understand the existing situation. However, in the past much of the data-collection process related to water resources has been concentrated on water supply aspects, on hydrology, and only recently and to a more limited extent on water quality monitoring. The collection of data on water demand levels and patterns of use has not attracted the attention of many national and international data-collection agencies in either developed or developing countries. For example, as late as the mid-1980s, for at least 25% of countries the data were more than 15 years old.[10] Analysis of water demand data, particularly with the objective of establishing trends, is virtually impossible. This deficiency was recognized by the International Conference on Water and the Environment, held in Dublin in 1992 (see WMO, 1992), where it was recommended that the public "have best feasible access to ... water-use data and information for those concerned in, or affected by, that water and their likely developments and demands."

The water demand assessment can be a good starting point for the identification of water resources management problems. Some of these problems can be solved by present water resources management projects and water resources management policies using known technology. Future problems connected with climate-induced changes are an important starting point for the identification of areas of research.

Water demand assessment as influenced by climate change is valuable because of the information it provides for policy making. Therefore, it is necessary to identify the projected regional and national water demand issues, to collect and assess relevant information, and to provide policy makers with a computer database that can be used in an interactive way for answering relevant questions.

5.6.3 Vulnerability of countries to changes in water availability

Climate-induced changes in availability of water coupled with changes in water demand may make some regions more vulnerable with respect to water resources (Nemec and Schaake, 1982; Nemec, 1986). Depending upon the relative demand for available water, a region or country may have a surplus of water or may face severe water shortages. Under climate change, when both water availability and water demand are affected, some countries that had a water surplus may experience water stress or may even face water scarcity. Which regions in the world are more vulnerable to water resources under the expected population growth and climate change? An answer to this question is provided by a recent study at the International Institute for Applied Systems Analysis (IIASA).[11] In this study, a country or region was classified under one of the following categories:

- Water surplus region.
- Water management problem region.
- Water stress region.
- Water scarce region.

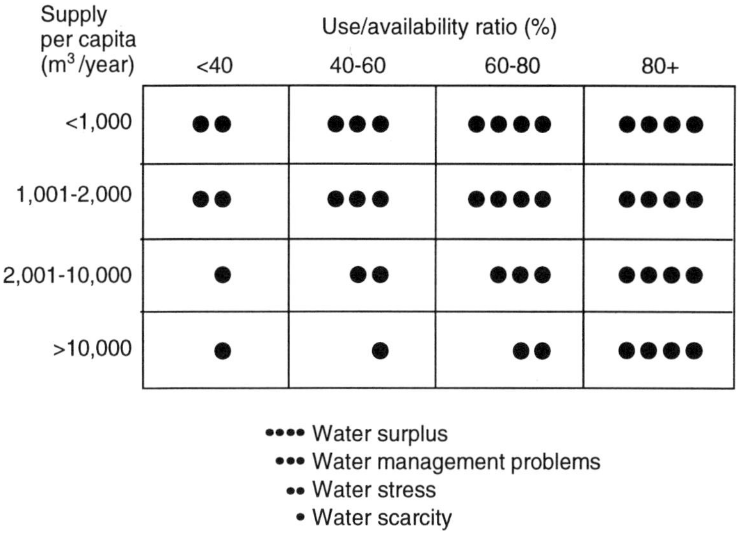

Figure 5.9. Joint availability and use criteria for identifying vulnerability of a region to water resources.

Criteria for each of these categories are shown in *Figure 5.9.* For each country, water availability and demand were estimated with and without climate change under the projected population for the year 2025. With respect to climate change, simple assumptions were made.[12]

- Per capita domestic and industrial consumption levels without climate change in the year 2025 would be at their respective 1990 level.
- Under the climate change scenario water supply in humid regions would increase by 10%, whereas in other regions it would decrease by 20% of the respective 1990 level.
- The domestic and industrial water demand per capita would increase by 5% of the 1990 level under the climate change.
- The irrigation water demand would increase by 10% under the climate change scenario.

The results of the study are summarized in *Table 5.5.* One conclusion that is inescapable is that if present trends of water demand continue, a majority of the world population will reside in regions with persistent water-related problems; the climate change will only make things worse. Without any climate-induced change, more than one-third of the world population will live in countries with acute water shortages. This situation is significantly worse than that during 1990, when only 6% of the total population

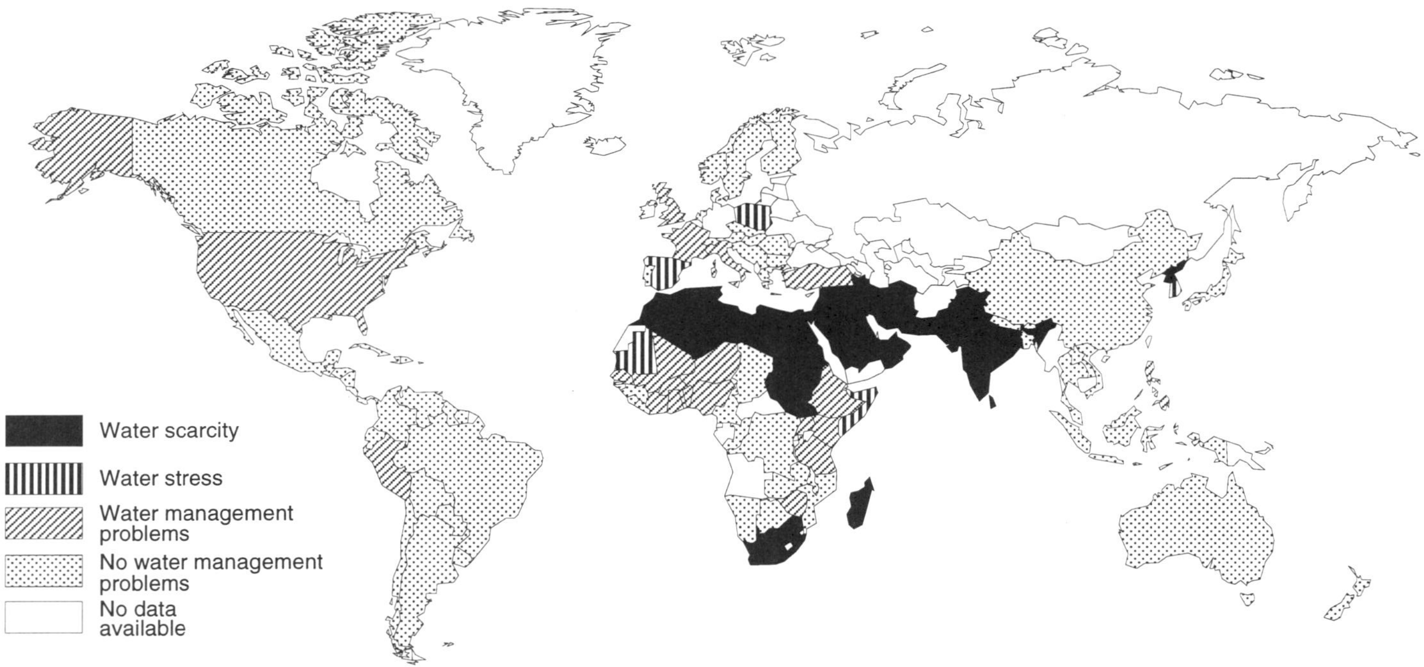

Figure 5.10. Distribution of regions of the world in the year 2025 by degree of vulnerability to water under climate change. Source: Strzepek *et al.* (1995).

Table 5.5. Distribution of world population by degree of vulnerability to water resources under alternate scenarios.

			Situation in 2025: no climate change		Situation in 2025: with climate change	
Degree of vulnerability	Population (millions)	% of total	Population (millions)	% of total	Population (millions)	% of total
Surplus	3,669	70	5,098	60	3,009	35
Management problems	1,320	24	468	6	643	7
Stress	154	3	369	4	1,801	21
Scarcity	136	3	2,513	30	2,995	37
Total	5,279	100	8,448	100	8,448	100

Source: Kulshreshtha (1993).

of the world faced water stress or water scarcity. One point of interest is that although climate change causes more strain on water resources, the situation has already gotten worse; even without the climate change, only 60% of the world population will be in a water surplus situation, and some 34% will face water stress or water scarcity.

The model presented by Kulshreshtha (1993) was re-estimated using information on the estimated change for various parts of the world by Strzepek *et al.* (1995) (*Figure 5.10*). These results suggest that the globe is further polarized in terms of water resources vulnerabilities. Currently, the UK and Poland are better off, but the situations in India and Algeria are worse. Under the climate change scenario the percentage of the population facing water stress rises dramatically, from 12% before climate change to 28% after climate change. Furthermore, the climate change may help those regions that may not need help at the expense of those that do.

5.6.4 Socioeconomic impacts

The degrees of vulnerability of various countries to climate change will have significant socioeconomic and political implications in those countries. A preliminary account of such changes is displayed in *Figure 5.11*. Climate change would affect all sectors of society. Some of these effects would be direct results of climate change. For example, Dudek (1989) reported that in California the net economic well-being generated by agricultural operations under climate change would be reduced by 14% to 17%. Similarly, Onyeji (1992) has reported a change in agricultural output of 6% to 99% for Egypt under different climate scenarios. Other economic impacts will be

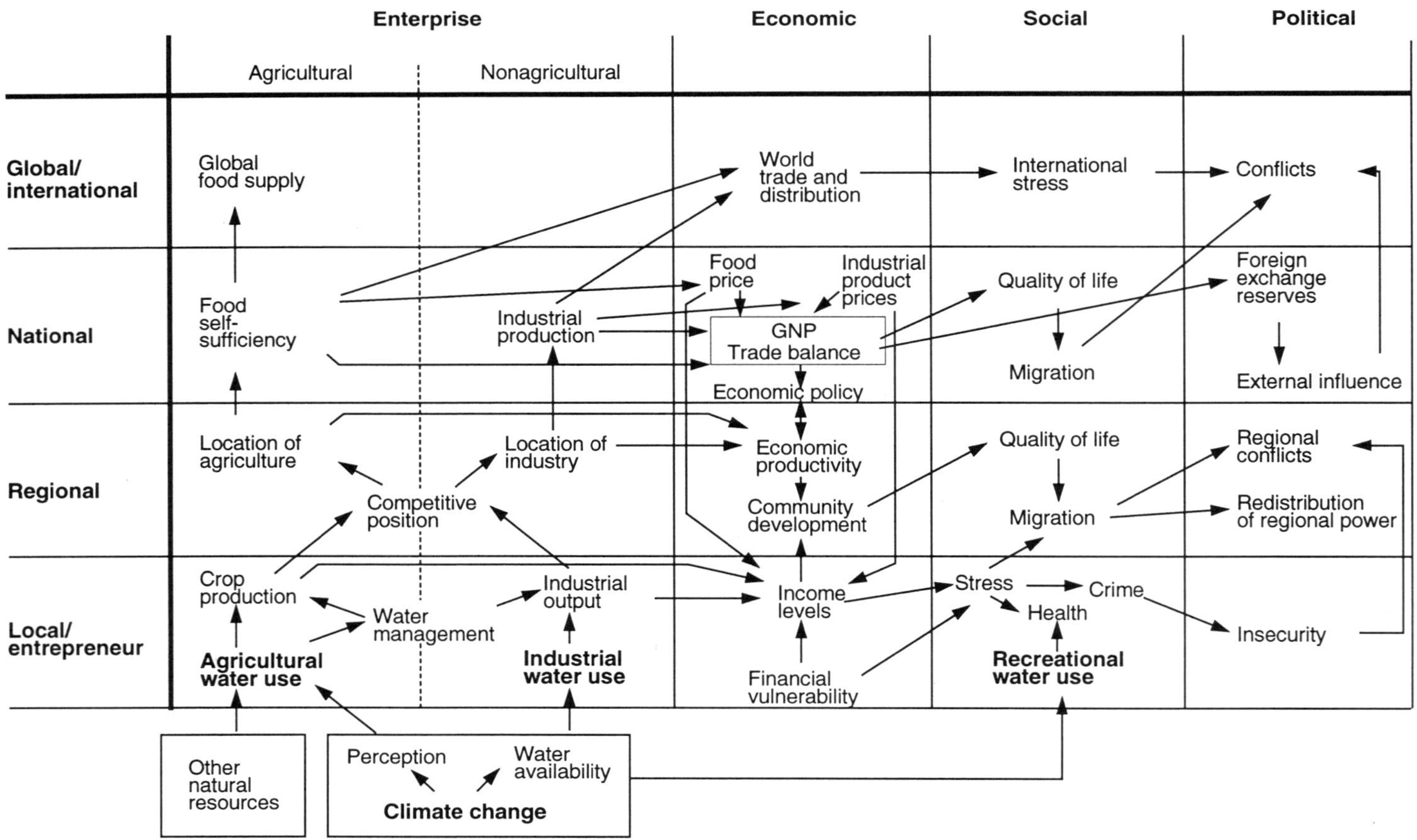

Figure 5.11. A preliminary list of socioeconomic/political impacts of climate change.

felt through the impact on nonagricultural production, through economic linkages. Extreme events, such as droughts or floods, would decrease the economic profitability of production in all sectors, but more so for agricultural production. This may eventually affect the competitive position of one region in relation to another, and may lead to interregional shifts in economic activity, followed by massive human migration. Adjustments in the regional pattern of agriculture in many countries could be triggered by climate change, which may eventually affect the agricultural and economic prospects at the regional or state level.

The impacts would also be felt at the national level, where food self-sufficiency may be threatened. If this gap is filled by imports, there are implications for the balance of payments, and therefore the vulnerability of the country to foreign influence. All these changes may result in more stress for the individual, and thus may have some health effects. The political effects of a changing climate may include human migration from one region to another, thereby weakening rural communities, causing a need for a redistribution of electoral boundaries, and leading at the global level to international conflicts. The conflicts among water users will not be confined to the international arena, but will also occur in various regions within a country. More conflicts will arise when a water resource is shared and a common management plan for the resource is not in place. Similar conflicts may also arise in situations where part of a country's water supply consists in an inflow of water from a neighboring country. Negotiations on these conflicts may well become the focal point of water resources management policies in the twenty-first century.

Climate change, through adjustments in various types of water use, will have other impacts that may be of high social significance. For example, the vulnerability of domestic water demand to climate change will be different in developed and developing countries. In developed countries the impact will be reflected in economic terms (Howe and Linaweaver, 1967). The increased water requirements for these countries can be met through technological change, for example, by building new water reservoirs or by reducing leakage and conveyance losses through reconstructing the water supply network.

In developing countries, however, there is a lack of capital, and this scarce capital has to be used to solve various social problems. A socially adequate water supply in these countries does not mean the coverage of all domestic water requirements, but rather the solution of the most critical situations. The developing countries, therefore, would be much more vulnerable to climate change than would the developed ones.

According to World Health Organization (WHO, 1980) studies, more than 83% of the population in developing countries are not adequately served by the public water resources supply. The organization of water management and the lack of capital investment funds are the central problems. The WHO, therefore, proposed a reduced objective, i.e., to supply water mainly through public stand posts. The expected climate change may adversely influence this minimum program in two ways: the potable water requirement will increase, but the scarce capital available for water management may even be reduced due to socioeconomic changes.

In assessing the potential impact of future climate changes on human activities, the fundamental subject of concern is the adaptive capacity of social and economic systems. Society in the past has adapted to similar changes. As climate change modifies human action, feedback will again occur as society adapts in a dynamic fashion (Warrick, 1989).

5.6.5 Integrated water resources management

Water resources management planners should be aware of the forthcoming challenge posed by climate change, and they should project and plan water resources in an appropriate manner. Such planning should be adaptable, robust, and resilient to withstand the possible future changes of climate. As the demand for dependable water supplies increases and supplies dwindle, the type of water resources development that will be needed under climate change will change as well. The simple withdrawal of water from streams or rivers will be replaced by the provision of water reservoirs to smooth out the irregularity in stream flow on the one hand, and inter-year variability in its use on the other. Measures will also have to be developed for proper management of ground water resources. According to Cook (1976), the ground water cycle is often so long that ground water can be taken as a nonrenewable resource; mining it at an increasing rate may leave future generations with water shortages. According to Williams (1989), possible water resources management responses under climate change may include the following: reallocating water from less valuable irrigated agriculture to municipal uses; changing agricultural methods; increasing incentives for integrated flood management; increasing incentives for watershed management; integrating ecosystem needs in water resources planning; and redesigning the operation of existing water projects. Kos (1986) has also suggested that in a multipurpose water resources system a transfer among the objectives may also produce a positive reaction to the impact induced by climate change, and thus may minimize the possible economic losses.

Most of the past assessment of climate change has concentrated on physical and biological aspects. The results of these studies need to be set within a framework of water resources management, agriculture, and economic analysis (Orlovsky and van Walsum, 1984). An adjustment in regional agricultural patterns may also take place in response to climate change, the impacts of which should be linked to more general approaches for developing optimal water resources management policies in the future (Skogerboe, 1982; Heady *et al.*, 1973).

The price of water may increase due to a reduction in water availability, which will further push the water demand down, causing additional socioeconomic problems. This water supply situation is complicated by the decreasing quality of water and the increasing need for water treatment. Under these conditions, the management of water quality may have to be given top priority in water policy making. It may be cheaper to solve the problem before it happens than to wait until it has major impacts on society.

Meo (1991) distinguished three types of strategies for adapting to the potential climate change: the no-action strategy, the stand-and-defend strategy, and the strategic-retreats strategy. If water resources management in the future is based on the no-action strategy, the outcome would be as described here. The stand-and-defend strategy includes structural and nonstructural measures to counter the impact of climate change. The strategic-retreats strategy includes measures for adapting to climate change impacts without directly countering them. A pertinent question is which of these strategies would make the most sense from a water resources management point of view? Complications are brought about by the fact that the environmental and socioeconomic systems are also closely interrelated at the local and regional levels, as suggested by Kairiukstis (1989); such linkages should be considered in the development of water resources management policies.

Water resources conservation will need to be elevated to the status of a major goal of water resources policy under the uncertainty created by climate change. Some kinds of water resources are renewable, such as surface water. As long as a reliable supply of renewable water exceeds demand there is no advantage to restricting the level of usage below water requirements. However, ground water can be taken to be a nonrenewable resource; increasing rates of ground water mining may lead to water shortage problems. In some regions that are supplied primarily by this kind of ground water, climate-induced changes may even worsen the situation. Optimal ground water use has to take into account not only the consequences of alternative courses of action in the near future, but also the effect of the climate-induced increase in water requirements. In such a case there is a strong need for conservation policy designed to protect the public interest under the hydrological and climatic uncertainties.

5.6.6 Demand management

With water supplies dwindling and water use increasing under climate change, many nations will have to institute proper demand management measures. This matter has been treated extensively by many researchers and has been summarized by the United Nations Department of Technical Cooperation for Development (UNDTCD, 1992). A properly developed demand management system can improve the efficiency of water use and reduce the need for massive investments in water supply infrastructure without bringing undue hardship on the users. The plan by its very nature cannot be generic, because measures need to be evaluated on their own merits for a given situation.

Under the looming fear of global warming, demand management measures would have to be devised to ensure that water demand is balanced with water availability; that acceptable quality of water is maintained; that water use during periods of drought is carefully managed; that seasonal water availability is matched with use; that inter-year variability in supply is managed through adjustments in water use levels; and that water is reused (Frankel, 1967). Each of these may require different strategies and, therefore, should be considered on a country-specific basis. In the following discussion, we have presented some general principles that might be applicable in some situations.

Management of Withdrawal Demand for Water through the Use of Economic Instruments

Water demand management is based on the premise that the behavior of the individual can be modified. Economic incentives (or penalties) can be provided to induce the desired behavior. However, one should note that economic methods are not the only means by which an individual's behavior can be changed. Through communicating with the users, one may be able to persuade them to alter their water use levels.

Among various economic methods, the institution of tariffs for water use is the predominant one. Tariffs can be uniform rate or block rate.

Under uniform rate tariffs, all water use is charged at the same rate. Under the block-rate system, increasing quantities of water may cost more (the increasing block-rates system) or less (the decreasing block-rates system). Based on the analysis of households in Saskatchewan, of these three pricing structures increasing block-rate pricing is the most consistent with water conservation.[13] Tariffs can also be devised on a seasonal basis to induce more water conservation during certain times of the year. However, before tariffs can be effective, the measurement of water use is an important prerequisite.

Management of In-situ Demand

Although management of withdrawal demand for water is obvious, under climate change *in-situ* water demand management may be just as important. Because under a warmer climate recreational activities are expected to increase, there may also be several environmental impacts. Overcrowded beaches have a decreased environmental and recreational value. Some recreational water uses (e.g., boating) can be considered part of the pollution in a system (e.g., noise generated and oil and gas leaks). These sources of pollution may increase in some regions due to the increased number of participants in water-related recreational activities. All these negative environmental impacts may grow in number and frequency with the expected climate change.

For bulk commodities, inland water transportation is the most economical means of transport. In transitional economies and in developing countries, unless the water depth in big rivers exceeds the required minimum, inland water navigation is hindered by the lack of capital investment that it often requires. The environmental impact of the navigation structures (e.g., canals) is often negative due to changes in the landscape value. Therefore, the combination of these environmental problems with the lack of capital would hinder the planning and building of large waterways.

Management of Indirect Water Demand

With dwindling water supplies and increased competition for water use, management of direct water use may not be sufficient. These measures may need to be supplemented by management of indirect water use. Electric power generation uses relatively large amounts of water. The most wasteful process in some countries is the direct use of electricity for heating, particularly at moderately low temperatures. Some of this waste can be reduced by measures that increase heating efficiency through heat recovery and better insulation of houses. Climate change may increase the temperature; consequently, summer cooling requirements may also grow. In the future, we may approach a situation where the seasonal demand for heating and cooling complement each other. Because in conventional heat production the real efficiency is extremely low, one could make a strong case for the introduction of some form of thermodynamic heating, e.g., the use of industrial waste heat, the combination of power and heat generation, or the use of thermally driven heat pumps.

5.6.7 Planning of irrigation systems

Because irrigation water demand is a significant use of water in many countries of the world, the planning of irrigation development is closely related to the planning of water resources systems. According to Holy and Kos (1981), the stochastic character of irrigation water requirements by irrigation systems is generally recognized, although it is not always observed in models of water resources systems in long-term planning.

In planning these systems, irrigation water requirements should be related to meteorological factors through an appropriate mathematical model (Kos, 1982a). The climate impact on agriculture might be caused either by a gradual change in the average climate or by a change in climatic risks. The two aspects produce different results. The first may result in a slow deterioration of the conditions of agricultural production; the second may cause changes in the frequencies of drought. These effects would not be uniformly distributed throughout the world. For example, in the Northern Hemisphere, the increase in irrigation water requirements would be higher through droughts than through the change in the mean temperature.

A major implication of climate change for irrigation systems would be the change in the design parameters. Some systems may have to be flexible and robust enough to withstand the increase of irrigation water requirements. An example of an estimation technique for irrigation system parameter estimation as implemented for the North Caucasus case study is presented in Chapter 12 of this monograph. The other issues that need to be considered by watershed non-point pollution models are impacts on soil erosion, surface and subsurface runoff pollution under various strategies of irrigation development, and climate scenarios.

5.6.8 International and interregional conflicts

Global climate change appears to be the most likely change that would affect international politics because of its wide scope (Gleick, 1989). Although the societal impacts of climate change will be almost the same for developing or developed countries, their respective abilities to adjust to the change are substantially different. In fact, the less developed countries are more vulnerable because they have fewer means to cope with climate change (see Karpe *et al.*, 1990). Also, the growing population in many developing countries represents a further stress on water resources and may lead to conflicts among water users both within the country and internationally. The situation will be worse for countries that share water resources, such as lakes, and for those where a large part of the water supply is obtained throughflows from other countries.

5.6.9 Legal and institutional development

Climate change would require changes in legal and administrative structures, both nationally and internationally. In various countries a need would emerge to develop a clear and concise water resources management policy and appropriate legal structures. Internationally, a need to develop management policies and supporting institutions for shared water courses would emerge. Mechanisms to settle disputes in such situations would also have to be devised.

Climate change may also bring about a major challenge for human resources development. To implement the new water resources management mandate, training of water resources managers, researchers, and policy makers will be required. Strengthening the training program to meet the needs of many developing countries will be a mammoth effort.

5.7 Conclusions

Climate change will have an extensive effect on water demand, both directly and indirectly, through changes in supply, year-to-year differences in availability, changes in water quality, or changes in intra-year distribution. Rising temperatures will increase the demand for urban domestic water use and for agricultural water use. Warmer water and altered intra-year distribution will also affect water quality and will worsen the existing balance between water supply and demand. Global climate change may also be instrumental in provoking disputes both within a country and internationally, and in altering international relations between developed and developing countries. These impacts would be felt through changes in water quantity and its use, as well as through the impact of climate change on the food supply.

The problems of future water supply–demand integration are central to climate impact assessment. The development of agricultural systems and their adaptive management strategies must be considered. These policies should be oriented to more resilient and robust agricultural systems that make sense in all situations, whether the climate changes or not.

Climate change creates another dimension in the uncertainty facing water resources management. Its effect would be realized both through the withdrawal type of water use and through increased demand for *in-situ* water use. The inability of countries to cope with these problems will have many socioeconomic and political impacts. Different nations may be affected by the climate change in different ways, and their reactions may differ. Therefore, as conditions evolve the prediction of future water demand must take place in the context of economic, demographic, and environmental development. The use of alternate scenarios of development and the use of integrated models (integrating water use with socioeconomic impacts) has merits. Integrated water management in the future must rise above disciplinary boundaries. There is an urgent need to train social scientists to work with natural scientists to seek a joint understanding of the future of mankind's climate and its likely impacts.

Notes

[1] According to Beck (1980) and McKean and Ericson (1975), the environmental, social, and climatological relations form an open system, created by the relation between man and nature and the relations within society.

[2] For an excellent treatment of the need for forecasts see Gardiner and Herrington (1986).

[3] Factors affecting demand of water for various uses have been reviewed by Kindler and Russell (1984).

[4] This is based on the situation in Canadian electric utilities power generation plants, excluding hydropower stations, as reported by Tate and Scharf (1985b).

[5] Penman's formula was derived for the semi-humid zone and has been recommended by organizations such as the World Meteorological Organization (WMO, 1966).

[6] Penman's equation for PET exists in a number of forms. The modification by the FAO is preferred here (see FAO, 1979). The formula is not discussed in detail, only the main parts that are important for the derivation of the model of irrigation water requirements are presented; more details are available in Kos (1982a).

[7] This assumption was discussed by Kos (1982a) and was found to be quite realistic.

[8] Such willingness-to-pay studies have been carried out for North American recreational sites. For a review of such studies see Kulshreshtha (1991), Merewitz (1966), and Gibbons (1986).

[9] An application of a simulation model for estimating the impacts of drought on farmers and on the region is provided by Kulshreshtha and Klein (1989).

[10] This information is based on the data on freshwater availability and use as reported by the World Resources Institute (1986).

[11] For details of this study, see Kulshreshtha (1993).

[12] These assumptions were necessary due to the lack of concrete evidence on the exact nature of the climate-induced impact on both water supply and water use levels.

[13] For details, see Brockman *et al.* (1987).

References

Anderson, Jr., R.J., 1982, The probability distribution of water inputs and the economic benefits of supplementary irrigation, *Water Supply & Management*, **6**(3):281–291.

Baumann, D.D., 1969, Perception and public policy in the recreational use of domestic water supply reservoirs, *Water Resources Research*, **5**(3):543–554.

Beck, M.B., 1980, Hard or Soft Environmental Systems, WP-80-25, International Institute for Applied Systems Analysis, Laxenburg, Austria.

Benedick, R.E., 1989, The Montreal ozone protocol: Lessons for global warming, in J.C. Topping, Jr., ed., *Coping with Climate Change*, The Climate Institute, Washington, DC, USA.

Brockman, J.L., Kulshreshtha, S.N., and O'Grady, K.L., 1987, *Municipal Water Use*, Saskatchewan Water Corporation, Moose Jaw, Saskatchewan, Canada.

Brown, L.R., 1989, What does global change mean for society? in R.S. DeFries and T.F. Malone, eds., *Global Change and Our Common Future: Papers from a Forum*, National Academy Press, Washington, DC, USA.

Budyko, M.I., 1989, Climatic conditions of the future, in World Meteorological Organization, *Conference on Climate and Water*, Volume 1, 11–15 September, at Helsinki, Finland, Valtion Painatuskeskus (VAPK), Helsinki, Finland.

Byrne, J.M., Barendregt, R., and Schaffer, D., 1989, Assessing potential climate change impacts on water supply and demand in Southern Alberta, *Canadian Water Resources Journal*, **14**(4):5–15.

Chen, R.S., and Parry, M., 1987, *Policy-Oriented Impact Assessment of Climatic Variations*, RR-87-7, International Institute for Applied Systems Analysis, Laxenburg, Austria.

Cicchetti, C.J., Smith, V.K., Knetsch, J.L., and Patton, R.A., 1972, Recreation benefit estimation and forecasting, *Water Resources Research*, **8**(4):840–850.

Cohen, S.J., 1985, Effects of climatic variations on water withdrawals in metropolitan Toronto, *Canadian Geographer*, **29**(2):113–122.

Cohen, S.J., 1987, Projected increases in municipal water use in the Great Lakes due to CO_2-induced climate change, *Water Resources Bulletin*, **23**(1):91–101.

Cook, E., 1976, Limits to exploitation of nonrenewable resources, in H.E. Daly, ed., *Economics, Ecology, Ethics*, W.H. Freeman & Co., San Francisco, CA, USA.

Cure, J.D., and Acock, B., 1986, Crop response to carbon dioxide doubling: A literature survey, *Agricultural and Forest Meteorology*, **38**:127–145.

David, E.L., 1968, Lake shore property values: A guide to public investment in recreation, *Water Resources Research*, **4**(4):697–707.

Dee, N., Baker, J., Drobny, N., Duke, K., Whitman, I., and Fahringer, D., 1973, An environmental evaluation system for water resource planning, *Water Resources Research*, **9**(3):523–527.

Dossat, R.J., 1978, *Principles of Refrigeration*, John Wiley & Sons, New York, NY, USA.

Dudek, D.J., 1989, Potential climate change effects on irrigated agriculture in California, in J.C. Topping, Jr., ed., *Coping with Climate Change*, The Climate Institute, Washington, DC, USA.

Dworkin, D.M., 1975, Water reuse: A flexible and efficient management alternative for municipal water supply, *Water Resources Research*, **11**(5): 607–615.

Edinger, J.E., Duttweiler, D.W., and Geyer, J.C., 1968, The response of water temperatures to meteorological conditions, *Water Resources Research*, **4**(5):1137–1143.

EPA (Environmental Protection Agency), 1989, *The Potential Effects of Global Climate Change on the United States*, Report to Congress, Washington, DC, USA.

FAO (Food and Agricultural Organization), 1979, Agrometeorological crop monitoring and forecasting, *Plant Production and Protection*, Paper No. 17, Food and Agricultural Organization, Rome, Italy.

Fleming, G., 1975, *Computer Simulation Techniques in Hydrology*, Elsevier, New York, NY, USA.

Frankel, R.J., 1967, Viewing water renovation and reuse in regional water resources systems, *Water Resources Research*, **3**(1):57–61.

Gardiner, V., and Herrington, P., 1986, The basis and practice of water demand forecasting, in V. Gardiner and P. Herrington, eds., *Water Demand Forecasting*, Geo Books, Norwich, UK.

Gibbons, D.C., 1986, *The Economic Value of Water*, Resources for the Future, Washington, DC, USA.

Giorgi, F., and Mearns, L.O., 1991, Approaches to the simulation of regional climate change: A review, *Reviews of Geophysics*, **29**(2):191–216.

Gleick, P.H., 1987, Regional hydrologic consequences of increases in atmospheric CO_2 and other trace gases, *Climatic Change*, **10**(2):137–161.

Gleick, P.H., 1989, Climate change and international politics: Problems facing developing countries, *Ambio*, **18**(6):333–339.

Griffen, R.C., and Chang, C., 1990, Pretest analysis of water demand in 30 communities, *Water Resources Research*, **26**(10):2251–2255.

Hanke, S.H., and de Maré, L., 1984, Municipal water demands, in J. Kindler and C.S. Russell, eds., *Modeling Water Demands*, Academic Press, New York, NY, USA.

Heady, E.O., Madsen, H.C., Nicol, K.J., and Hargrove, S.H., 1973, National and international models of water demand, land use, and agricultural policies, *Water Resources Research*, **9**(4):777–791.

Hekstra, G.P., 1990, Sea-level rise: Regional consequences and responses, in N.J. Rosenberg, ed., *Greenhouse Warming: Abatement and Adaptation*, Resources for the Future, Washington, DC, USA.

Holy, M., and Kos, Z., 1981, Water resources planning for irrigation systems, *Water Supply and Management*, **5**:195–200.

Howe, C.W., and Linaweaver, Jr., F.P., 1967, The impact of price on residential water demand and its relation to system design and price structure, *Water Resources Research*, **3**(1):13–32.

Hufschmidt, M.F., 1965, Field level planning of water resource systems, *Water Resources Research*, **1**(2):147–163.

Jacoby, H.D., 1990, Water quality, in P.E. Waggoner, ed., *Climate Change and U.S. Water Resources*, John Wiley & Sons, New York, NY, USA.

Kairiukstis, L., 1989, Ecological sustainability of regional development: Background to the problem, in L. Kairiukstis, A. Buracas, and A. Straszak, eds., *Ecological Sustainability of Regional Development*, International Institute for Applied Systems Analysis, Laxenburg, Austria.

Karpe, H.J., Otten, D., and Trinindade, S.C., eds., 1990, *Climate and Development*, Springer-Verlag, Berlin, Germany.

Kerr, R.A., 1990, New greenhouse report puts down dissenters, *Science*, **249**(3):481–482.

Kindler, J., and Russell, C.S., eds., 1984, *Modelling Water Demand*, Academic Press, London, UK.

Kos, Z., 1970, Simulation models of water supply for irrigation in Water Resources Systems, *ICID Bulletin*, **19**(1):7–52.

Kos, Z., 1982a, *Stochastic Water Requirements for Supplementary Irrigation in Water Resources Systems*, RR-82-34, International Institute for Applied Systems Analysis, Laxenburg, Austria.

Kos, Z., 1982b, Operation of water resources systems in Czechoslovakia, in Z. Kaczmarek and J. Kindler, eds., *The Operation of Multiple Reservoir Systems*, CP-82-S3, International Institute for Applied Systems Analysis, Laxenburg, Austria.

Kos, Z., 1986, Design of water management systems: General report, *Seminar on Water Management Systems*, UN ECE, Bratislava, Czechoslovakia.

Kulshreshtha, S.N., 1991, Estimation of value of water for water-based recreation in Saskatchewan, *Canadian Water Resources Journal*, **16**(3):1–16.

Kulshreshtha, S.N., 1993, *World Water Resources and Regional Vulnerability: Impact of Future Changes*, RR-93-10, International Institute for Applied Systems Analysis, Laxenburg, Austria.

Kulshreshtha, S.N., and Klein, K.K., 1989, Agricultural drought impact evaluation model: A system's approach, *Agricultural Systems*, **30**(1):81–96.

Kulshreshtha, S.N., and Spriggs, J.D., 1982, *Water use in Saskatchewan: Present and future levels*, Saskatchewan Drought Study Element No. 11, Prairie Farm Rehabilitation Administration, Regina, Canada.

Loucks, D.P., French, P.N., and Taylor, M.R., 1982, Water resources and environmental planning using interactive graphic, *Water Supply & Management*, **6**(4):303–320.

Manabe, S., and Wetherald, R.T., 1987, Large-scale changes of soil wetness induced by an increase in atmospheric carbon dioxide, *Journal of the Atmospheric Sciences*, **44**(8):1211–1235.

McKean, J.R., and Ericson, R.K., 1975, Quality of life and environmental social overhead impact, *Water Resources Research*, **11**(6):789–790.

Mearns, L.O., Katz, R.W., and Schneider, S.H., 1984, Extreme high temperature events: Changes in their probabilities with changes in mean temperatures, *Journal of Climate and Applied Meteorology*, **23**(12):1601–1613.

Meo, M., 1991, Policy-oriented climate impact assessment: The Tennessee Valley Authority and Apalachicola Bay, *Global Environmental Change*, **1**(2):124–138.

Merewitz, L., 1966, Recreational benefits of water resources development, *Water Resources Research*, **2**(4):625–640.

Middleton, D.J., 1981, It all depends on clean water, Environmental Quality News, *The Conservationist*, **36**(2).

Mintzer, I.M., 1989, Developing country energy strategies: An overview of the climate implications of energy strategies, in J.C. Topping, Jr., ed., *Coping with Climate Change*, The Climate Institute, Washington, DC, USA.

Morton, F.I., 1983, Operational estimates of areal evapotranspiration and their significance to the science and practice of hydrology, *Journal of Hydrology*, **66**(1/4):1–76.

Murdock, S.H., Hamm, R.R., Albrecht, D.E., and Parpia, B., 1986, *An Analysis of the Effects of Socio-demographic Factors of Daily Per Capita Residential Water Use in Texas City*, Draft report, Texas Water Resources Institute, College Station, TX, USA.

Nemec, J., 1986, Assistance to design and operation of water management systems by activities of the World Meteorological Organization, *Seminar on Water Management Problems*, UN ECE, Bratislava, Czechoslovakia.

Nemec, J., 1988, Implications of a changing atmosphere on water resources, in World Meteorological Organization, *The Changing Atmosphere: Implications for Global Security*, Conference proceedings, World Meteorological Organization, Geneva, Switzerland.

Nemec, J., and Schaake, J., 1982, Sensitivity of water resource systems to climate variations, *Hydrological Sciences Journal*, **27**(3):327–343.

O'Connor, D.J., 1967, The temporal and spatial distribution of dissolved oxygen in streams, *Water Resources Research*, **3**(1):65–79.

Onyeji, S.C., 1992, *A Socio-economic Analysis of Integrated Climatic Change Impacts in Egypt*, A report prepared for the US Environmental Protection Agency, International Institute for Applied Systems Analysis, Laxenburg, Austria.

Orlovsky, S.A., and van Walsum, P.E.V., 1984, Water Policies in Regions with Intense Agriculture, WP-84-40, International Institute for Applied Systems Analysis, Laxenburg, Austria.

Parry, M.L., 1986, Some implications of climate change for human development, in W.C. Clarke and R.E. Munn, eds., *Sustainable Development of the Biosphere*, Cambridge University Press, New York, NY, USA.

Parry, M., and Carter, T.R., 1989, The assessment of the effects of climatic variations on agriculture: Aims, methods, and summary of results, in M.L. Parry, T.R. Carter, and N.T. Konijn, eds., *The Impact of Climatic Variation on Agriculture, Volume I*, Kluwer Academic Publishers, Dordrecht, The Netherlands.

Parry, M., and Jiachen, Z., 1991, The potential effect of climate changes on agriculture, in J. Jäger and H.L. Ferguson, eds., *Climate Change: Science, Impacts and Policy*, Cambridge University Press, New York, NY, USA.

Parry, M., Carter, T.R., and Konijn, N.T., 1989, *The Impact of Climatic Variations of Agriculture. Volume 1: Assessments in Cool Temperate and Cold Regions. Volume 2: Assessments in Semi-Arid Regions*, Kluwer Academic Publishers, Dordrecht, The Netherlands.

Penman, H.L., 1954, *Evaporation Over Parts of Europe*, Proceedings IAHS General Assembly, Publication No. 38, International Association of Hydrological Sciences, Rome, Italy.

Priazhinskaya, V.G., 1989, An assessment of climate change impacts on water resources management in irrigated zones, *Conference on Climate and Water*, Volume 2, 11–15 September, at Helsinki, Finland, Valtion Painatuskeskus (VAPK), Helsinki, Finland.

Riebsame, W.E., and Magalhaes, A.R., 1991, Assessing the regional implications of climate variability and change, in J. Jäger and H.L. Ferguson, eds., *Climate Change: Science, Impacts and Policy*, Cambridge University Press, Cambridge, UK.

Rind, D., Goldberg, R., and Reudy, R., 1989, Changes in climate variability in the 21st century, *Climate Change*, **14**(1):5–37.

Rodda, J.C., Downing, R.A., and Law, F.M., 1976, *Systematic Hydrology*, Newness-Butterford, London, UK.

Scott, M.J., Rosenberg, N.J., Edmonds, J.A., Cushman, R.M., Darwin, R.F., Yohe, G.W., Libertrau, A.M., Hunsaker, C.T., Bruns, D.A., DeAngelis, D.L., and Hales, J.M., 1990, Consequences of climate change for the human environment, *Climate Research*, **1**(1):63–79.

Seuna, P., 1977, *On the Hydro-meteorological Factors Affecting Irrigation*, Vesihallitus, Helsinki, Finland.

Shiklomanov, I., 1990, Water consumption, water availability and large-scale water projects in the world, in *The Impact of Large Water Projects on the Environment*, UNESCO, Paris, France.

Skogerboe, G.V., 1982, Planning framework and modelling of water quality from rain-fed agricultural lands, *Water Supply & Management*, **6**(5):359–374.

Skogerboe, G.V., *et al.*, 1982, Development process for improving irrigation water management, *Water Supply & Management*, **6**(4):329–342.

Stevens, J.B., 1967, Recreation benefits from water pollution control: A further note on benefit evaluation, *Water Resources Research*, **3**(1):63–64.

Strzepek, K., Nieman, J., Somlyódy, L., and Kulshreshtha, S.N., 1995, A global assessment of national water vulnerabilities: Sensitivities, assumptions, and driving forces, International Institute for Applied Systems Analysis, Laxenburg, Austria.

Swaminathan, M.S., 1991, Agriculture and food systems, in J. Jäger and H.L. Ferguson, eds., *Climate Change: Science, Impacts and Policy*, Cambridge University Press, New York, NY, USA.

Tate, D.M., and Scharf, D.N., 1985, *1981 Industrial Water Use Survey Tables, Volume II*, Environment Canada, Ottawa, Canada.

Turnovsky, S.J., 1969, The demand for water: Some empirical evidence on consumers' response for a commodity uncertain in supply, *Water Resources Research*, **5**(2):350–361.

United Nations, 1978, *Register of International Rivers*, Pergamon Press, Oxford, UK.

UNDTCD (United Nations Department of Technical Cooperation for Development), 1992, *Legislative and Economic Approaches to Water Demand Management*, United Nations, New York, NY, USA.

Warrick, R., 1989, Report on session IV: Expected impacts, in R. Fantechi and A. Ghazi, eds., *Carbon Dioxide and Other Greenhouse Gases: Climatic and Associated Impacts*, Commission of the European Communities, Brussels, Belgium.

Warrick, R., and Bowden, M., 1981, Changing impacts of drought in the Great Plains, in M. Lawson and M. Baker, eds., *The Great Plains, Perspectives and Prospects*, Center for Great Plain Studies, Lincoln, NE, USA.

Williams, P., 1989, Adapting water resources management to global climate change, *Climatic Change*, **15**(1–2):83–93.

Woodwell, G.M., 1990, The effects of global warming, in J. Leggett, ed., *Global Warming: The Greenpeace Report*, Oxford University Press, New York, NY, USA.

WHO (World Health Organization), 1978, Malaria control: A recognized strategy, *WHO Chronicle*, **32**(6):226–230.

WHO (World Health Organization), 1980, A challenge to developing countries: The international drinking water supply and sanitation decade, 1981–1990, *WHO Chronicle*, **34**(9):327–331.

WMO (World Meteorological Organization), 1966, *Measurement and Estimation of Evaporation and Evapotranspiration*, WMO Report No. 201, TP 105, Technical Note No. 83, Geneva, Switzerland.
WMO (World Meteorological Organization), 1992, *International Conference on Water and the Environment: Development Issues for the 21st Century*, 26–31 January, at Dublin, Ireland, Report of the conference, WMO, Geneva, Switzerland.
World Resources Institute, 1986, *World Resources 1986*, Basic Books Inc., New York, NY, USA.

Chapter 6

Climate Change and Water Resources Planning

Zdzislaw Kaczmarek, Zbigniew Kundzewicz, and Valentina Priazhinskaya

6.1 The Issue: From Climate to Water Resources

Water resources engineers and decision makers have always been confronted with random properties of hydrological phenomena and uncertain projections of demographic and economic conditions, which influence the reliability and efficiency of water systems. The addition of one more uncertainty – the issue of climate change – seriously complicates the practice of water resources planning. However, water resources planners are prepared to deal with variability of natural processes and various types of uncertainties. Water resources projects are planned and managed so that water will be available for various kinds of users in sufficient quantity and good quality, and so that human lives and economic activity will be protected from the harmful effects of extreme hydrological situations. Most hydraulic structures are implemented to cope with floods and droughts, i.e., to mitigate negative social and economic consequences of the natural variability of water resources. The climate change expected to be caused by increased emissions of carbon dioxide and other greenhouse gases is still highly uncertain in terms of its effect on regional precipitation, storm events, potential evapotranspiration (PET), and other characteristics usually used as input data for hydrological models. Any effect climate change has on these characteristics used for input data will affect water resources planning, because

Z. Kaczmarek et al. (eds.),
Water Resources Management in the Face of Climatic/Hydrologic Uncertainties 168–193.

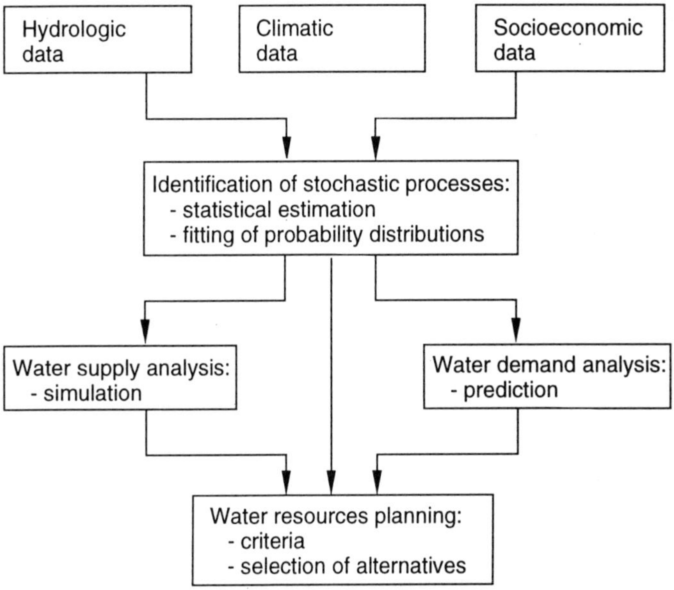

Figure 6.1. Diagram of water management methodology.

the results of calculations using these models, in turn, form a database for water resources investigations, and, in particular, for planning future water investments.

For hundreds of years, people have adapted their habits and economic activities to what they assumed were the natural climatic and hydrological conditions. Implicit in this assumption was that these conditions were stable. Despite the knowledge that, through dams, diversions, and water intakes, man has altered natural hydrological regimes, the fundamental assumption of stationarity of key hydrological processes still dominates the planning and designing of water resources development projects. This assumption may no longer be valid, because projected global environmental disturbances can cause serious alterations in the stochastic properties of hydrological time series. Moreover, the long time scales usually associated with large-scale regional water projects make them particularly sensitive to anthropogenic climate change.

The commonly accepted water resources management methodology (*Figure 6.1*) is based on the assumption that the probability distributions of water supply in the next decades can be evaluated on the basis of recorded hydrological data. This is certainly not true in the case of nonstationary climatic and hydrological processes, because information contained in the past records is of little value for estimating present or future probabilistic structure of these processes. Georgakakos and Yao (1993), discussing hydrological input to water management systems, state that "existing data records are atypical of future input realizations due to natural or anthropogenic causes (e.g., global climate changes). On such occasions, probabilistic input characterizations are inappropriate even in the average sense."

Also, water demand values must be forecast in a way that is compatible with assumed climate scenarios. Transformation of the structure and characteristics of the water requirements for the population and various branches of the economy should be expected, in particular due to a change in agricultural water demands. For example, a study by Peterson and Keller (1990) on the impact of climate on irrigation in the USA shows that for various climate scenarios the potential net water requirement for agriculture increases as the climate warms, because of the extended growing season, regional shifts in crops, and increased PET. These conclusions are, however, questioned by other scientists because of the CO_2 fertilization effect on actual evaporation and the possible increase in regional precipitation. Very few studies are reported in the literature on the possible implications of global warming on urban and industrial water demands. Water resources planners usually assume increasing demand due to population growth and economic development, but in many cases their projections do not match actual water demand. It is even more difficult to predict changes in water demand caused by climatic perturbations, because of the many possibilities for water conservation (Kaczmarek and Kindler, 1989).

As was said above, the issue of climate change adds a new dimension to the old problem of water resources planning for uncertain future water supply and demand. There are two crucial questions:

- What will be the scale of possible disturbances in water resources management caused by changes in global geophysical processes?
- How adjustable are the water resources systems affected by these changes, and what might be the costs of necessary adaptive measures?

The first question may be answered by improving forecasts of climatic and hydrological phenomena, as well as of climate-dependent water requirements. This is discussed in other parts of the book. To answer the second question some vision of future water resources management policy should be formulated. To some extent, we may learn something from the past experience by looking, for instance, at the performance of water systems during periods of long-lasting extreme climatic situations.

Water balance elements – runoff, evaporation, and catchment storage – vary with time; the extremes of their temporal variations manifest in floods and droughts. When planning in a stationary stochastic environment, decisions are usually made with due regard to water supply fluctuations. Most of the existing water resources projects are able to withstand severe stress because of their conservative design and the surplus of water in "normal" climatic conditions. In many cases an ample margin of safety assures the users of an acceptable water supply in extreme situations. An investigation of the past water resources management policy for a number of drought periods, carried out recently in an industrial region in southern Poland, lead to the conclusion that even a 20% to 30% decrease of water supply had a negligible effect on most municipal and industrial water users. In Sweden, with its abundant water resources, the overall water use has decreased substantially over the last two decades due to technological improvements and increased environmental concern. However, adapting to drought conditions may be much more difficult in regions where water supply and use are closely matched. Riebsame (1988), discussing several cases of both successful and failed adjustments to extreme climatic events, comes to the conclusion that "This record of mixed success in social adjustment should concern us in the face of future climate change." The assessment of present water resources systems' vulnerability to extreme situations is a good starting point for predicting their performance in a nonstationary environment.

6.2 Water Resources Planning Techniques in an Uncertain Climatic Environment

Conflicts are inherent in regional water resources management, particularly in regions with scarce water resources. The risk of conflicts will be enhanced if decision makers are faced with increased uncertainty about future water supply and demand. Cohon and Revelle (1980) identify four groups of objectives important for rational and sustainable planning: economic efficiency, environmental quality, equity considerations, and reliability or risk. Many methods and techniques have been applied to deal with the multi-objective character of regional water resources problems where conflicts have arisen. A specific category of methods is aimed at solving conflicts in an uncertain environment, and it may be of particular interest for climate impact assessment.

Thousands of papers have been published in the last 30 years on various applications of systems analysis in the study of water resources. It is outside the scope of this book to discuss them in detail. The crucial question is how existing methodology can be used for solving problems raised by the issue of climate change. Before answering this question, a short review of the fundamental concepts of the systems approach may help to set the stage for further consideration. According to Quade (1978), the essential role of systems analysis is to find objective answers to a number of questions crucial for decision making through the following process: (a) identification of objectives and constraints; (b) selection of a feasible alternative solution; (c) evaluation of positive and negative consequences of these alternatives, with due account given to existing uncertainties; and (d) comparison of alternative solutions according to predetermined performance criteria.

This general approach is applicable both to stationary and nonstationary climatic and hydrological conditions. The problems are how to identify objectives (for instance, future water demands in agriculture); how to evaluate various alternatives (e.g., by stochastic optimization techniques); and what kind of performance criteria may be informative enough for making decisions in the face of global warming. The simplest procedure is to assume future supply and demand scenarios and to apply standard methods of analysis to them. As has been shown in the previous chapters, constructing scenarios is not an easy task; it requires broad knowledge of future atmospheric processes and of future relations between climate and water resources. Similarly, as in the case of climate modeling (IPCC, 1992), two types of water resources scenarios should be distinguished: equilibrium scenarios and transient scenarios. The first assumes certain supply and demand conditions in a given region for, for example, carbon dioxide doubling; the second predicts a time-dependent change. Very few reported studies have used the second approach.

Most of the mathematical techniques used in water resources planning and operation take into account, either implicitly or explicitly, the random variability of hydrological processes. Examples are stochastic optimization, used for finding the "best" operational rule of reservoir operation, and planning methods based on simulated time series of river discharges. Consequently, scenarios of future water supply should represent not only changes in mean characteristics, but also changes in standard deviations, correlation coefficients, and all other parameters necessary to describe stochastic properties. Unfortunately, this requirement is often neglected in hydrological papers that deal with climate impact assessment. The need for a comprehensive description of anthropogenic changes in probabilistic structures of water supply processes is important not only for the planning of the systems's extension, but also for the operation of water resources structures. This can be illustrated by the following example.

Assume that the current control of a system of storage reservoirs is implemented with the help of a computer program based on stochastic optimization. In this case, the input data usually include some information on the current state of the system (e.g., the available storage, the inflow the previous day, etc.), and on parameters of the stochastic inflow process estimated on the basis of past hydrological records. If there is a reason to believe that climatic and land surface processes over the catchment will undergo changes, then operational decisions made on the basis of "historical" statistics may be far from optimal. Some additional assumptions are needed in the control procedures to adjust parameters of the inflow processes to the current climatic situation.

Analyzing the concept of uncertainty, Kundzewicz and Somlyódy (1993) observed, "The meaning of this term may range from the state of being unknown, indefinite, indeterminate, on one extreme, to minor uncertainties about the value of perfectly certain parameters." In the analysis of climate impact, the level of uncertainty depends on the temporal disaggregation, on the spatial scale of analysis, and even on the type of climatic characteristic of interest. Situations in which the water resources analyst is in a position to evaluate changes in numerical values of process parameters and then to use the altered values for his or her model, are ideal. In most cases such assessment must be based on purely subjective premises, or alternative methods of analysis should be applied. Georgakakos and Yao (1993) present a control approach where input variables are only expected to belong to certain sets. The objective is to determine sets of control actions guaranteeing that the water resources system under consideration will remain within desirable bounds. Other possibilities are offered by the fuzzy sets theory (Kacprzyk and Orlovski, 1987), but their applications to water resources problems are still very limited (Kindler, 1988).

In conclusion, most of the existing methods and techniques used for water resources planning and optimal operation can be adapted to nonstationary climatic and hydrological conditions. The basic concept of randomness of supply characteristics combined with quasi-deterministic prediction of future water requirements should be supplemented by uncertainties entered at several levels of analysis. The scenario approach may help in the formalization of impact assessments, but it usually introduces some kind of subjectivity in decision making. It can be treated, for instance, by the Bayesian statistical decision approach (Fiering and Rogers, 1989). This approach may sometimes appear to lead to unreasonable choices; however, the combined effects of the assumed prior (or expert) probabilities and the calculated regrets may overcome this tendency.

6.3 Criteria of Water Systems Performance

There currently is general agreement that the classic criteria for evaluating the performance of water resources systems (e.g., cost versus benefit) do not sufficiently describe a system's efficiency, even when its environment is assumed to be stable. Interest is increasingly focused on the details related to the episodes of unsatisfactory system performance – recurrence, duration, severity, and consequences of system failures. To assess various aspects of the performance of a technical or a natural system, it is necessary to use multiple criteria characterizing the possible alternatives, scenarios, and policies. In general, the term "satisfactory system performance" pertains to a combination of several attributes; if all variables of concern remain within the permissible limits, the system is said to perform satisfactorily. There is certainly some degree of arbitrariness and subjectivity in assigning the permissible range of satisfactory states of the system. For the sake of simplicity, assume that there is a single variable deciding the system's performance. This is adequate for many practical water resources situations. The system performs satisfactorily in the time instant t, if the variable describing the system performance belongs to the set of satisfactory states:

$$x(t) \in \{S(t)\} \quad . \tag{6.1}$$

According to the formalism originating in mechanical engineering, the satisfactory system performance is tantamount to the fulfillment of the inequality:

$$\text{load} < \text{resistance} \quad . \tag{6.2}$$

The applicability of this notation goes far beyond the structural mechanics. Examples of this analogy are the following inequalities, which describe the satisfactory performance of a water system:

- Flow rate in the river < maximum admissible flow.
- Concentration of a chemical < respective quality standard.
- Water supply > water demand.

The measures of different aspects of unsatisfactory performance of a water resources system may pertain either to time-related characteristics (duration and frequency), or to amplitude-related characteristics. The latter indices are maximum instantaneous entry into the state of unsatisfactory performance and the cumulated entry into the state (e.g., maximum water deficit and accumulated deficit, respectively). The ultimate criteria, combining both of the above types of characteristics, measure severity and consequences of the unsatisfactory performance. Characteristics of reliability have recently been given a more systematic form via a number of criteria: risk, reliability, resilience, vulnerability, and robustness (Fiering, 1982; Hashimoto *et al.*, 1982; Kundzewicz, 1989). Some of these concepts are discussed below. It should be noted that despite the existence of many references on water resources performance criteria, there is no general consensus as to the definition.

Among the basic notions of the mathematical theory of reliability are the following:

- The reliability function, i.e., the probability that the system will not enter the unsatisfactory state during a given time interval (or until the time instant t).
- The availability function, i.e., the probability that the system will act satisfactorily at the time instant t.

The classic notion of reliability in water resources applications refers to the probability of satisfactory system performance in a single time instant or time interval of interest (typically, the design period). The latter interpretation has been used by Duckstein and Plate (1985) under the name *mission reliability.* Other definitions of reliability used by Kritskiy and Menkel (1952) include the following:

- Occurrence reliability – the ratio of the number of years in which the system entered the satisfactory state to the total number of years considered.
- Temporal reliability – the amount of time a system remained in the satisfactory state divided by the total time considered.
- Volumetric reliability – the portion of the volume of water supplied during the time period considered with the rate equal to or less than demanded divided by the demanded volume.

In addition to temporal reliability as defined above, Kindler and Tyszewski (1989) have used *annual reliability indices*, defined as the relative frequency of the demand being satisfied in each particular year. These indices take into account the annual cycle of hydrological events. Even at very high levels of temporal reliability, there might be unsatisfactory episodes to the extent that irreversible changes occur and the system cannot recover under most plausible hydrological situations.

As demonstrated above, there are several concepts labeled with only one term, reliability. Even more ambiguity accompanies the concept of *resilience*, which in principle should represent the system's ability to recover from a state of unsatisfactory system performance to a state of satisfactory system performance. In particular, a resilient system should be able to accommodate temporary environmental shocks. The idea was introduced in ecology by Holling (1985) to describe systems that can survive intensive stress. A great number of alternative indices of resilience were considered by Fiering (1982) and by Hashimoto *et al.* (1982), among others. The definition often used in the context of water resources states that the resilience criterion is the conditional probability of a system's recovery from failure. In the case of a single time interval, this criterion reads:

$$\text{Resilience} = \text{Prob}(S|F) = 1 - \text{Prob}(F|F) \quad , \tag{6.3}$$

where S means "success" and F means "failure" in the preceding time period. For several time intervals (e.g., months) the resilience index is the probability of the logical product:

$$\text{Resilience} = \text{Prob}(S|F_0 \times S_2|F_1 \times S_3|F_2 \times \ldots) \quad , \tag{6.4}$$

where S_i means a "success" in the ith time interval, and F_{i-1} means "failure" in the $(i-1)$th time interval. In a more practical sense, the resilient system should be able to operate under a range of external conditions with relatively low losses.

The notion of vulnerability is typically associated with the system's susceptibility to negative external influences, for instance floods, as that vulnerability relates to the likelihood of economic and social consequences of the system's unsatisfactory performance. There is no generally accepted definition of water system vulnerability to climatic and hydrological variability and trends. A number of empirical indicators have been proposed by Gleick (1990), such as the ratio of storage volume in a given catchment to the average annual runoff, or the ratio of basin consumptive depletion to total basin annual mean renewable supply. These indicators, calculated on the basis of present climatic conditions, show great differences among 21 US water resources regions. Their numerical values are less important than are the relative characteristics for various parts of the country. In a similar way, Brouwer and Falkenmark (1989) evaluated vulnerability of water supply in

European countries to expected doubling of carbon dioxide concentration in the atmosphere.

A convenient instrument for evaluating the water resources systems performance criteria, in particular those related to temporal aspects of the unsatisfactory performance, is the renewal theory. The concept of its application in the continuous time framework has been sketched in an article by Kundzewicz (1989), where a detailed description can be found. The term *renewal* means a recovery, or a comeback of the system from the state of unsatisfactory performance to the state of satisfactory performance. Criteria based on this concept render a convenient framework, embracing most of the measures of reliability and resilience reviewed above.

The crucial question here is how the performance criteria developed for classic water resources investigations can be applied in the changing climatic environment. The renewal theory lends itself well to applications in the nonstationary case. In fact, such cases have been used extensively in the areas of their earlier applications, where the possibility of accounting for the aging process was incorporated (Baecher *et al.*, 1980). Consideration of a nonstationary reliability scheme is not new in engineering practice (for example, aging of technical objects, assessment of insurance costs, etc.), but is rather rare in dealing with complex regional systems. Other measures, such as resilience and robustness, must be adjusted to nonstationary conditions before they can be applied in studies of climate impact.

Such adjustments should not be difficult when two equilibrium climate scenarios – e.g., the "historical" one and the alternative scenario for the $2\times CO_2$ conditions – are investigated to assess the implications of climate change on regional water resources management. In such cases, the performance criteria should be evaluated independently for both scenarios, and the results should be compared to draw respective conclusions. For example, in a study implemented for the Kariba reservoir on the Zambezi River (Kaczmarek, 1990), the reliability and resilience indices for hydropower production were calculated and were compared for recorded inflow data and for assumed changes in inflow parameters. The average mean annual inflow to the reservoir is equal to 1,458 m^3/s, and the standard deviation $\sigma(Q) = 578$ m^3/s. Assumed changes in the mean and standard deviation are $\pm$ 10% and $\pm$ 20%. The reliability criterion has been defined as the ergodic probability of storage level being in the upper part of storage space during the whole year. The resilience index has been calculated by means of formula (6.3), where "failure" means an empty reservoir (less than 20% of maximum storage). Probability values of storage levels were calculated by means of a stochastic storage model of the Kariba reservoir. Results of calculations

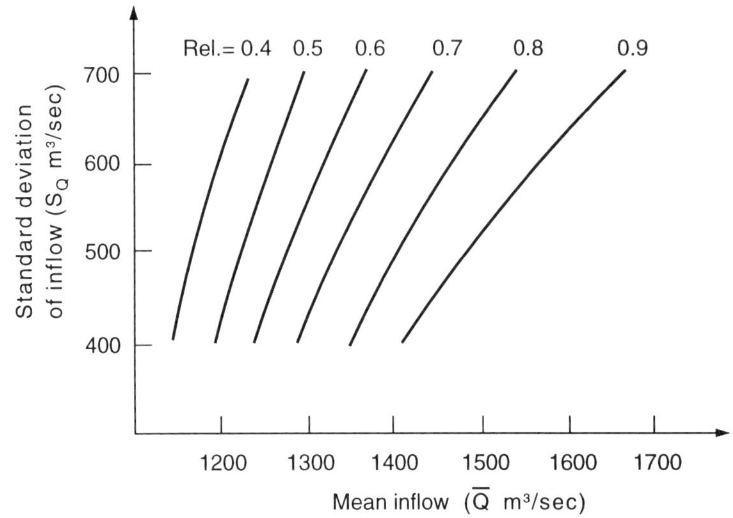

Figure 6.2. Reliability (Rel.) of the Kariba Dam operation.

are shown in *Figures 6.2* and *6.3*. As can be expected, both reliability and resilience indices increase with the increase of the assumed average inflow. The impact of changes in the standard deviation is insignificant, particularly for the resilience criterion.

It seems that the criteria discussed in this chapter are a versatile means of describing various aspects of the system's performance. However, in some cases these indices may be insufficient, and the set of adopted criteria would depend on the particular situation. Assume, for instance, that a key parameter that helps to define the net benefit of the operation of a water resources system is changing, and that this change causes a sharp decrease in net benefit. If the change is abrupt, a large drop in benefit can be observed because the system cannot adapt to the new situation. Only after some time and after having passed the minimum value of the parameter, the system may reach the steady new level of net benefit. In the case of a gradual change, the response and the resulting decrease of benefit may be less dramatic. It is difficult, however, to describe quantitatively the response to the speed of change with the criteria described above. Fortunately, the expected climate change will be relatively slow in comparison with the design horizon of most of the large-scale water resources projects, so there will be enough time for system adaptation. However, this requires overcoming the tendency of conservative and delayed political decision making.

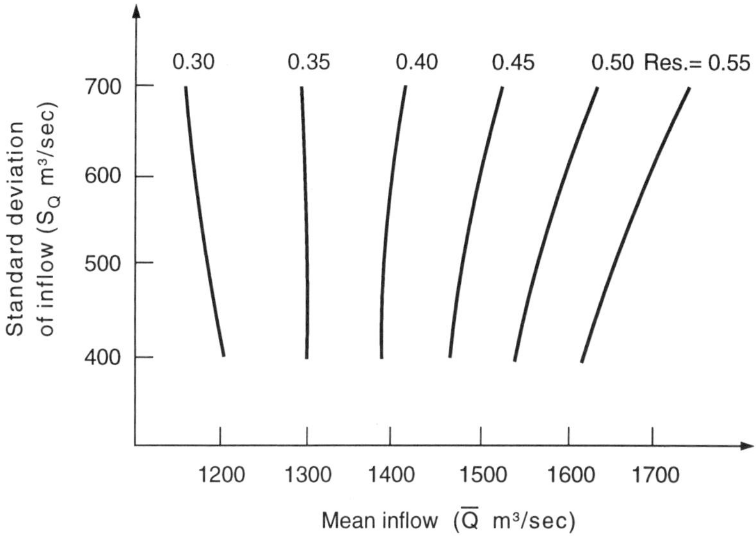

Figure 6.3. Resiliency (Res.) of Kariba Dam operation.

Finally, it seems worthwhile to stress that it is not the non-stationarity of hydrological processes that is most bothersome, but rather the uncertainty in predicting future climatic trends. One can cope with non-stationarity with somewhat more complicated methods, but it may be far more difficult to identify the characteristics of uncertainties accompanying the water resources planning process.

6.4 Lessons from Water Resources Impact Studies

The growing interest in the possible consequences of anthropogenic climate change on regional water resources has given rise to a wealth of scientific papers on the sensitivity of physical hydrology to climatic input variables. A great number of numerical results are reported in the literature on the application of hydrologic models for assessing changes in runoff, evaporation, and soil moisture for assumed regional climate scenarios. Much less information is available on the possible effects on managed water resources systems. Many general statements on the threats to water resources that may be caused by expected shifts in regional precipitation and temperature patterns can be found in the literature. However, most of the analyses to date focus on the supply aspects of the problem, neglecting economic and societal consequences of expected climate evolution.

One of the reasons water resources analysts are very cautious in providing decision makers with substantial information on possible effects of climate change is the complexity of water systems where all of the interactions and feedbacks are difficult to identify and assess. Moreover, climate is not the only attribute of these systems that is likely to undergo transformation over the next decades. As a result, most of the known water resources impact studies are devoted to isolated elements of the system (e.g., to the operation of a single reservoir with a decreased inflow), rather than to comprehensive analysis of the system as a whole. This limits the ability to convince higher-level decision makers of the importance of the problem and of the need for action to be taken to mitigate potential negative consequences of climate change.

Several chapters of this book are devoted to results of regional case studies, with the guiding principle being to reflect, if possible, the physical, technical, and economic aspects of water resources problems of concern. The lessons that can be drawn from these regional experiments are summarized in the concluding chapter of the monograph. In this section, we present some of the concepts and results of earlier water resources impact studies as they developed and were described in the literature during the last two decades.

Probably the first attempt to assess possible consequences of climate change on water resources was undertaken by Schwarz (1977) in a paper prepared for the US National Academy of Sciences. Although his conclusions were not supported by the results of any numerical calculations, they reveal the extensive experience that water resources planners have in coping with natural climatic and hydrological temporal variability. Assuming that the magnitude and timing of possible regional climate changes are largely unknown (which was the situation at the time the paper was written), Schwarz suggested three steps of analyses:

- To review individual cases on the basis of previous response of water resources systems to climate variations, supported by subjective reactions of local water resources managers.
- To select certain performance criteria and speculate on the effects that changes in supply processes might have on each of them.
- To observe the effects of these variations on performance criteria using generated streamflow with varying statistical parameters as a substitute for climate scenarios.

Results of the second approach led Schwarz to the following important conclusions. The decrease in streamflow may have a significant effect on the yield from reservoirs and on system reliability. Pressure for a water system's expansion would be created if shortages were to occur repeatedly, but the effectiveness of interbasin connections would not change very much. The quality of raw water probably would be insignificantly affected, except where large reservoirs are drawn to very low levels. The increase of variance in streamflow may severely affect yield from unregulated rivers, and may have a moderate effect to no effect on reservoir yields, depending on storage capacity in relation to drainage area. Pressure for expansion would be obvious, particularly for additional reservoir construction. The effectiveness of interbasin connections would increase, particularly if the spatial correlation of runoff is small. The increase of runoff skewness has little if any significance for a system's performance, except for the yield from unregulated streams, because the number of days of low flow may increase relative to the few periods of very high flow. The role of streamflow persistence may be significant or severe for yields from reservoirs with limited long-term storage. Also, ground water yields could be severely affected for long periods of time. As a result, the pressure for new investments would mount over time and increase the likelihood for adaptive actions. The speed with which climate change occurs could influence the system's expansion policy, but long-term changes will generally be ignored by existing institutions.

It is interesting to note that this set of conclusions, drawn about 15 years ago on the basis of a relatively simple methodological approach, generally has not been called into question by the results of later studies. Most of the specific regional impact assessments based on sophisticated models and optimization techniques end with a number of very general statements and recommendations similar to those presented by Schwarz. The reason may be that the uncertainty factor accompanying the issue of global warming plays a crucial role in most of the existing assessment experiments.

Another set of rather general conclusions about the possible effects of global climate change on regional water resources is discussed in the previously mentioned paper by Gleick (1990). Based on five subjectively chosen vulnerability criteria, Gleick evaluates the expected impact of climatic variations on various water resources regions in the USA. The results can be summarized as follows:

- Regions with small storage compared with supply may be particularly vulnerable to a climate-induced reduction in water and to wet and dry extremes.

- Climate changes that affect the quantity of water would have the most noticeable effects in regions with high consumption relative to available renewable supply.
- Climate change that reduces water availability in regions with a high proportion of hydroelectricity relative to total electricity would have serious economic and environmental impacts.
- Regions where there is a high ratio of ground water overdraft to total ground water withdrawal are especially vulnerable to climate change, if it causes a decrease in water availability.
- Current streamflow variability suggests increased vulnerability to extreme hydrologic situations – floods and droughts.

In addition, Gleick stresses that qualitative vulnerability of water resources to climate change can be exacerbated by nonrationed system design, by operational inflexibility, or by legal and institutional constraints.

Gleick's approach was criticized by Klemeš (1990), who expressed the opinion that a reasonably accurate assessment of future climate is "beyond the reach of our knowledge." Furthermore, he says that "the only useful message from the prospect of climate change is to adopt more robust and flexible solutions to current problems." It seems that such a pessimistic view is shared by many politicians and water resources decision makers, who delay any active response until more convincing arguments are available. This discussion among two leading scientists shows the difficulties of getting a unified opinion on possible active responses to the climate-induced changes in the current water resources policy.

Of the numerous site-specific or sectoral impact studies in the literature, we shall mention only some representative examples. Schwarz and Dillard (1990) have reported results of interviews with top-level engineers, managers, and planners of large urban systems in different hydrological and climatological regimes, for example, Indianapolis, New Orleans, New York, and Tucson in the USA. The representative of the Indianapolis Water Company expressed the opinion that longer droughts and reduced annual precipitation could affect safe yield. Also, warmer average temperatures might increase demand and evaporation from the surface of reservoir. What was particularly interesting was his view that to initiate any adaptive measures climate change would have to become manifest by either a "measurable trend over at least 5 years or an unprecedented drought."

Managers of the water supply system for New Orleans, much of which lies below sea level, expressed an opinion that supply would be greatly affected by a rise in sea level, because of increased salinity. The possible decrease of runoff would be insignificant because the flow of the Mississippi River is much larger than the intake. Asked about the response actions, the managers said that action would be taken if the responsible federal agency (in this case, the US Army Corps of Engineers) would "recognize the coming climatic change and their local effects."

The New York water resources planners worry about the increased salinity of the Delaware estuary (for details, see Chapter 8). But they also voice the concern that the system is sensitive to climate-induced changes in water demand, and state that a rise in sea level of 25 cm or more and a rainfall decrease of 5% to 10% below the historic minimum (which is rather unlikely) would harm the system. At the same time the New York City water resources managers have taken a conservative view that before taking any action, "near unanimity among scientific and professional bodies" is needed.

In the case of Tucson, those responsible for the water supply system state, "Planning for adverse impacts of climatic change would likely need good scientific evidence sanctioned by the Central Arizona Water Conservation District and the Arizona Water Resources Department."

As these few examples show, experienced water resources managers are generally aware of possible changes in stochastic properties of hydrological and climatic processes that may adversely affect the planning or operation of water supply systems. They are are, however, cautious in undertaking any real, active response until the negative consequences of climate change materialize, or at least are definitely proved by scientific findings. In our opinion, this stage of understanding will not be reached before the end of this century.

Several water resources impact studies focused on design and operation of storage reservoirs, which are the most important tools for coping with natural and anthropogenic streamflow variability. Klemeš (1985) examined the results of the operation of two midsize reservoirs for a simulated series of monthly flows corresponding to assumed changes in precipitation and PET for two basins in the USA. He concludes that, should the climate become substantially drier, the reduction in streamflow cannot be compensated for by building reservoirs. Some doubts about this statement arise, however, in the case of storage capacities substantially lower than the average yearly inflow to the reservoir, where a great amount of water may be lost for practical use during the flood periods. Reliability indices, measured in terms of the time during which the actual release is equal to or greater than the target value, were evaluated by Klemeš for various climate scenarios. The results show that reliability differences due to changes of PET over the catchment were much lower than those due to precipitation change. This confirms the findings of other researchers who conclude that runoff is more sensitive to precipitation than to changes in temperature (see Chapter 1). In his final comment, Klemeš concludes, "As far as the future conditions are concerned, there is little that a water manager can do other than to refrain from developing a water resource up to its last drop and keep a reasonable margin of safety."

Other studies (Fiering and Rogers, 1989; Kaczmarek, 1990; Lettenmaier *et al.*, 1990) attempted to better understand the vulnerability, resilience, and robustness of reservoirs under climatic uncertainty. Some of the results on the reliability of meeting the target release for a given storage capacity were shown earlier (see *Figures 6.3* and *6.4*). Another study shows that reservoir reliability was not seriously affected by changes in the climate-induced variation in runoff, if the storage capacity was larger than 50% of the mean annual inflow. It was also observed that adjustments made in the reservoir operating policy generally did not help in coping with the consequences of climate change. All of these conclusions should, however, be treated as only preliminary results, and must be re-examined after more empirical data are collected.

Two strictly regional water resources impact studies seem to be very informative. The first one concerns the drought that has been occurring in the Sahel since 1968 (Sircoulon, 1991). The drought extends from Mauritania to Sudan, covering an area of several million square kilometers. The countries affected by the drought have a total population of over 40 million inhabitants; this number is likely to increase to more than 100 million by the year 2025. The drought in the Sahel is exceptional for its persistence, its extent (areal coverage), and its severity. The regional rainfall deficits reached 20% to 40%. The total water yield in the francophone Sahel decreased by 43% compared with the previous period. The available data show a high sensitivity of the Sahelian water balance to changes in climatic characteristics; this confirms results of other studies showing an increase of sensitivity and elasticity indices with a decrease of the annual runoff coefficient. The case of the drought in the Sahel may teach the world community about the possible consequences of climate non-stationarity. It raises a number of scientific questions about regional atmospheric and land surface processes (e.g., the problem of persistence), and should open a new phase of discussions on the need for concerted international action aimed at combating possible effects of regional climatic catastrophes.

The second regional impact study that is undoubtedly worthy of consideration concerns the water resources implications of present and projected climatic conditions in the Great Lakes basin (Sanderson, 1987). The

Canadian Climate Centre provided the impact analysts with a projection of climate conditions in the region, based on the output of the Goddard Institute for Space Studies (GISS) general circulation model. The average temperature increase is approximately 4.5°C, and the yearly sum of precipitation is projected to increase slightly in some parts of the region and to decrease in others. The water resources impact analysis focused on navigation and hydropower generation in the basin. First, some hydrological analyses show a possible reduction in lake levels of 30 to 80 cm due to decreased catchment runoff and increased evaporation. The GISS model also predicts a dramatic decline in ice-cover duration and, consequently, an increase in the length of the shipping season.

Using economic models, the effects of climatic and hydrological changes were converted into the costs of operation for Great Lakes shipping companies and the costs of the production of energy. The estimated shipping costs increased by approximately 30%, and the possible loss of energy for the Canadian hydro-generation plants may exceed 4,000 GWh of power generation. Combining the expected reduction of energy demand in warmer winters with this reduction in electricity generation may result in substantial annual savings of money for consumers. Researchers involved in the impact studies stressed that all the hydrologic and economic models used in the analyses are subject to serious uncertainties. Therefore, the above-mentioned economic forecasts should be taken with necessary caution.

In this section we have described a range of approaches used and the results obtained, in an attempt to identify water resources implications of the enhanced greenhouse effect when physical, technical, economic, and societal aspects are taken into account. Experience is still limited. As can be seen from one of the studies presented, water resources decision makers will not undertake any active response unless credible analysts provide them with necessary scientific information and take part of the responsibility for decisions with costly consequences. The crucial problem in decision making is still the high level of uncertainty in forecasting future environmental conditions. However, it should not be forgotten that future non-climatic factors are also uncertain. Klemeš (1990) refers to his own experience with the operating policy of a dam built in the first half of this century near Brno in the Czech Republic, for which the main objective was changed five times within 20 years. The dam was designed for irrigation water supply; during the Second World War it was used mainly for power generation; after the war its operating policy was subordinated to flood control; in the 1950s, the main use was drinking-water supply; and 10 years later the new priority was recreation.

Water resources impact assessments based on a set of various climate scenarios will obviously give different results. An analyst responsible for the implementation of a given study should first of all extract those findings that are common to all or most of the scenarios. In many cases, the calculated model results may differ in the magnitude of the effect but agree in direction or general tendency of changes. If, for instance, a model predicts droughts more severe than at present for most of the assumed scenarios, then for decision making it is not so important to know exactly how large the change will be.

An educated decision maker might also like to know which uncertainties will be reduced in a few years due to expected new scientific findings, and which are not likely to be reducible in the near future. Schneider (1989) expressed an opinion that "Uncertainties that are easily reducible ... might encourage waiting before implementing policy, whereas uncertainties that are irreducible or difficult to reduce might suggest acting sooner." Although this is probably only partially true, the analysts of impact studies should inform water resources managers about the nature of the scientific evidence. This will be of considerable help in deciding what kind of adaptive measures should be taken and when the investing process should start.

6.5 Conclusions: What Rational People Should Do

Planning for an uncertain future is a risky task; however, water resources analysts are accustomed to designing hydraulic structures in a stochastic environment and planning for highly uncertain future demands. The only new factor that is bothersome now is non-stationarity or trends in water supply processes that previously were usually assumed to be stable. Some guidance is needed in the methodology of impact studies and in the respective research programs. Because to date the experience in designing water systems for various climate scenarios is limited, the following suggestions should be treated as being preliminary.

Probably the simplest, and in many cases the most effective, method is to base decisions on the opinions of experienced water managers – the experts in the field. One should clearly distinguish between the sensitivity of hydrological processes and the sensitivity and vulnerability of water resources systems to changes in climate. In the first case, the physical laws of conservation of mass and energy determine the linkages between atmospheric phenomena and water balance; in the second case, the relation between "load" and "resistance" (or supply and demand) decides. Let us take as an example two urban water supply systems, one with demand closely matching the available resources and another with supply largely exceeding the present water requirements and those expected for the future. In the first case a careful analysis of possible consequences of a climate-induced decrease of supply is necessary; in the second case the vulnerability of the system to supply variations will be negligible. Looking at the problem from the reliability point of view, one must conclude that the probability of meeting demand may vary greatly in the first case, and in the second case supply will be always higher than the demand. In many cases, practical conclusions may be reached by soliciting the judgements of water specialists, without the need to undertake expensive and time-consuming investigations.

These situations do not appear very often, because in most densely populated regions the ratio of water use to supply is high enough to create problems. For these areas, water resources projections and planning studies are done routinely from time to time, independent of the issue of climate change. In Poland, for instance, national master water plans were elaborated in the mid-fifties by the Polish Academy of Sciences; between 1968 and 1972 this plan was reformulated in the framework of a United Nations Development Program (UNDP) project for the Vistula River basin; and it was again improved in the late seventies. Each time the hydrological and new water demand data were collected and predictions concerning future water requirements were made. Although substantial changes in the predicted input values could be observed in the subsequent studies, the main conclusions on the location and sizing of key elements of the Polish water resources infrastructure schemes remain relatively unchanged for all versions of investment programs.

As can be seen from the earlier considerations, the expected climate change introduces some new problems for the evaluation of the behavior of water systems in conditions resembling those of the future. The main difficulties are caused by a sequence of uncertainties characteristic of climate impact studies. This, in turn, requires that simulation models, optimization procedures, and other methods of analysis be applied to a range of climatic and hydrological scenarios. Often, another obstacle to the incorporation of the issue of climate change into the planning process is the difference of approaches between water resources analysts and the real-world decision makers. Usually, decision making is a complex hierarchical process in which not all of the participating actors may be ready to accept the importance of environmental changes that will only become apparent after several decades. Paradoxically, it currently may be easier to negotiate transnational agreements and protocols for addressing problems of global interest, such as the Montreal Protocol on Ozone Convention and the Climate Convention signed in Rio de Janeiro in 1992, than it is to convince national governments and local decision makers of the need for active responses to expected global threats.

The role of scientists and engineers, however, is to provide politicians and decision makers with the necessary information about the future, independent of their current attitude toward the global climate change problem. Methodological guidelines for assessing impact of climate change can be found in Carter *et al.* (1992) and in a number of papers dealing specifically with water resources. These suggestions may be summarized in several steps typical of an impact analysis. The first step in conducting any water resources impact analysis is to define precisely the objectives and specific goals of the study. For instance, in the cited work of Carter *et al.* (1992), one may find the warning that "an assessment of the future hydrological impacts of climate change in a river catchment has quite different requirements for data and expertise if the goal is to estimate the capacity for power generation, than if it is to predict changes in agricultural income as a result of changes in the availability of water for irrigation." The assessment goals are closely linked to the study area, its economic and societal context, the time horizon, and the data requirements.

The second step of analysis is to provide all the necessary information on the present water resources situation in the region: characteristics of current water supply and its natural variability, critical objectives of regional water management, data on water uses, description of regional infrastructure of water resources, administrative and legislative aspects of decision-making procedures, etc. This baseline representation should include all the climatic information necessary for evaluating not only the climatic and hydrological "norms," but also their temporal and spatial variability.

The third step is to embrace past and current predictions of future behavior of all non-climatic factors that may have an impact on regional water resources management. This includes, in particular, population projections, assessment of future water requirements in various sectors of the economy, planned investments in the water sector, etc. Although most of the above predictions are far from being exact, and although they incorporate some level of uncertainty, it is unusual in the current practice to present them in the form of alternative scenarios.

Projecting non-stationarity of future climatic conditions is the fourth step of the procedure. Such predictions are usually formulated as alternative climate scenarios. These alternative scenarios should contain all the variables that will be used later for assessing changes in water supply and demand characteristics. The methodology of building climate scenarios is discussed in the first chapter of this book. For the needs of impact assessment, the scenarios have to meet the following conditions:

- Data should be mutually consistent, comprising combinations of changes in different variables.
- Spatial compatibility should be preserved to assure that regional changes are compatible with global atmospheric processes.
- Predicted data must be suitable as inputs to hydrological and other impact models.

As was mentioned in Chapter 1, there are four types of scenarios usually applied for water resources impact studies: arbitrary adjustments of "historical" climatic characteristics, temporal analogues including paleoclimatic data, spatial analogues, and scenarios constructed in various ways on the basis of global atmospheric circulation models. The choice of approach lies with the water planners.

In the fifth step of water impact study, the analyst must decide on the methods he or she wants to use to transfer climatic information into water resources decision variables. A great range of models and mathematical techniques have been developed for this purpose in the last 30 years. Hydrological models with climatic input characteristics are discussed in Chapter 2. Simulation techniques, which are widely applied in the generation of streamflow time series, may be adjusted to climate change conditions by changing parameters of stochastic hydrologic models in a way consistent with the assumed scenarios of climatic variables. Finally, a number of optimization techniques based on deterministic and stochastic programming can serve as tools for evaluating the best design and operational characteristics of water resources systems for various hydrological and climatic situations. Performance criteria described earlier in this chapter will help in the comparison of results obtained for alternative climate scenarios.

It seems that it is not the lack of hydrological models and evaluation techniques that causes difficulties in assessing the possible effects of climatic non-stationarity on water resources. Most of the existing methodological tools can be applied for changed conditions, as they are used now in very different climatic regions. For the decision-making process, the critical question is what practical conclusions can be drawn from a set of results based on uncertain and differentiated scenarios. If the range of possible solutions is not too broad, then some kind of subjective expert judgement may help. If, however, the alternative results (concerning the required capacity of the water reservoir, for example) are highly dispersed, the decision-making process may be more difficult and may raise dramatic controversies. The only approach that has been advocated in such circumstances is to put decisions in the Bayesian framework.

The possibility of using the Bayes theorem in decision making under climate change has been discussed *inter alia* by Fiering and Matalas (1990). This approach may be summarized as follows:

- There are various hypotheses concerning possible changes of climate parameters. Prior probability distributions are assigned to each of these hypotheses; such assignments may be assumed on a wholly subjective basis.
- If new information becomes available, the prior probabilities $p(i)$ are modified by means of the Bayes formula:

$$p(i/Y) = p(i) * p(Y/i) \quad , \tag{6.5}$$

  where $p(Y/i)$ is the conditional probability of information Y if the hypothesis i is true.
- A regret matrix is formed, containing penalties associated with the decisions based on the erroneous choice of climate scenario.
- Updated prior (or posterior) probabilities are used to minimize expected regret.

The main difficulty in applying the Bayes theory to studies of climate impact is that new information on changes in climatic characteristics cannot be expected in the next few decades. An alternative feasible solution is that this information will concern current greenhouse gas emission values, provided that each emission scenario leads to a certain climate scenario, as was discussed in the IPCC Scientific Assessment (1992), for example. If the observed global emission proceeds according to a so-called business-as-usual scenario, then CO_2 doubling and a global temperature rise of 3.0°C to 4.0°C can be expected in the middle of the next century. In contrast, if the mitigation action proposed by the Climate Convention is successful, then the entire warming process may be significantly delayed. It may be added that Waggoner and Revelle (1990) make an important observation on the Bayesian decision-making process by saying, "So long as the prior probabilities are nearly constant and the differences in regret are not too great, other criteria are needed for decisions."

These alternative decision-making criteria may be based on intercomparison of various performance indices – reliability, robustness, and resilience of water resources systems – calculated for a set of alternative climate scenarios. In this case, the decision maker will be provided with numbers showing the possible consequences of each scenario. The final choice of action will stay with him or her, depending on a subjective attitude to various forecasts and the calculated indices of system performance. We conclude this chapter with a thought from Waggoner (1990): "So long as the future remains unsure, and that seems a long time ahead, rational people will make decisions that solve present problems and make water supply robust, resilient, and flexible for any future."

References

Baecher, G.B., Pate, M.E., and De Neufville, R., 1980, Risk of dam failure in benefit-cost analysis, *Water Resources Research*, **16**(3):449–456.

Brouwer, F., and Falkenmark, M., 1989, Climate induced water availability changes in Europe, *Environmental Monitoring and Assessment*, **13**(1):75–98.

Carter, T.R., Parry, M.L., Nishioka, S., and Harasawa, H., 1992, *Preliminary Guidelines for Assessing Impacts of Climate Change*, Intergovernmental Panel on Climate Change (IPCC), Geneva, Switzerland.

Cohon, J., and Revelle, C., 1980, Approaches and techniques for resolution of conflict situations in regional water management, in Proceedings of IIASA Conference on Criteria, Conflicts, Uncertainty, and Institutions in Regional Water Management, International Institute for Applied Systems Analysis, Laxenburg, Austria.

Duckstein, L., and Plate, E.J., 1985, A system framework for hydrologic reliability with application to the design of structures, *Proceedings of the Hamburg Symposium*, International Association of Hydrological Sciences (IAHS) Publication No. 147, IAHS Press, Wallingford, UK.

Fiering, M.B., 1982, Alternative indices of resilience, *Water Resources Research* **18**(1):33–39.

Fiering, M.B., and Matalas, N.C., 1990, Decision making under uncertainty, in P.E. Waggoner, ed., *Climate Change and U.S. Water Resources*, John Wiley & Sons, New York, NY, USA.

Fiering, M.B., and Rogers, P., 1989, Climate change and water resources planning under uncertainty, Report to the US Army Engineers Institute for Water Resources, Fort Belvoir, VA, USA.

Georgakakos, A.P., and Yao, H., 1993, New control concepts for uncertain water resources systems, *Water Resources Research*, **29**(6):1505–1516.

Gleick, P.H., 1990, Vulnerability of water systems, in P.E. Waggoner, ed., *Climate Change and U.S. Water Resources*, John Wiley & Sons, New York, NY, USA.

Hashimoto, T., Stedinger, J.R., and Loucks, D.P., 1982, Reliability, resiliency and vulnerability criteria for water resources system performance evaluation, *Water Resources Research*, **18**(1):14–20.

Holling, C.S., 1985, Resilience of ecosystems: Local surprise and global change, in W.C. Clark and R.E. Munn, eds., *Sustainable Development and the Biosphere*, Cambridge University Press, Cambridge, UK.

IPCC (Intergovernmental Panel on Climate Change), 1992, *Climate Change: The IPCC Scientific Assessment*, J.T. Houghton, B.A. Callander, and S.K. Varney, eds., Cambridge University Press, Cambridge, UK.

Kacprzyk, J.W., and Orlovski, S., eds., 1987, *Optimization Models Using Fuzzy Sets and Possibility Theory*, D. Reidel Publishing Company, The Netherlands.

Kaczmarek, Z., 1990, Impact of Climatic Variations on Storage Reservoir Systems, WP-90-20, International Institute for Applied Systems Analysis, Laxenburg, Austria.

Kaczmarek, Z., and Kindler, J., 1989, The impacts of climate variability and change on urban and industrial water supply and wastewater disposal, in *Proceedings of Conference on Climate and Water*, Vol. 2, Helsinki, Finland.

Kindler, J., 1988, Design of water resource system for imprecisely estimated water demands: Problem of resources allocation, Reports of the Warsaw Technical University, Water Resources and Sanitary Engineering, No. 2, Warsaw, Poland [in Polish].

Kindler, J., and Tyszewski, S., 1989, Multicriteria evaluation of decision rules in the design of a storage reservoir, in *Proceedings of Symposium on Closing the Gap Between the Theory and Practice*, International Association of Hydrological Sciences (IAHS) Publication No. 180, IAHS Press, Wallingford, UK.

Klemeš, V., 1985, Effects of hypothetical climate change on reliability of water supply from reservoirs, in *Sensitivity of Water Resources Systems to Climate Variations*, World Meteorological Organization, WCP-98, Geneva, Switzerland.

Klemeš, V., 1990, Sensitivity of water resource systems to climatic variability, in *Proceedings of the Canadian Water Resources Association 43rd Annual Conference*.

Kritskiy, S.N., and Menkel, M.F., 1952, *Water Management Computations*, Gidrometeoizdat, Leningrad, USSR [in Russian].

Kundzewicz, Z.W., 1989, Renewal theory criteria of evaluation of water resource systems: Reliability and resilience, *Nordic Hydrology*, **20**:215–230.

Kundzewicz, Z.W., and Somlyódy, L., 1993, Climatic Change Impact on Water Resources: A System View, WP-93-30, International Institute for Applied Systems Analysis, Laxenburg, Austria.

Lettenmaier, D.P., Bretman, K.L., and Vail, L., 1990, Robustness of a multiple-use reservoir to seasonal runoff shifts associated with climatic change, in Y.Y. Haimes and E.Z. Stakhiv, eds., *Risk-Based Decision Making in Water Resources*, American Society of Civil Engineers, New York, NY, USA.

Peterson, F., and Keller, A.A., 1990, Irrigation, in P.E. Waggoner, ed., *Climate Change and U.S. Water Resources*, John Wiley & Sons, New York, NY, USA.

Quade, E.S., 1978, An introduction to systems analysis: What it is and what it does, *Archiwum Automatyki i Telemechaniki*, **24**(4) [in Polish].

Riebsame, W.E., 1988, Adjusting water resources management to climate change, *Climatic Change*, **13**(1):69–97.

Sanderson, M., 1987, Implications of climate change for navigation and power generation in the Great Lakes: Summary of Great Lakes Institute Reports, Environment Canada, Ottawa, Canada.

Schneider, S.H., 1989, The greenhouse effect: Science and policy, *Science*, **243**(February 10):771–780.

Schwarz, H.E., 1977, Climatic change and water supply: How sensitive is the Northeast? *Climate, Climatic Change and Water Supply*, US National Academy of Sciences, Washington, DC, USA.

Schwarz, H.E., and Dillard, L.A., 1990, Urban water, in P.E. Waggoner, ed., *Climate Change and U.S. Water Resources*, John Wiley & Sons, New York, NY, USA.

Sircoulon, J., 1991, Climate, water and development, in J. Jäger and H.L. Ferguson, eds., *Climate Change: Science, Impact and Society*, Cambridge University Press, Cambridge, UK.

Waggoner, P.E., and Revelle, R.R., 1990, Summary, in P.E. Waggoner, ed., *Climate Change and U.S. Water Resources*, John Wiley & Sons, New York, NY, USA.

Part II

Case Studies

Chapter 7

Introduction to Case Studies

Kenneth M. Strzepek and Zdzislaw Kaczmarek

7.1 Introduction

The following chapters of this book seek to establish linkages between the general methodology of climate change impact assessment and a certain number of real water resources problems around the world. In the majority of regional climate impact studies a water resources system can be treated as an "exposure unit," components of which are sensitive to changes in precipitation, air temperature, and other climatic characteristics. In most cases these systems are complex structures, composed of several natural and man-made water supply sources and a number of water users. Generally speaking, it must be assumed that all of the system's elements may be influenced by a number of climatic and non-climatic factors. The interdependence of the water system's components suggests that integrated regional impact assessment is probably the best way to understand the complex character of these influences. An ideal case study should present an analysis of the impact of climate change on regional water resources that combines all possible interactions, feedback, and possible response actions, and takes into account uncertainties accompanying the climate change issue.

There are no universally applicable methods for solving water resources problems under changing environmental and economic conditions. If differences in the level of current water stress, main goals of management, and expected future economic and political conditions are disregarded, the resulting analysis will likely be a failure. One can, however, identify some common methodological features that are typical for any climate impact

Z. Kaczmarek et al. (eds.),
Water Resources Management in the Face of Climatic/Hydrologic Uncertainties 197–201.

study in the water resources field. In the next section we will discuss some of them.

7.2 Key Methodological Issues

Each climate impact case study describes an actual region and the site-specific environmental and socioeconomic disturbances caused by non-stationarity of climatic processes. Because a case study may be representative of a certain class of water resources problem, it can be informative to other water managers. Some lessons drawn from results of past regional assessments were briefly presented in Chapter 6. It is our hope that the set of case studies included in this book will help the water resources community to better understand the threats and opportunities linked to the expected global warming. Toward this goal each regional analysis should fulfill some conditions:

- The study area characteristics and the goals and constraints of regional water management should be clearly defined and described.
- Key methodological approaches used for the analysis should be formulated in a way that allows readers to apply this methodology in their own analyses.
- Assumptions used for creating baseline climatic and hydrologic conditions, as well as alternative scenarios, must be presented.

In view of the uncertainties associated with the climate change issue, no single methodology can be considered better than the others. The use of various methodological approaches may be necessary, depending upon the information sought.

For long-term water resources planning it must be acknowledged that it is not just climate that may change; most of the non-climatic factors will vary in time, as well. Because we are most interested in the incremental effects of climatic variations, it is particularly important to distinguish the role of various influencing factors in each case study. This permits a comparison of the results of various climate impact studies and allows assessment of the sensitivity of water resources systems to climatic forcing alone.

Another important methodological issue is the definition of baseline conditions and the future characteristics of those climatic variables that are crucial for a given study. The baseline data are usually calculated from "historical" hydrologic and meteorological records, but the periods of observations often differ from one study to another. Also, the climate change scenarios used for impact analysis are highly differentiated in their origin, spatial and temporal resolution, and magnitude of changes. At present it does not seem feasible to unify on global scale all the background assumptions needed for such an analysis. Scientists still differ in their opinions on future emission of greenhouse gases, on the effects of changed chemical composition of the atmosphere on regional climates, and on methods for constructing scenarios. The approaches vary from using subjectively prescribed changes in temperature and precipitation, to using temporal and spatial climatic analogues, to using results obtained by means of a number of general circulation models (GCMs) with differentiated parameterization. Therefore, a minimum requirement for those water resources analysts publishing climate impact results should be to describe as clearly as possible all the basic data and assumptions concerning the current climatic and hydrologic conditions, as well as the future scenarios used in the study.

Each climate/water resources impact analysis is based on a number of models, from simple regression-type relations to sophisticated catchment models and optimization schemes. There is no reason that the same models must be used in areas with different environmental and socioeconomic conditions; most models were developed to solve specific regional water resources problems in the best possible way. The experience gained by authors of climate impact case studies should be shared with other interested water analysts through descriptions of the necessary details of the models used and, even more important, descriptions of the successes and failures in their application (unfortunately, sharing failures is not very common in scientific literature). Therefore, a good case study should contain as much information as possible on the models and techniques used in the calculations, including subjective assumptions, necessary simplifications, and uncertainties.

The main characteristics of the seven case studies presented in this book are listed in *Table 7.1*. The case studies analyze climate change impacts on a wide range of water resources elements, including reservoir management, hydropower, irrigation, municipal and industrial use, and water quality and instream flows for environmental protection.

Four case studies focus only on water quantity issues, two focus only on water quality issues, and one addresses both water quantity and water quality issues. Geographically, they include one global case study, two case studies in North America, one in Western Europe, one in Central Europe, one in Eastern Europe, and one in Africa. The areas studied ranged in size from small catchments in Belgium to the entire Upper Zambezi in Africa.

Table 7.1. Summary of case study characteristics.

	Impact focus		Scenario		Scale			Water resources impacts assessed					
Chapter	Quality	Quantity	GCM	Prescribed	Global	Regional	Local	Runoff	Reservoirs	Hydropower	Irrigation	Municipal & industrial	Environ.
8	*	*	*			*		*	*			*	
9	*		*		*		*		*				*
10	*		*	*			*	*	*				*
11		*	*			*	*	*	*	*			
12		*		*		*		*	*		*		
13		*	*				*	*	*				
14		*	*			*		*	*				

Chapter 8. Effects of Climate Change on Water Resources in the Delaware River Basin
D.M. Wolock, G.J. McCabe, Jr., G.D. Tasker, and M.E. Moss

Chapter 9. Impact of Climate Change on Sensitivity of Lake Statification: A Global Perspective
G. Meyer, I. Masliev, and L. Somlyódy

Chapter 10. Effects of Climate Change on Water Quality in the Central Valley of California
G.K. Meyer, G.T. Orlob, and C. Jokiel

Chapter 11. Impact of Climate Change on the Lake Kariba Hydropower Scheme
K.A. Salewicz

Chapter 12. Climate Change and Water Resources Management Strategy: The Terek River Basin
V. Shnaydman and J. Niemann

Chapter 13. Climate Change and Streamflow and the Sizing of Flow Control Reservoirs: Two Catchments in Belgium
D. Gellens

Chapter 14. National Assessment: Poland
Z. Kaczmarek and J. Kindler

Five of the case studies used GCM-based climate scenarios, one used only prescribed scenarios, and one used both GCM and prescribed scenarios.

The case studies have one feature in common: they all analyze a "managed system." All case studies include a reservoir or systems of reservoirs in the analysis. Reinforcing the focus of this book – water resources management in the face of climatic and hydrologic uncertainties – in six of the seven case studies, some form of modeling of climate change impacts on runoff was performed, although the modeling approaches were different in each of the six cases. These common features provide a means for comparison and understanding of certain aspects of the interrelation between climate and the water resources system.

Although climate is a global-scale problem, water resources management is a regional or local concern with distinct issues across varying hydroclimatic zones. Therefore, the case studies were chosen to not only have features in common to allow comparison, but also to span a wide range of hydro, climatic, and economic environments to provide the reader with a breadth of problems and alternative techniques for analysis. In Chapter 15, which follows the case studies and concludes the book, we will attempt to synthesize the issues and results of the case studies and present some insights into lessons learned from them.

Chapter 8

Effects of Climate Change on Water Resources in the Delaware River Basin

David M. Wolock, Gregory J. McCabe, Jr., Gary D. Tasker, and Marshall E. Moss

8.1 Introduction

In 1988, in response to concerns about the potential effects of CO_2 on water resources, the US Geological Survey began an interdisciplinary study of the sensitivity to climate change of water resources in the Delaware River basin (Ayers and Leavesley, 1988). The objective of the study was to evaluate the sensitivity of critical water resources in the basin to the potential effects of long-term changes in temperature, precipitation, and sea level.

8.2 The Delaware River Basin

The Delaware River basin, which lies along the eastern coast of the USA, covers an area of about 30,000 km^2 (*Figure 8.1*). The basin has a humid, temperate climate with a mean annual temperature of about 12°C and mean annual precipitation of 1,200 mm. Within the basin, soils range from thick, sandy loam soils to thin clay soils, and topography varies from low-relief coastal areas in the southern part of the basin to the Appalachian Mountains in the northern part of the basin.

Z. Kaczmarek et al. (eds.),
Water Resources Management in the Face of Climatic/Hydrologic Uncertainties 202–224.

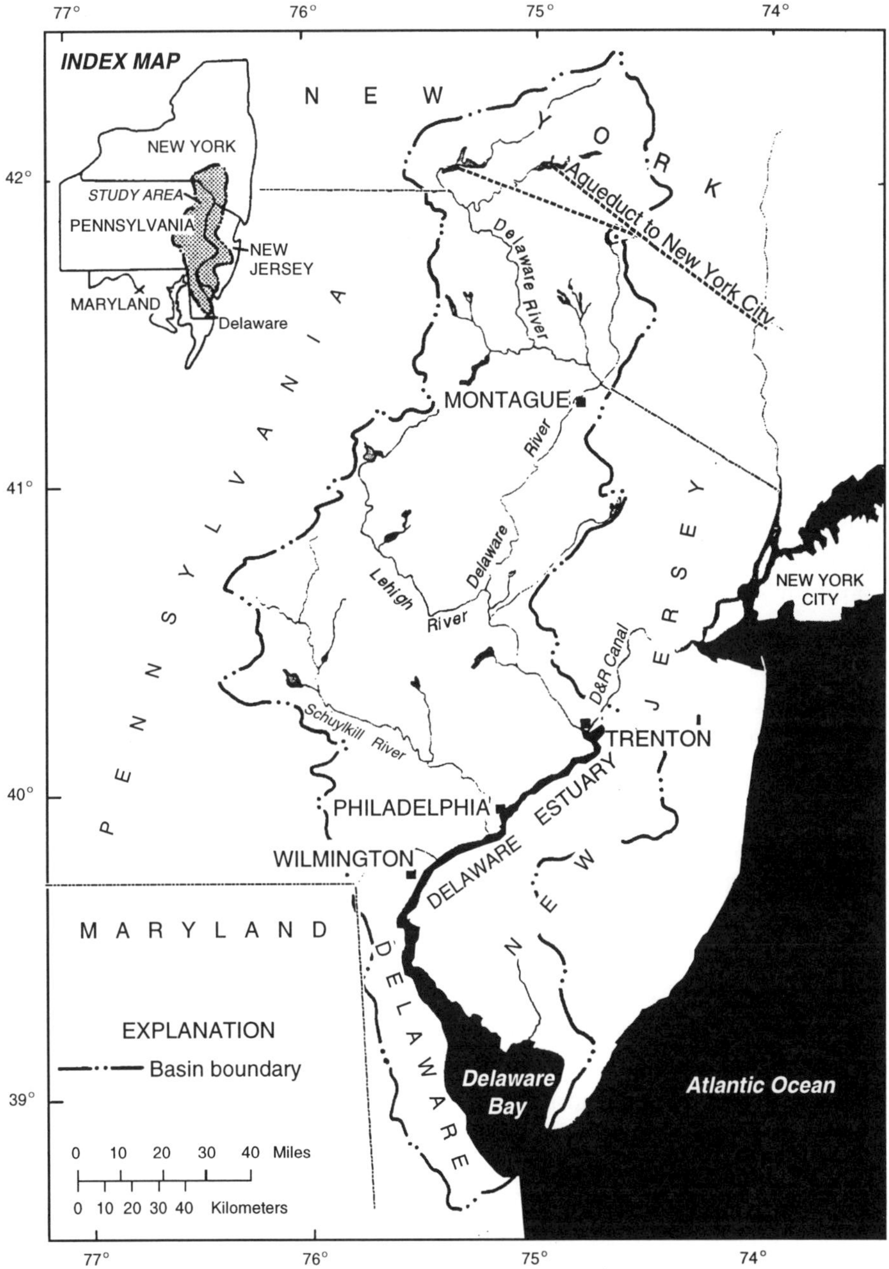

Figure 8.1. The Delaware River basin.

The Delaware River basin provides water for about 7 million people living within its boundaries (Albert, 1987). A major population center is located in and around Philadelphia, Pennsylvania. In this area water supply is withdrawn both directly from the Delaware River and from a regional aquifer system.

The Delaware River basin also provides water for an additional 13 million people living primarily in two areas outside the basin. The first is New York City, which is connected by an aqueduct system (*Figure 8.1*); the second is a region of New Jersey, which is connected to the basin by the Delaware and Raritan Canal (D & R Canal, *Figure 8.1*).

Management of water resources in the basin is centered on two critical components: the water storage in reservoirs used to supply water to New York City and the position of the salt front in the Delaware River estuary relative to water supplies in the Philadelphia area. The position of the salt front in the estuary is closely related to flow rates in the river: when river flow is low, the salt front migrates farther upstream. The relation between streamflow and salt-front position results in conflicts between the interests of the Philadelphia area and those of New York City. Currently, releases from the New York City reservoirs (and other mid-basin reservoirs) are required to maintain specified minimum flows at two points in the river, Montague and Trenton, New Jersey (see *Figure 8.1*). These minimum target flows are large enough to keep saline water downstream of the water supply sources in the Philadelphia area.

The magnitude of the target flows in the river (and therefore of the required releases from the New York City reservoirs) depends on the position of the salt front in the estuary and the actual storage level in the reservoirs. The storage level in the reservoirs defines the drought status of the basin (*Figure 8.2*), which along with the position of the salt front determine the required minimum flows (for an example, see *Table 8.1*). Larger target flows and reservoir releases are required when the position of the salt front moves upstream, but the target flows and required releases are decreased as the actual storage in the reservoirs decreases. During several droughts in the period from 1950 to 1970, New York City water managers refused to make the releases needed to maintain river flows sufficient to prevent upstream migration of the salt front. Their actions resulted in legal proceedings and even Supreme Court decrees. Existing reservoir operating procedures are the result of many years of negotiation and compromise between New York City and downstream users, such as those in the Philadelphia area (Albert, 1987).

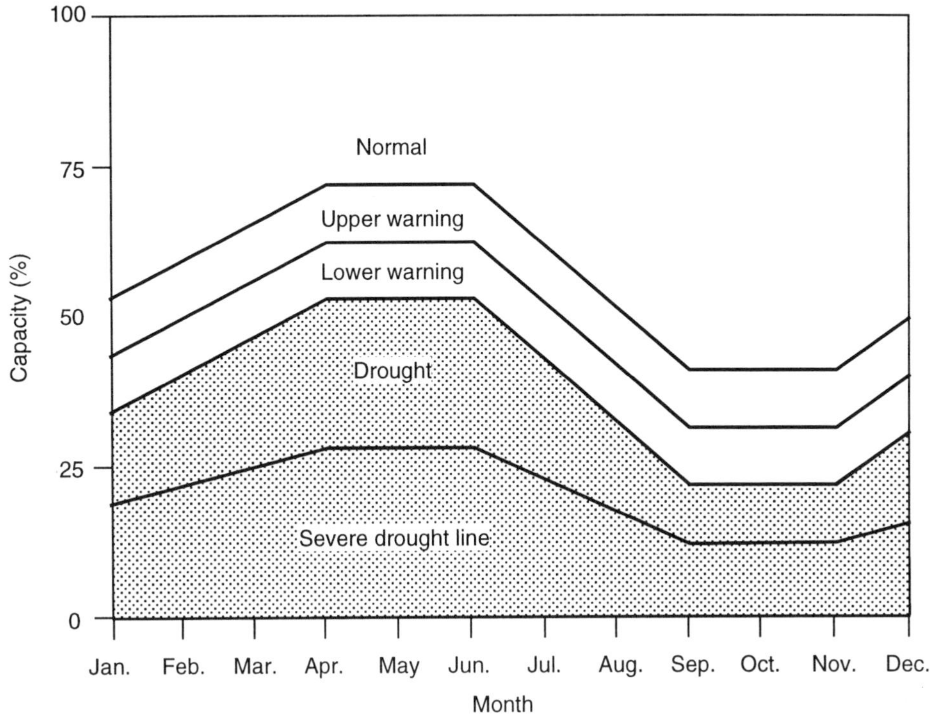

Figure 8.2. Drought status based on storage conditions in the New York City reservoirs.

Table 8.1. Flow targets at Trenton, New Jersey, for the months May through August.

New York City reservoir condition	Flow requirement (m^3/sec)
Normal	85
Warning	
Upper and lower	76
Drought	
Salt-front location	
Upstream of kilometer 149	82
Kilometer 140 to 149	76
Downstream of kilometer 140	71

This chapter addresses the potential effects of climate change on the two critical and conflicting water resources components in the Delaware River basin. It examines the sensitivity of New York City reservoir storage and the movement of the salt front in the Delaware River estuary to long-term changes in temperature, precipitation, and sea level. The objective of the study is to determine how climate change might affect the water resources conflict between New York City and the Philadelphia area, and whether changes in water resources management policy could alleviate some of the possible detrimental effects of climate change on water resources.

8.3 Scenarios of Climate Change

There is great uncertainty about the effects of an increase in atmospheric CO_2 on regional climate; therefore, several prescribed climate changes from current average conditions (1895 to 1988) were used in this study. The prescribed climate changes included combinations of an increase in temperature of 0°, 1°, 2°, 3°, and 4°C, and changes in precipitation of –20%, –10%, 0%, +10%, and +20%. These changes in temperature and precipitation represent the range of GCM projections reported in the literature (Schneider *et al.*, 1990). GCM projections were not used directly, because there is an inherent scale problem in interpolating projections from GCMs to areas the size of the Delaware River basin. GCM nodes are spaced on a grid of about four degrees latitude by five degrees longitude, and the size of the Delaware River basin measures only about three degress latitude by one degree longitude. Thus, the basin is smaller than the distance between model nodes, and any projection derived from such spatially coarse models should be questioned.

A wide range of upstream shifts in the mean monthly position of the salt front in the Delaware estuary was examined to determine the effects of increases in sea level on the water resources of the Delaware River basin; the magnitudes of these shifts were 0, 5, 10, 15, 20, 25, and 30 kilometers. Prescribed shifts were used because the effects of a rise in sea level on the movement of the salt front are complex and largely unknown at this time. Rough estimates, however, suggest that a 1-meter rise in sea level (which is more than twice the current estimates of sea-level rise for $2\times CO_2$ conditions) will cause about a 15-kilometer upstream shift of the mean monthly position of the salt front (Robin, 1986; Titus, 1986).

8.4 Hydrological Model of the Delaware River Basin

A three-component hydrological model of the basin was developed to examine the potential sensitivity of water resources in the Delaware River basin to increasing atmospheric CO_2. The three components are shown in *Figure 8.3*: (1) a multisite, monthly precipitation and temperature generator; (2) a modified Thornthwaite water balance model; and (3) a basin operations model. The precipitation and temperature generator was used to create time series of monthly temperature and precipitation values for each of five subregions in the basin (*Figure 8.4*). The resulting time series of monthly temperature and precipitation were input into a water balance model to simulate monthly streamflow in each of the five subregions. Finally, the monthly flows generated by the water balance model were routed through a basin operations model, which simulated the operation of reservoirs, diversions, and water use. Predictions from the basin operations model included storage in the New York City reservoirs and the position of the salt front in the Delaware River estuary. A quantitative assessment of the effects of climate change was made by tracking how often the New York City reservoirs were in severe drought conditions (defined here as storage level being at least one-half the distance from the upper limit of drought status to being completely empty; see *Figure 8.2*) and how often the salt-front position was at or above kilometer 160, a point in the river close to the sources of Philadelphia-area water supplies.

8.4.1 Precipitation and temperature generation

Monthly temperature and precipitation values were generated to simulate gradual climate changes for the next 100 years. The time series of temperature and precipitation were generated independently, because historical data indicated no significant correlation between these climate characteristics.

Time series of monthly temperature and precipitation were generated for the five subregions of the basin using a five-site Markov model (Matalas and Wallis, 1976). The values were computed by randomly generating serially correlated and cross-correlated residuals from long-term monthly mean values of temperature and precipitation (1895 to 1988). A square-root transformation was used to normalize the skewed precipitation residuals.

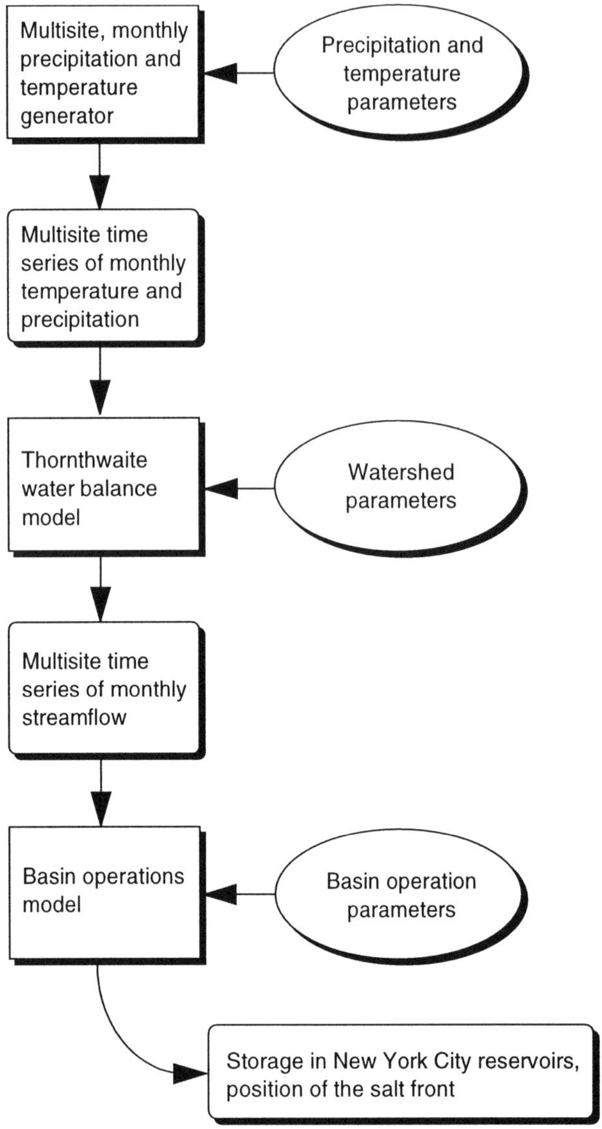

Figure 8.3. Hydrological model components.

In the five-site Markov model, the temperature or precipitation deviation $z_{t,j}$ at time t and site j is written as follows:

$$z_{t,j} = c_j + a_{j,j} z_{t-1,j} + \sum_{i=1}^{j} b_{j,i} e_{t,i} \quad , \tag{8.1}$$

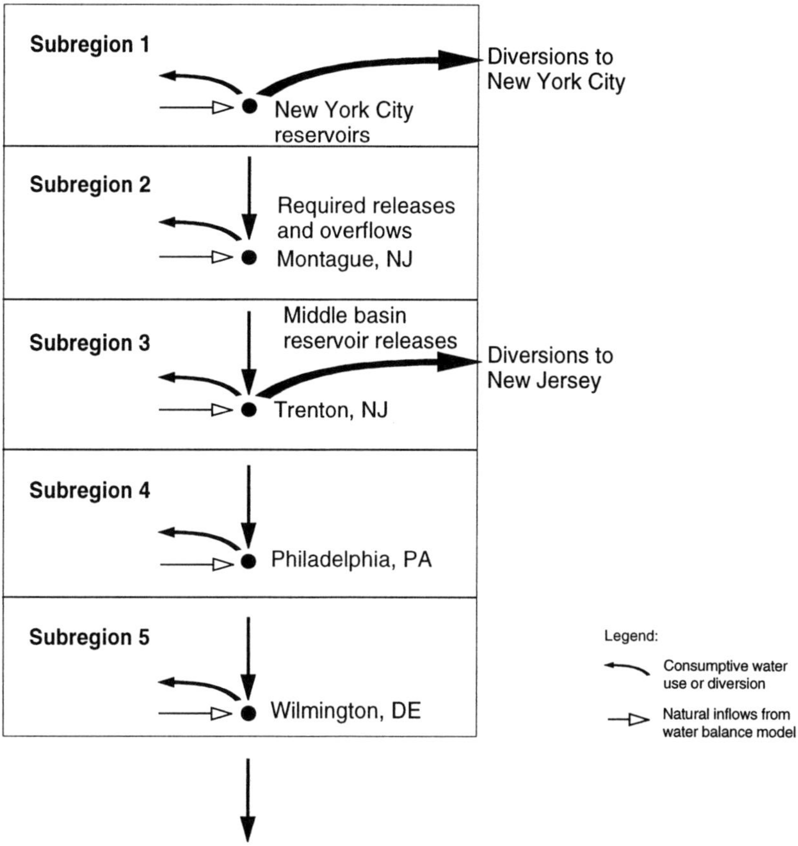

Figure 8.4. The basin operations model.

where c_j is $Z_j(1-a_{j,j})$; $a_{j,j}$ is the lag-one correlation of z; Z_j is the mean of z; $e_{t,i}$ is a standard normal variable; and $b_{j,i}$ are elements of a 5-by-5 lower triangular matrix, B, given by the solution of $BB^T = M_o - AM_oA^T$. The matrix A is a 5-by-5 diagonal matrix whose (j, j)th element is $a_{j,j}$; and M_o is a 5-by-5 matrix of covariances of z.

The values of $a_{j,j}$ and c_j were estimated from precipitation and temperature records in the basin. The effect of uncertainty in the estimates of $a_{j,j}$ and c_j was included in the simulations by using the method of Stedinger and Taylor (1982). Values of $e_{t,i}$ were produced by a random-number generator (Press *et al.*, 1986) to generate noisy 100-year time series of monthly precipitation and temperature. Multiple, unique random-number sequences were used to produce multiple, unique noisy time series to evaluate the effects of natural climate variability on water resources.

The Markov model was used to generate gradually changing temperature and precipitation time series. Thus, temperatures were linearly increased over the 100-year simulations, and precipitation was linearly increased or decreased. The total amount of change in temperature and precipitation over a 100-year simulation was equal to the prescribed climate changes.

The degree of realism of the Markov model was checked by comparing historical data with the statistical characteristics of simulated monthly temperature and precipitation (assuming no climate change). The analysis showed that the simulated climate was realistic, including the persistence of long-term droughts.

8.4.2 Modified Thornthwaite water balance model

A monthly time-step water balance model was used to simulate streamflow in each of the five subregions given the time series of temperature and precipitation and given some watershed parameters. The model was a modification of the Thornthwaite water balance model, which is a water budget procedure that accounts for soil moisture, evapotranspiration, water deficit, snowmelt, and surface runoff (Thornthwaite and Mather, 1955; Tasker, 1990).

The inputs to the water balance model are temperature, day length, and precipitation (*Figure 8.5*). The state variables are snowpack storage, soil moisture storage, and basin storage, which includes ground water and surface water storage. The model parameters are soil moisture holding capacity and the parameters that affect the volume of snowmelt and direct and delayed runoff. The outputs from the model are the three sources of runoff (snowmelt, direct runoff, and delayed runoff) and evapotranspiration. Parameter values were chosen from previous applications of water balance models in the Delaware River basin and by subjective calibration by comparing model output to observed monthly flows for several streamflow gauging stations in the basin. After calibration, monthly streamflow predicted by the water balance model closely matched observed streamflow when historical records of monthly precipitation and temperature were used as inputs to the model (*Figure 8.6*). For the simulations described in this study, outputs from the Markov precipitation and temperature generator were used as inputs to the water balance model, resulting in simulated runoff.

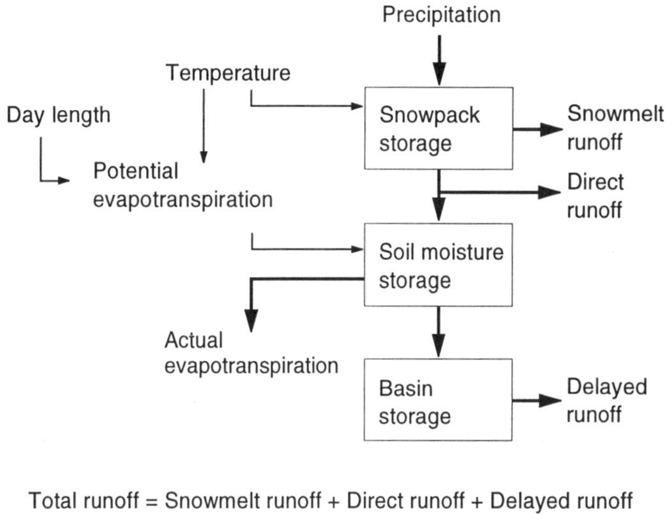

Figure 8.5. The water balance model. Bold arrows show potential water paths; light arrows show other conceptual relations.

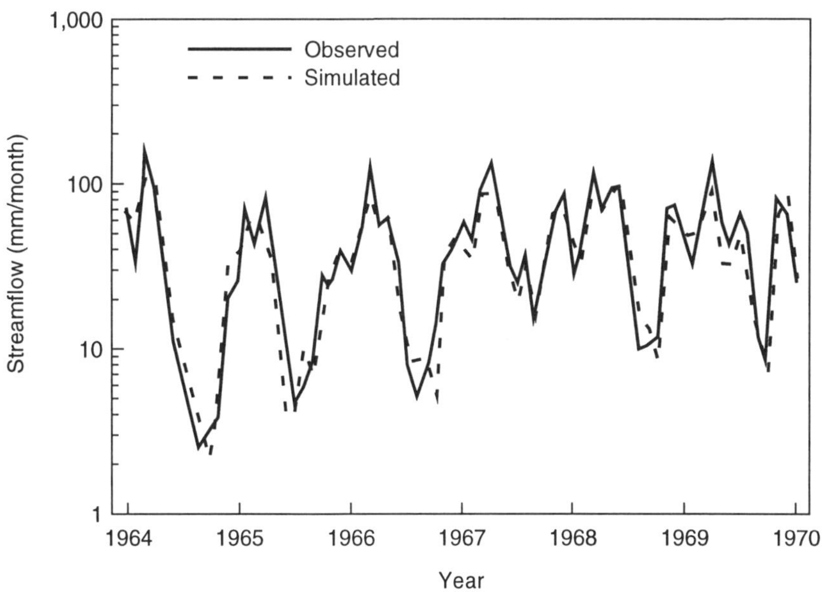

Figure 8.6. Simulated and observed streamflow in the Delaware River basin.

8.4.3 Basin operations model

The basin operations model (see *Figure 8.4*) predicted storage in the New York City reservoirs and the position of the salt front in the estuary using the following information: estimates of unregulated streamflow simulated by the water balance model for each of the five basin subregions, consumptive water use statistics, reservoir storage capacities, operating procedures for reservoir releases and diversions (such as those in *Figure 8.2* and in *Table 8.1*), and an empirical relation between river flow and the position of the salt front in the estuary. The length of time that the New York City reservoirs were in severe drought (see *Figure 8.2*) and the amount of time that the salt front was at or above kilometer 160 (near the Philadelphia-area water supplies) were computed for each 100-year simulation. These two aspects of water resources in the basin will be referred to as a New York City water crisis and a Philadelphia water crisis, respectively.

The only inputs to the basin operations model are the unregulated streamflow output values from the water balance model for each of the five subregions. The state variables are storage in the New York City reservoirs; storage in other basin reservoirs; streamflow at Montague, New Jersey; streamflow at Trenton, New Jersey; streamflow at Philadelphia, Pennsylvania; streamflow at Wilmington, Delaware; and the position of the salt front in the estuary. The model parameters are prescribed levels of consumptive water use, constants of the equations that represent the operating procedures, and constants of the empirical equation that relates the position of the salt front in the estuary to streamflow at Trenton, New Jersey. The basin operations model predicts diversions to New York City and New Jersey based on values of the state variables and the operating procedures.

To investigate the sensitivity to climate change of New York City reservoir storage and the position of the salt front in the Delaware River estuary, 100-year simulations of prescribed gradual changes in precipitation and temperature were input to the water balance model, resulting in simulated unregulated streamflow for the five subregions of the basin. These projections of unregulated streamflow were input to the basin operations model, resulting in 100-year simulations of storage in the New York City reservoirs and the position of the salt front in the estuary. For each 100-year simulation, the percentage of time that the New York City reservoirs were in severe drought (New York City water crisis) and the length of time that the position of the salt front in the Delaware River estuary was near the Philadelphia-area water supplies (Philadelphia water crisis) were calculated.

8.5 Results and Discussion

8.5.1 Effects of changes in temperature and precipitation

Results indicate that the New York City water resources are more sensitive to the range of prescribed changes in temperature and precipitation used in this study, and thus are more prone to crisis, than are the Philadelphia water resources (*Figure 8.7*). When precipitation was unchanged from the current average (1895 to 1988), a Philadelphia water crisis never occurred, even with a gradual 4°C temperature increase. In contrast, keeping precipitation constant the frequency of a New York City water crisis increased from 4% to 11% with temperature increases from 0°C to 4°C.

The effect of temperature increase on water crises was affected by the prescribed change in precipitation. *Figure 8.7* shows the effect of increasing temperatures on water crises when precipitation was simultaneously increased gradually by 20%. The frequency of a New York City water crisis was not affected by increases in temperature when precipitation was gradually increased; with a 20% increase in precipitation, a New York City water crisis occurred 3% of the time regardless of the prescribed temperature changes.

The sensitivity of water resources to temperature and the consequent risk of water crises are amplified when the warming is coupled with a gradual decrease in precipitation (see *Figure 8.7*). When precipitation is gradually decreased by 20%, the frequency of a Philadelphia water crisis ranges between 0.5% and 1.7% for temperature changes of 0°C and 4°C, respectively. Thus, it is only when a decrease in precipitation occurs that a Philadelphia water crisis is predicted by the model. The frequency of a New York City water crisis becomes very sensitive to increases in temperature when precipitation is decreased. With a gradual 20% decrease in precipitation, the frequency of a New York City water crisis increased from 27% to 42% for temperature increases of 0°C to 4°C. Given a precipitation decrease of 20% and increases in temperature from 0°C to 4°C, the range in the frequency of a New York City water crisis is 15%, compared with a range of 7% when precipitation remained unchanged. Note that the slopes of the lines showing the relationship between a New York City water crisis and temperature change are nearly constant, indicating that the sensitivity of New York City water resources to temperature changes is almost linear (see *Figure 8.7*). The slopes of the lines illustrate the greater sensitivity of New York water resources to temperature changes compared with the sensitivity of Philadelphia water resources. The slopes of the lines also show

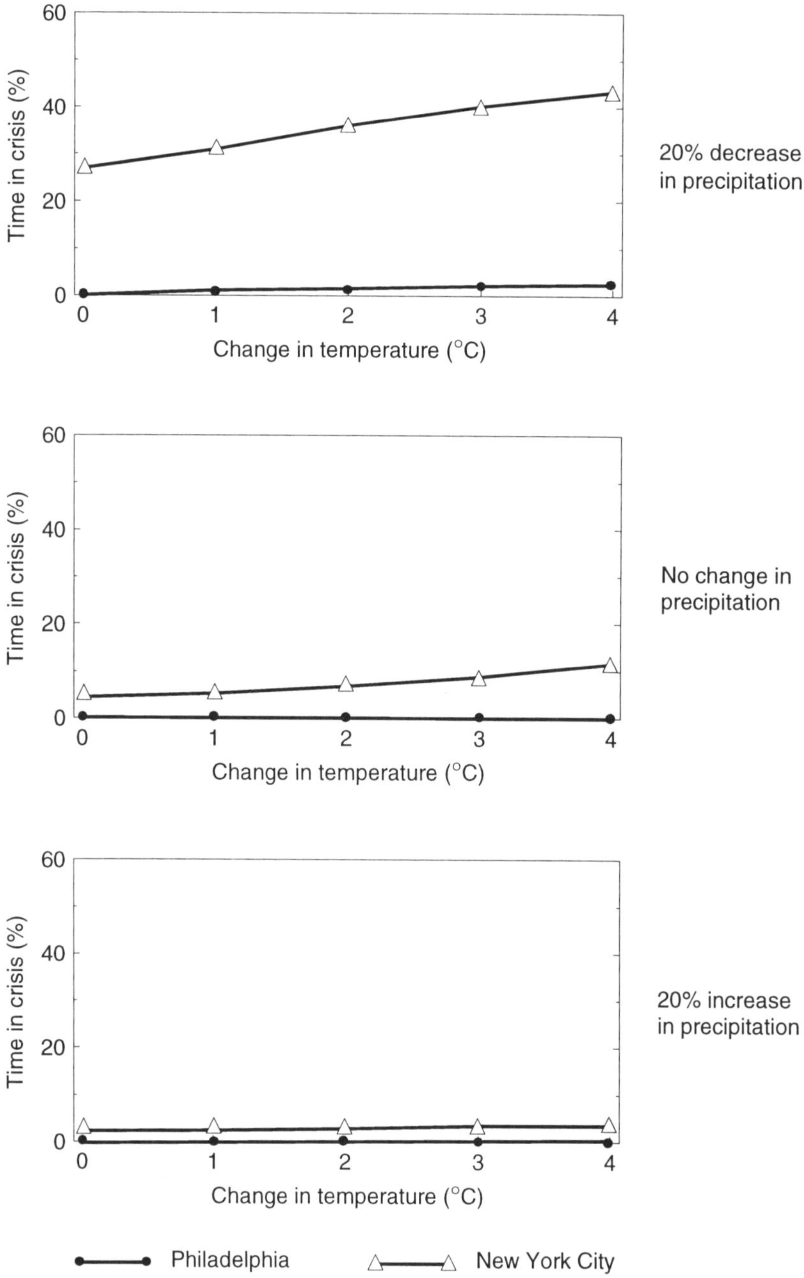

Figure 8.7. Sensitivity of water resources to changes in temperature with selected changes in precipitation.

the increased sensitivity of both New York City and Philadelphia water resources to temperature with a simultaneous decrease in precipitation.

The results also indicate that water resources in New York City are more sensitive to precipitation changes than are Philadelphia water resources (*Figure 8.8*). In contrast to the linear effect of temperature on water crises, the effect of precipitation changes on water crises is nonlinear. The slopes of the lines show that the sensitivity of water resources to precipitation change is not constant; the slopes increase as the value of percentage change in precipitation decreases from +20% to –20%.

Figure 8.8 also shows that the sensitivity of water resources to precipitation changes and the consequent risk of water crises increases as the temperature increases. With a 0°C warming, the range of the frequency of a New York City water crisis is between 3% and 27% for precipitation changes of +20% to –20%. With a 4°C warming, however, the range of the frequency of a New York City water crisis is between 3% and 42% for precipitation changes of +20% to –20%.

Water resources in New York City and Philadelphia are more sensitive to changes in precipitation than to changes in temperature for the ranges evaluated in this study. If the range in temperature changes were increased and the range in precipitation changes were decreased, however, the relative sensitivity of water resources to changes in temperature and precipitation might be different from what is reported here.

The greater sensitivity of water resources to precipitation changes than to temperature changes can be understood in terms of the effects of those changes on runoff. The water balance model indicated that runoff was more sensitive to the prescribed changes in precipitation than to the prescribed changes in temperature (*Figures 8.9* and *8.10*). The figures show runoff for subregions one and five (see *Figure 8.4*), the most northerly and southerly subregions, respectively. The slopes of the lines in *Figure 8.9* that show the sensitivity of runoff to changes in temperature are less steep than are those in *Figure 8.10*, which show the sensitivity of runoff to changes in precipitation.

8.5.2 Effects of a rise in sea level

The effects of a potential rise in sea level on the water resources of the Delaware River basin were examined by inducing prescribed shifts in the mean monthly position of the salt front in the Delaware River estuary and observing changes in the percentage of time New York City and Philadelphia were in water crisis situations.

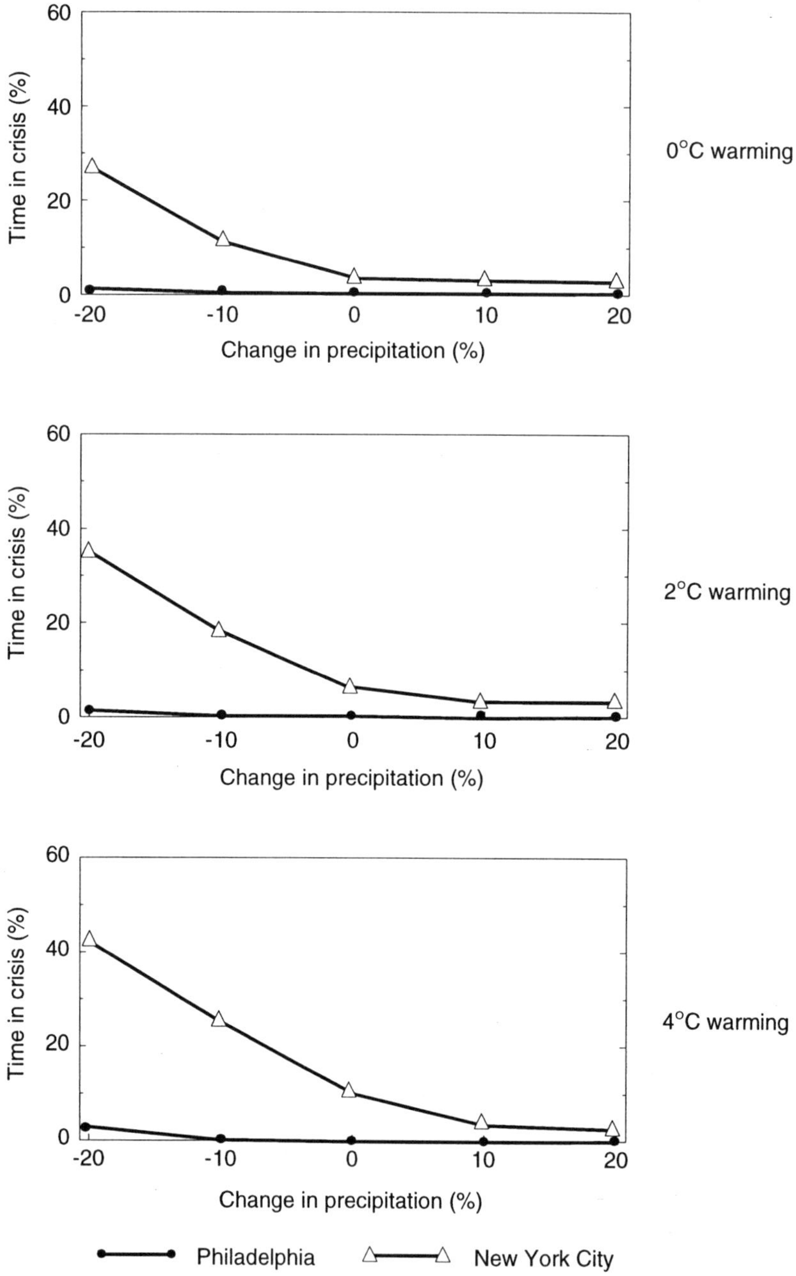

Figure 8.8. Sensitivity of water resources to changes in precipitation with selected changes in temperature.

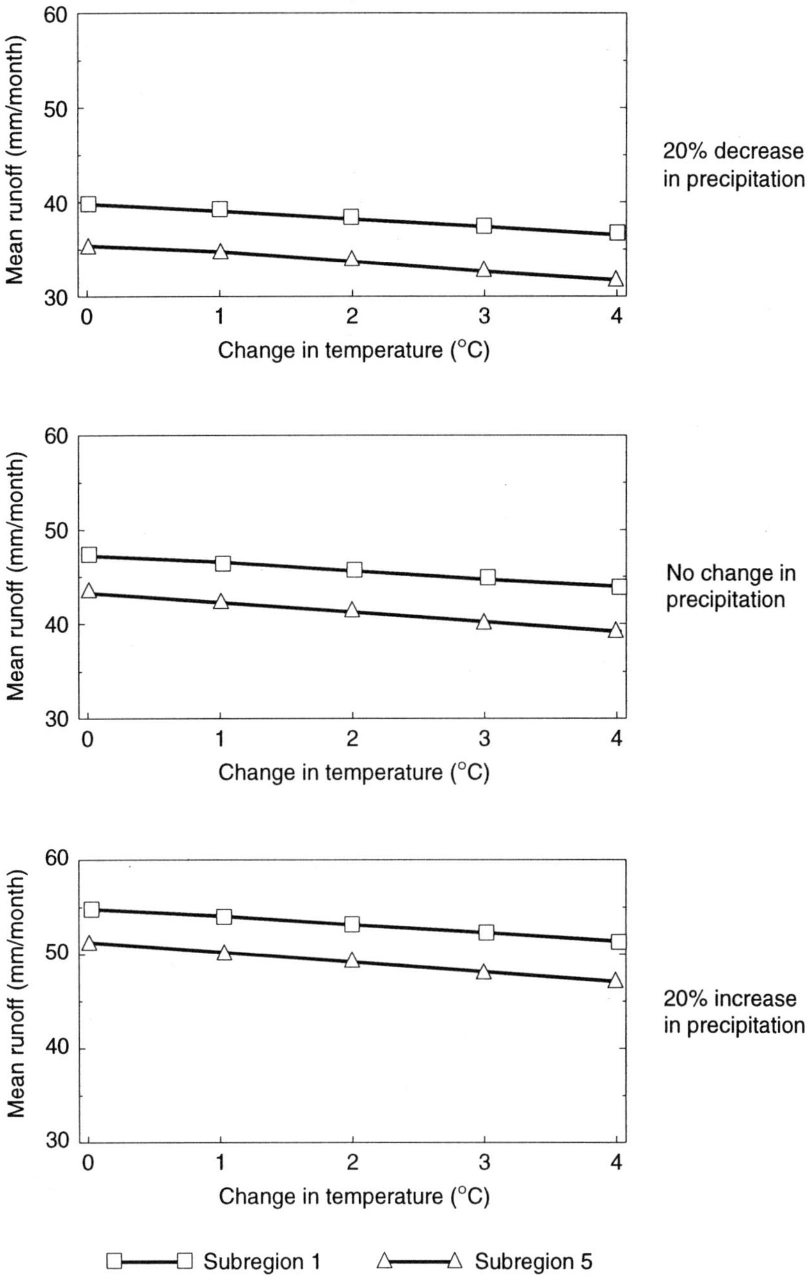

Figure 8.9. Sensitivity of runoff in subregions one and five to changes in temperature with selected changes in precipitation.

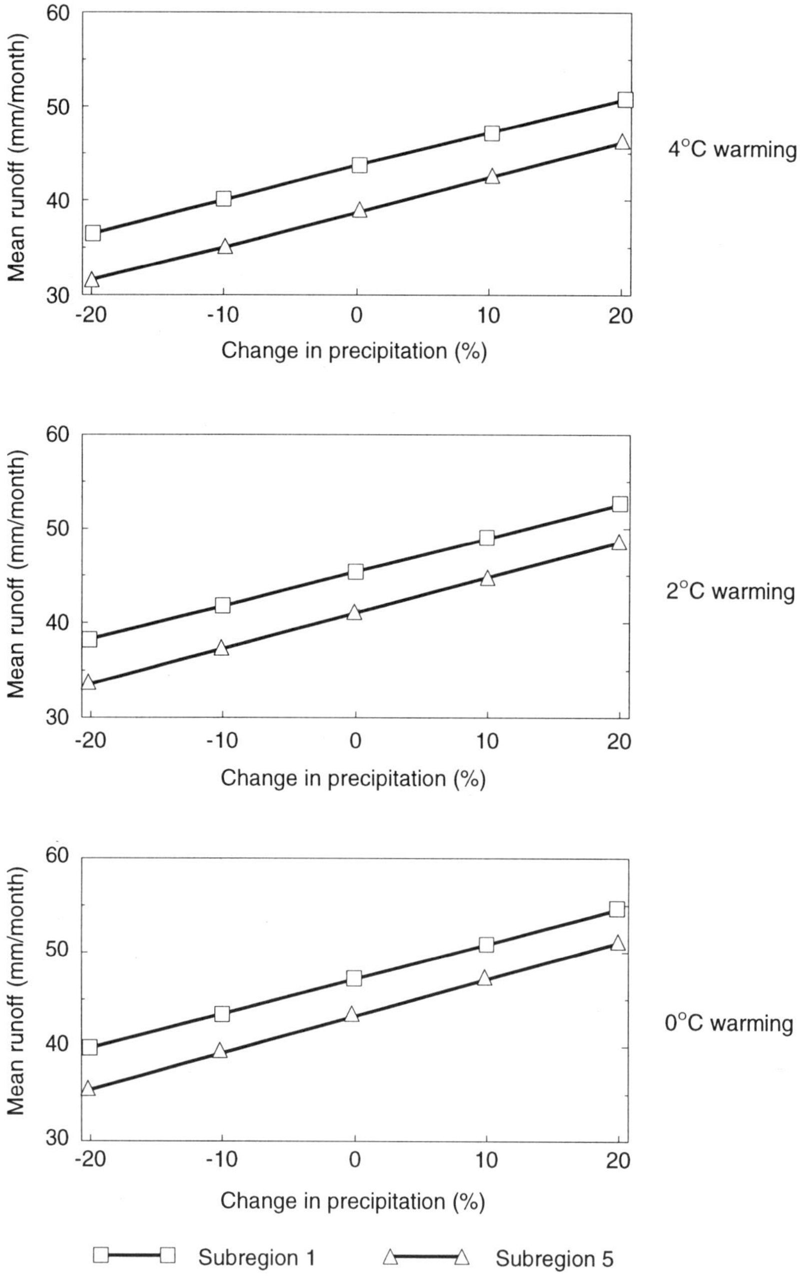

Figure 8.10. Sensitivity of runoff in subregions one and five to changes in precipitation with selected changes in temperature.

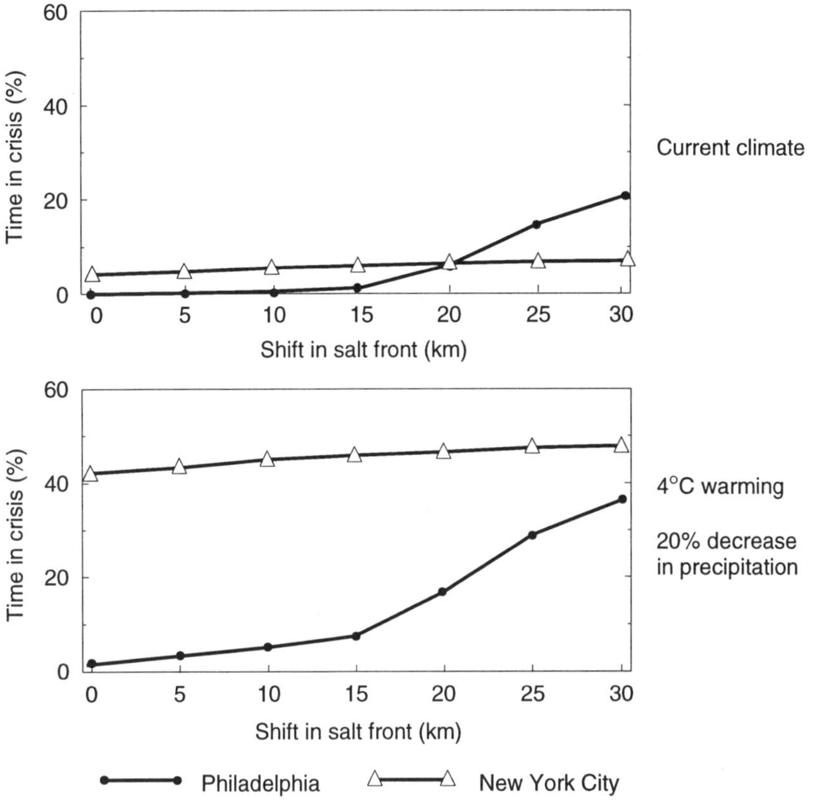

Figure 8.11. Sensitivity of water resources to shifts in the position of the salt front for two selected climate change scenarios.

The effect of a shift in the mean monthly position of the salt front was examined for two scenarios of changes in precipitation and temperature (*Figure 8.11*): current average climate conditions (1895 to 1988), and an extreme climate change, specifically a 4°C temperature increase and a 20% decrease in precipitation. In both cases, an upstream shift in the mean monthly position of the salt front had a greater effect on a Philadelphia water crisis than on a New York City water crisis. For current average climate conditions, an upstream shift in the mean monthly position of the salt front (the assumed effect of a rise in sea level) had almost no effect on the frequency of a New York City water crisis, as evidenced by the gentle slope of the sensitivity line. A shift in the position of the salt front has some effect on storage in the New York City reservoirs because target flows (for an example, see *Table 8.1*) and required releases are determined in part by the position of the salt front.

The effect of an upstream shift in the mean monthly position of the salt front on a Philadelphia water crisis was not apparent until an upstream shift of 20 km or more was prescribed. The results for the scenario with a 4°C warming and a 20% decrease in precipitation were similar. The effect of an upstream shift in the mean monthly position of the salt front on a New York City crisis was small compared with its effect on a Philadelphia water crisis. In the scenario representing a 4°C warming and a 20% decrease in precipitation, the effects of an upstream shift in the mean monthly position of the salt front in the estuary became apparent for a smaller upstream shift (5 km) than in the scenario representing current average climate conditions (20 km).

8.5.3 Effects of management practices

A number of model simulations were performed to determine if the water resources' sensitivity to climate was affected by the basin operating procedures. The specific aspect of basin management evaluated in this study was the use of target flows at Montague and Trenton, New Jersey, to determine required releases from the New York City reservoirs. The sensitivity of New York City and Philadelphia water crisis frequencies to changes in temperature and precipitation was determined for current target flows and for a number of modified target flows. Target flows were decreased by prescribed amounts. It was expected that with decreased targets the New York City reservoirs would be required to release less water; therefore, the frequency of a New York City water crisis was expected to decrease, potentially without an increase in the frequency of a Philadelphia water crisis.

Two climate change scenarios were used to evaluate the effects of decreasing the target flows on the sensitivity of water crisis frequency to changes in precipitation and temperature (*Figure 8.12*): current average climate conditions (1895 to 1988), and a 4°C increase in temperature and a 20% decrease in precipitation. *Figure 8.12* shows that the frequency of a New York City water crisis is decreased by decreasing the target flows. When targets are decreased by more than 40%, the frequency of a Philadelphia water crisis increases.

The effects of decreasing the target flows are dramatic for the drastic climate change scenario (*Figure 8.12*). Decreasing the target flows by 40% effects a decrease in the frequency of a New York City water crisis from 42% to 28%. Surprisingly, there is no adverse effect on Philadelphia when the

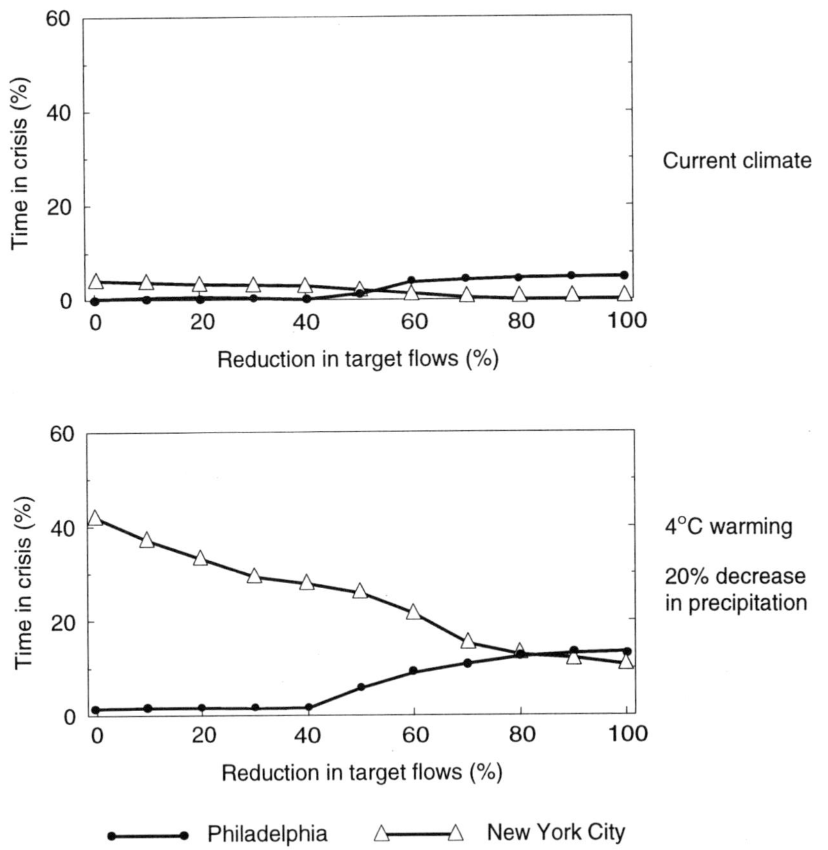

Figure 8.12. Effects of changing the target flows on water resources for two selected climate change scenarios.

targets are decreased by this amount. In fact, the frequency of a Philadelphia water crisis actually decreases from 1.7% to 1.2% as the target flows are decreased by 40%. This result may seem counterintuitive, because the larger target flows would be expected to keep the salt front downstream in the estuary. The probable explanation for the result is that by decreasing target flows, the New York City reservoirs release less water, thereby keeping water in reservoir storage (as evidenced in the decreased frequency of a New York City water crisis). The conserved water then is available at times when it is needed to maintain target flows in the Delaware River. Therefore, current target flows at Montague and Trenton, New Jersey, are unnecessarily large; decreasing the targets by 40% effects a decrease in the expected frequency of a New York City water crisis and actually decreases

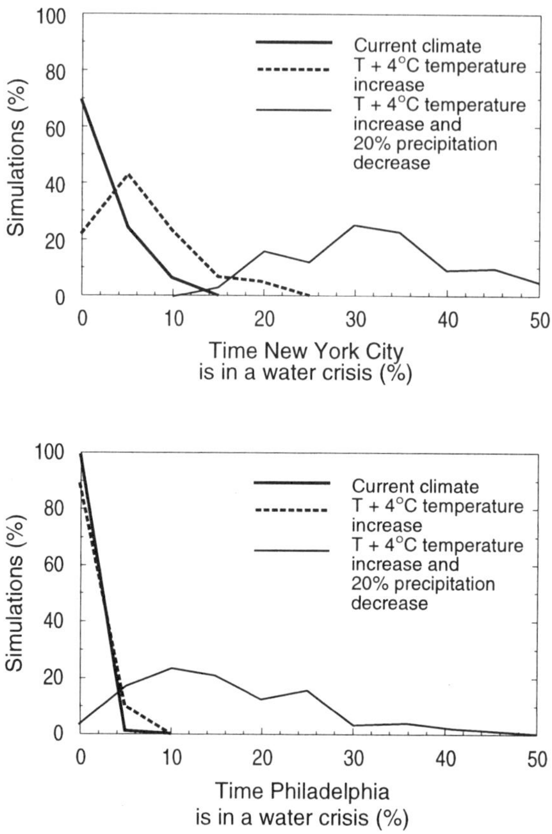

Figure 8.13. Effect of natural variability and climate change on the water resources.

the expected frequency of a Philadelphia water crisis if temperature increases and precipitation decreases.

8.5.4 Effects of natural climate variability

Natural climate variability is a major factor affecting the detection of the effects of climate change on water resources in the Delaware River basin. For a given expected future climate (that is, a specified climate change scenario), natural variability creates the potential for a wider range of climate conditions. Each scenario was evaluated for 100 unique simulations by using 100 different random-number sequences in the stochastic precipitation and temperature equations. The resulting frequency distributions for the percentage of time that the water crises occurred, derived from the 100 simulations for each of several temperature- and precipitation-change scenarios, are presented in *Figure 8.13*. Each scenario had a number of possible water crisis frequencies; for example, the figure shows that the frequency of a New York City water crisis varied from 0% to 15% of the time for a temperature change of 0°C and no change in precipitation. When temperature was increased 4°C, the range of the frequency of a New York City water crisis was from 0% to 25% of the time. When a 4°C temperature increase was coupled with a 20% decrease in precipitation, the frequency of a New York City water crisis ranged between 10% and 50% of the time. The overlap in these distributions implies that natural climate variability can mask the effects of projected long-term climate changes. Even if accurate projections of long-term changes in mean temperature and precipitation can be made, a wide range of drought frequencies is possible because of unpredictable random variability in precipitation and temperature. Both New York City and Philadelphia water crisis probability distributions showed substantial overlap among the various prescribed climate change scenarios.

8.6 Conclusions

The effects of potential climate change on water resources in the Delaware River basin are uncertain. First, uncertainty in GCM projections of potential changes in future precipitation is propagated into water resources projections because of the sensitivity of the water resources to precipitation changes. Second, natural variability in climate is substantial. Even if projections of long-term changes in temperature and precipitation were certain, natural variability in climate is unpredictable and of such magnitude that it could overwhelm any long-term trends in precipitation and temperature.

Despite the uncertainties in projecting the effects of climate change on water resources, several conclusions about the sensitivity of water resources in the Delaware River basin can be drawn. If current reservoir operating procedures are followed in the future, changes in temperature and especially precipitation will affect the frequency of a New York City water crisis more than it will affect the frequency of a Philadelphia water crisis. Also, the expected frequencies of water crises under adverse climate changes can be reduced by modifying the operating procedures used in the basin.

References

Albert, R.C., 1987, *Damming the Delaware*, The Pennsylvania State University Press, University Park, PA, USA.

Ayers, M.A., and Leavesley, G.H., 1988, Assessment of the potential effects of climate change on water resources of the Delaware River basin: Work plan for 1988–1990, US Geological Survey Open-File Report 88-478, USGS, West Trenton, NJ, USA.

Matalas, N.C., and Wallis, J.R., 1976, Generation of synthetic flow sequences, in A.K. Blewas, ed., *Systems Approach to Water Management*, McGraw-Hill, New York, NY, USA.

Press, W.H., Flannery, B.P., Teukolsky, S.A., and Vetterling, W.T., 1986, *Numerical Recipes*, Cambridge University Press, Cambridge, UK.

Robin, G., 1986, Changing the sea level: Projecting the rise in sea level by warming of the atmosphere, in B. Bolin, B.R. Döös, J. Jäger, and R. Warrick, eds., *The Greenhouse Effect, Climate Change, and the Ecosystem*, John Wiley & Sons, New York, NY, USA.

Schneider, S.H., Gleick, P.H., and Mearns, L.O., 1990, Prospects for climate change, in P.E. Waggoner, ed., *Climate Change and U.S. Water Resources*, John Wiley & Sons, New York, NY, USA.

Stedinger, J.R., and Taylor, M.R., 1982, Synthetic streamflow generation 2: Effect of parameter uncertainty, *Water Resources Research*, **18**(4):919–924.

Tasker, G.D., 1990, Some hydrological impacts of climate change for the Delaware River basin, in R.M. Khanbilbardi and T.C. Gooch, eds., *Optimizing the Resources for Water Management: Proceedings of the 17th Annual National Conference of the Water Resources Planning and Management Division*, 17–21 April, at Fort Worth, TX, USA, American Society of Civil Engineers, New York, NY, USA.

Thornthwaite, C.W., and Mather, J.R., 1955, The water balance, *Climatology*, **8**(1):1–104.

Titus, J.G., 1986, The basis for expecting a rise in sea level, in C.J.H. Hull and J.G. Titus, eds., *Greenhouse Effect, Sea-level Rise, and Salinity in the Delaware Estuary*, EPA 230-05-86-010, Environmental Protection Agency, Washington, DC, USA.

Chapter 9

Impact of Climate Change on Sensitivity of Lake Stratification: A Global Perspective

Gabriela K. Meyer, Ilya Masliev, and László Somlyódy

9.1 Introduction

Water-related impacts of climate change are discussed extensively in the literature, as well as in this book. As indicated in Chapter 3, water quality effects have not yet been systematically analyzed, although the related uncertainties are perhaps even greater than in other fields of hydrology and water resources management.

Water temperature is one of the few exceptions to the general uncertainty: methodologies and models are available for consistent studies for surface water (see Chapter 4). Water temperature is a crucial indicator regarding the impact of climate change. It is a physical parameter characterizing water quality directly, which itself has an indirect impact through a number of processes, particularly if lakes and reservoirs are considered.

The question of how freshwater lake temperatures respond to changed atmospheric conditions was addressed by Hondzo and Stefan (1991). Three dimictic lakes located near 45° N latitude and 93° W longitude in the north-central USA were selected. Historical records were used to specify a "warm" (future) climate scenario (the summer of 1988) and a "normal" (past) condition (the year 1971). A vertically one-dimensional hydrothermal model was used to analyze spatial and temporal changes in temperature, as well

Z. Kaczmarek et al. (eds.),
Water Resources Management in the Face of Climatic/Hydrologic Uncertainties 225–273.

as changes in components of the thermal energy budget and in the stratification pattern. The study indicated that the water temperature within the epilimnion varied considerably, and there was an increase in surface mixed layer depths. The present chapter addresses the same question. However the context, and thus the approach, is quite different. Instead of a detailed analysis of a small number of specific lakes, our primary aim is to raise a broader question, i.e., what is the regional sensitivity of lake stratification patterns to future climate change depending on latitude and longitude? One possible way to study this issue would be to evaluate past and future thermal behavior for a large number of lakes of various morphologies and locations by a proper heat budget model. However, such an approach is not practical because the data and labor requirements would be tremendous.

For this reason we decided to follow a three-step procedure. The first step incorporates the definition and analysis of a number of hypothetical lakes (from shallow to deep) to identify sensitive regions of the globe by systematically changing the location of these hypothetical lakes in terms of latitude. In the second step, real lakes are selected from the sensitive regions identified in step one, still maintaining a number of simplifying assumptions related to morphology, inflow, and outflow, etc. (see below). Base-case simulations use monthly averaged historical climatic data; future conditions are obtained from a general circulation model (GCM). Both scenarios depend on latitude and longitude. The third step is beyond the scope of our present effort: it is the detailed analysis of the set of lakes from the second step using specific data and resolving the simplifying assumptions.

In this chapter, we first offer a detailed problem formulation, followed by an outline of the approach to be employed. Section 9.4 summarizes properties of databases to be used for generating historical and future climate scenarios. Section 9.5 summarizes major features of thermal models, which will serve as basic tools for our study. A simplified method is outlined in Section 9.6, which is then used for the specification of hypothetical lakes. The identification of sensitive regions and the selection of real lakes are the subjects of Section 9.7. Results obtained and a discussion of these results are given in Section 9.8. Section 9.9 presents a summary and conclusions of this study.

9.2 Problem Formulation

The heat balance in a lake is a result of heat absorption, dissipation, and distribution within the entire water body. Heat absorption takes place at the surface of the water. The low thermal conductivity of water prevents heat redistribution into deeper layers; heat redistribution is possible almost entirely due to mechanical energy, with wind being the most important driving force. It is therefore expected that any changes in meteorologic and climatic conditions can affect the thermal household of the system.

The surface heat transfer of a lake depends on several components, such as net short wave (solar) radiation, net long wave (atmospheric) radiation, water surface reflective radiation, evaporative loss, and conductive loss (see Orlob, 1983; Octavio *et al.*, 1977). Net solar radiation depends on the clear sky solar radiation, a function of latitude, time of the year, and cloudiness. Net long wave radiation is dependent on air temperature, with clcudiness and air humidity also playing a role. Water surface reflective radiation is determined by the surface water temperature. The evaporative heat flux is a function of the water pressure (or specific humidity), wind speed, and surface water temperature. An additional component, conductive loss, is also influenced by the air temperature.

The brief review above shows that a change in climate would affect the heat balance of lakes primarily through factors such as air temperature, cloudiness, relative humidity, and wind speed. Among these air temperature is pivotal. Its potential increase would result in higher atmospheric radiation and modifications in reflective radiation, evaporation losses, and heat conduction. The likely end result of all these changes would be an increase in the surface water temperature. For the lakes studied by Hondzo and Stefan (1991), the average air temperature for the period from May to August was approximately 3°C higher in 1988 than it was in 1971, leading to a similar increase in the average epilimnion temperature (the difference between 1977 and 1988 conditions fluctuated between –4°C and +10° C, with approximately only two months with colder conditions in the future scenario). The changes in hypolimnetic temperatures were much smaller and somewhat inconsistent. A comparison of simulations for the "warm" and "normal" summers showed that higher back radiation (due to higher epilimnion temperatures) compensated for the increase in atmospheric radiation. Conductive heat transfer turned out to have a negligible effect, while evaporative heat loss increased by about 40%.

The heat balance of a lake obviously depends on transformation processes in the water body. Deep lakes and reservoirs are subject to a phenomenon called stratification. Stratification occurs due to heat flux differential along the depth, resulting in the formation of a stable density profile with denser water layers below less dense ones (Wetzel, 1983). Lakes and reservoirs experience seasonal changes in their degree of stratification (see, e.g., Octavio *et al.*, 1977). The characteristic degree of stratification depends on climatic factors and on morphology, and in the case of reservoirs, on the throughflows, as well. In temperate zones, in early spring a lake is isothermal with temperatures typically around 4°C. As the season progresses, the lake begins to warm and differential absorption of heat induces stratification, forming the epilimnion and hypolimnion. Later, during cooling periods, if the surface temperature drops below the temperature of the water beneath, convective mixing occurs through density instabilities due to the formation of denser water at the surface. Pronounced water circulation and mixing can eventually destroy thermal stratification, leading to water overturn, particularly when surface waters reach 4°C, i.e., maximum water density. Wind mixing contributes to the maintenance of a mixed upper layer, a situation especially enhanced when the water temperatures are low due to the lower density gradients in the cold water ranges.

An alteration of the magnitude and timing of the climatological factors that drive stratification thus can change the stratification pattern and possibly the lake turnover, both of which strongly influence water quality and the behavior of aquatic ecosystems. Under stratified conditions the deeper water is isolated from the atmosphere, impeding oxygenation of these layers. Furthermore, decomposition of organic matter at the bottom can deplete oxygen and create anoxic conditions that can foster the growth of anaerobic microorganisms. Lakes with overturn periods temporarily experience a mixing of the whole water column. It is evident that if there is a significant change in the length of the stratified period and the number of overturns, material cycling can be drastically altered. Under extreme conditions, deep lakes may behave like shallow ones, or vice versa.

The objective of this chapter is to study whether such changes can occur and if so, in which regions of the world. A related question is to what extent would duration of periods of ice cover and stratification change? In other words, our interest is the evaluation of the effects of global climate change on regional sensitivity of lake stratification.

This issue is closely related to geographical classification used in limnology (see, e.g., Wetzel, 1983). The thermal behavior of lakes changes with latitude and altitude locations, depending on the relation of the seasonal variability of water temperature and the point of maximum density (4°C). Accordingly, six distinctive categories can be identified:

- Amictic polar lakes, with permanent ice cover.
- Cold monomictic lakes in the latitude range between 60° and 80° N/S, which circulate fully during the summer.
- Dimictic (temperate) lakes in the latitude range between 40° and 60° N/S, which have two periods of overturn in a year (most of the northern Eurasian and North American lakes belong to this group).
- Warm monomictic lakes of the subtropical type (20° to 40° N/S), which show circulation only during the winter.
- Oligomictic lakes, with little or no circulation located close to the equator at low latitudes.
- Polymictic water bodies in latitudes close to the equator, with frequent periods of overturn due to high altitudes and intense cooling at night.

Another type of obvious distinction exists between shallow lakes and deep lakes. For shallow lakes, the depth is not sufficient for a hypolimnion to form and stratification can only be maintained temporarily. The mixing depth of a lake depends primarily on wind speed. According to Schwoerbel (1987), 1 m to 2 m mixing depth corresponds to 2 m/sec wind speed, and 6 m to 12 m mixing depth corresponds to 10 m/sec. The mixing depth required for a lake to become thermally stratified varies greatly with site-specific features, such as surface area, the volume-to-depth relation, the prevailing wind direction, the fetch, and the protection provided by surrounding topography and vegetation.

The geographical classification also depends on morphology and other local conditions. A dimictic or monomictic lake should be deep enough to develop a hypolimnion (in contrast, a shallow lake can be considered polymictic). Altitude also plays a role: the behavior of standing waters located at higher altitudes is similar to that of standing waters at lower altitudes located at higher (colder) latitudes. Low latitudes lead to exceptions in the above categories, primarily due to the strong influence of oceanic climate (Wetzel, 1983).

Having defined the two interrelated classification systems (geographical and geomorphological), it may be of interest to know whether a shift in classes along the latitude will be the result of climate change. Similarly, we can study whether depending on location a lake can change from being "deep" to being "shallow" (i.e., if because of climate change it no longer develops a hypolimnion). If so, this would have serious implications from the viewpoint of material cycling and water quality, as noted earlier.

9.3 Approach

The regional analysis of climate change on lakes and the approach adopted stem directly from the formulation of objectives.

First, climate data from databases were retrieved to formulate a realistic, systematic analysis. The available data were a longitudinally averaged historical data set, long-term monthly average data for a large number of stations (of different latitudes and longitudes), and predictions of a future changed climate from GCMs (for details see Section 9.4).

Next, two hydrodynamic models were compared to select the one most appropriate for the analysis. These models were the model developed by the Institute for Water and Environmental Problems (IWEP), Siberian Branch of the Russian Academy of Sciences, in Barnaul, Russia, and the hydrothermal component of the Water Quality of River-Reservoir Systems model (WQRRS) developed for the US Army Corps of Engineers. As shown in Section 9.4, the similarity of the simulations performed by the two models leads to the use of the IWEP model, because its running time is shorter and it is capable of simulating periods of overturn and ice cover.

To produce the systematic runs whereby the behavior of water bodies ranging from shallow to deep is assessed, hypothetical lakes were generated, all having a surface area of 100 km^2 and parabolic bathimetry. The selection of the depth of shallow, medium-depth, and deep lakes was performed using a simplified two-layer version of the Massachusetts Institute of Technology (MIT) wind-mixing model (Octavio *et al.*, 1977). The model is computationally efficient and well-suited for the purpose (see Section 9.6).

Next, latitudes where lakes are sensitive to air temperature change were identified using the IWEP model, hypothetical lakes, and the historical longitudinally averaged monthly data. Real water bodies (lakes and reservoirs) were then chosen within the identified sensitive latitudes. The morphology of these lakes is characterized in a simplified fashion on the basis of real surface area pertaining to real mean depth.

Present climate scenarios were formulated using real-station monthly averaged climatic data (for each water body the closest station was used). Future scenarios, in which a doubling of atmospheric CO_2 is assumed, were derived from the GCM of the Geophysical Fluid Dynamic's Laboratory (GFDL) of the National Oceanographic and Atmospheric Administration (NOAA).

9.4 Data Used

As outlined earlier, three main sets of meteorological data were needed. One contains monthly and longitudinally averaged historical data; this was used to test overturn sensitivity at different latitudes. A second set contains historical observations of weather stations located at various latitudes and longitudes. This data set was applied to produce present (base-case) scenarios for the real lakes analyzed, which are located in various sites within belts sensitive to climate change. The third data set is the GFDL model output, used to derive global climate change (future) scenarios for the real lakes analyzed. A detailed description of these data sets follows.

9.4.1 Longitudinally and monthly averaged data

The climatic data used to generate hypothetical lakes (Section 9.6) and belts of the globe sensitive to climate change (Section 9.7) were those compiled by Telegadas and London (1954). These data contain summer and winter data for different longitudes and latitudes compiled from 30 years of data prior to 1954, and include air temperature, relative humidity, wind speed, cloudiness, and pressure. These data were further averaged longitudinally and within 10° latitude ranges, and were converted to monthly values by interpolation. Sinusoidal interpolation was used for temperature, relative humidity, and cloudiness; quadratic interpolation was applied for wind speed; no interpolation of pressure data was necessary. Latitudinal changes of major parameters are illustrated in *Figures 9.1* to *9.5*.

9.4.2 Historical observations

Historical observations of climate stations located close to the real lakes studied were obtained from the IIASA database developed by Leemans and Cramer (1991). This database is a compilation of the following weather records:

- World Weather Records, 1940–1951 (Weather Bureau, 1959).
- Klimadiagramm-Weltatlas (Walter and Lieth, 1960–1967).
- Selected Global Climatic Data Set for the Vegetation Sciences (Müller, 1982).
- Selected Climate Tables for the World (UK Meteorological Office, 1966, 1972, 1973, 1978, 1980, 1983).

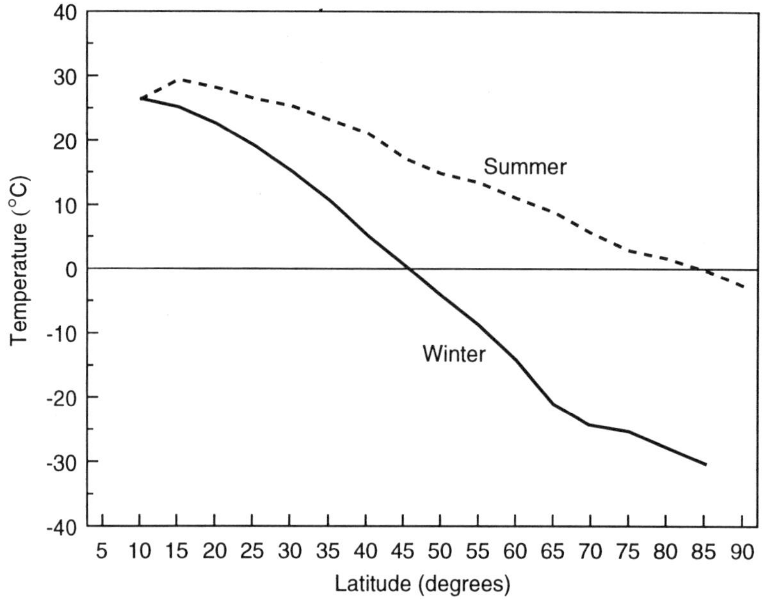

Figure 9.1. Historical longitudinally averaged temperature.

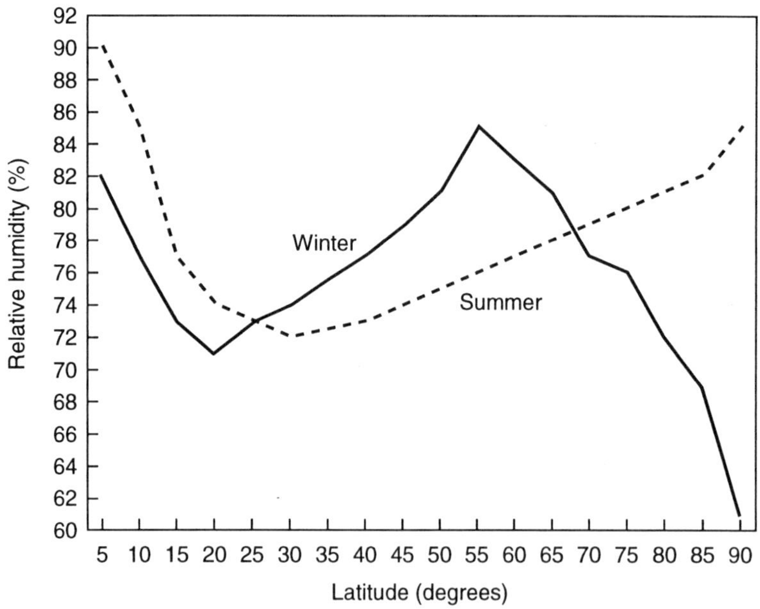

Figure 9.2. Historical longitudinally averaged relative humidity.

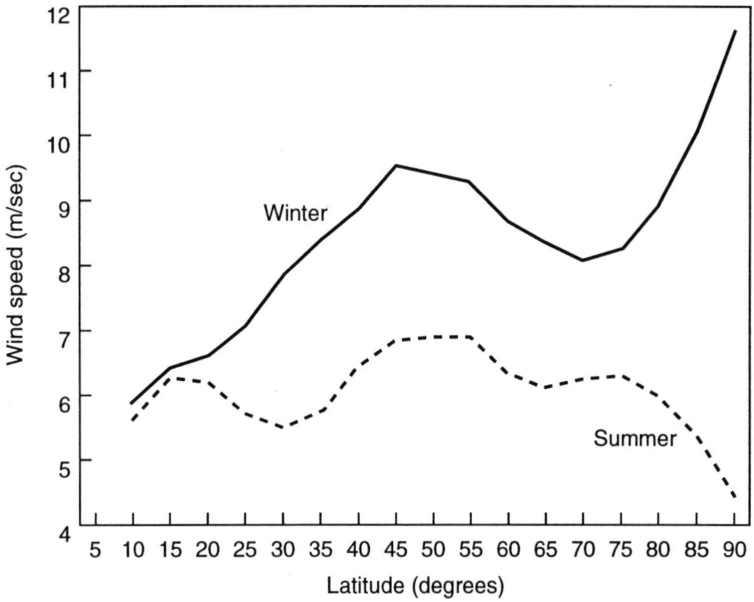

Figure 9.3. Historical longitudinally averaged wind speed (height = 13 m).

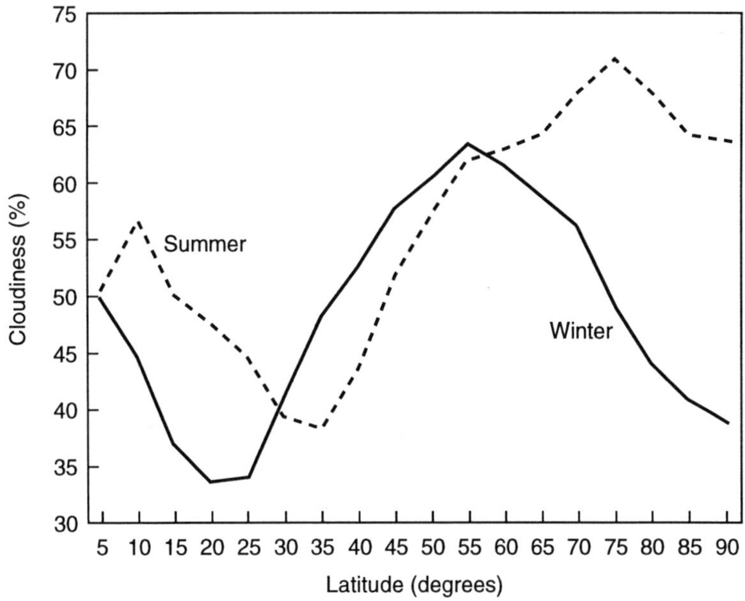

Figure 9.4. Historical longitudinally averaged cloudiness.

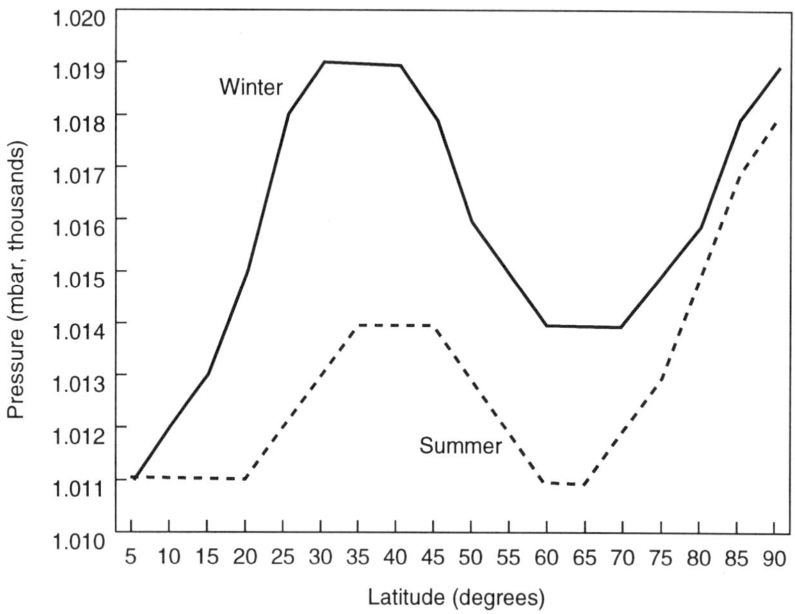

Figure 9.5. Historical longitudinally averaged pressure.

Leemans and Cramer integrated all these records into a single database within one format that covers the following variables: mean daily temperature, mean daily maximum temperature (average of daily maximum temperatures from the historical records); mean daily minimum temperature (average of daily minimum temperatures from the historical records); absolute maximum temperature; absolute minimum temperature; mean relative humidity; mean precipitation; maximum precipitation; minimum precipitation; maximum precipitation in 24 hours; mean number of days with precipitation greater than 2.5 mm; mean duration of sunshine; mean quantity of radiation; mean potential evapotranspiration; mean wind speed; mean predominant direction of the wind (in degrees); soil moisture; and monthly total radiation transmittance.

These records are monthly averages over the period from 1941 to 1960 (some stations have data for a shorter time period within the same period). This period was chosen by Leemans and Cramer as being representative of what they defined as "current climate." All the station data were checked separately for coding errors by evaluating extreme values and averages, and by comparing each station with other stations in the immediate

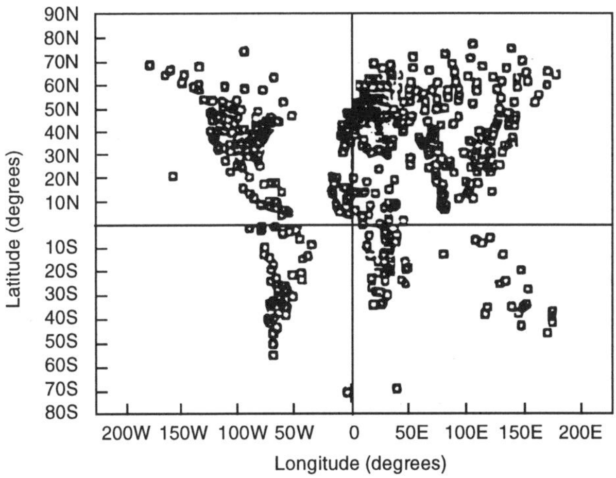

Figure 9.6. Selected climate monitoring stations (for details see Leemans and Cramer, 1991).

surroundings. The longitude and latitude for each station (location) were also checked.

Unfortunately, data on cloud cover were lost from the database. To resolve this problem the interpolated cloudiness data from Leemans and Cramer (1991) were used (interpolated to a 0.5° latitude × 0.5° longitude grid). Implementation of the original station data with cloudiness was done using the value of the grid point nearest to each of the stations. It was decided that this method was sufficiently accurate given the short distances between grid points.

A subset of stations was then selected. The selection was made by choosing only those stations that had the following variables: mean daily temperature, mean relative humidity, mean wind speed, mean predominant direction of the wind, and cloudiness.

Figure 9.6 shows the location of this subset of stations. A total of 540 stations were available.

9.4.3 GCM data

GCM data from the IIASA database were obtained. It was decided to use the simulation results of the GFDL model (Manabe and Stouffer, 1980). According to the analysis done by Kalkstein (1991), GFDL predictions differ from existing conditions by a maximum of 3°C in North America, Europe,

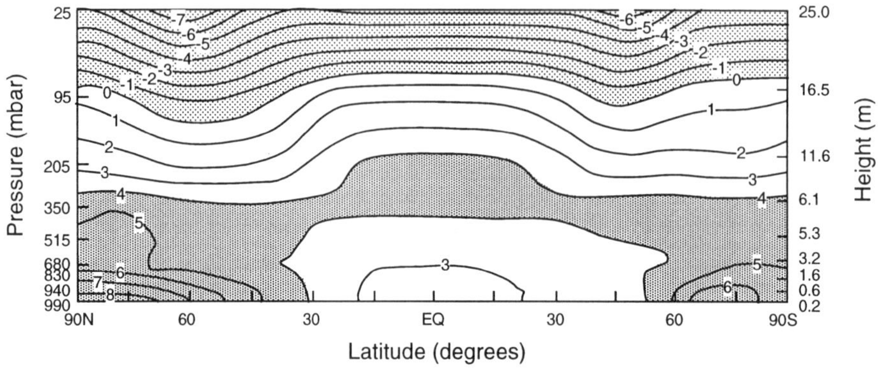

Figure 9.7. Longitudinally averaged predictions of air temperature differences from the GFDL model. Source: Wetherald and Manabe (1988).

and western Eurasia. In tropical Asia, Australia, Africa, and South America, the GFDL model predicts colder temperatures than exist in reality.

The variables obtained from the GFDL model include monthly mean precipitation and temperature. Both variables are given for present conditions ($1\times CO_2$) and for a future scenario in which a doubling of CO_2 is assumed ($2\times CO_2$). The variable used for the analysis was the temperature for both scenarios, because it is the most important variable from a hydrophysical point of view and it is available from our database.

Analysis of GCM output reveals that, given the relatively coarse resolution of the GCMs (four degrees latitude by five degrees longitude, or about 500×500 km for the GFDL model), within the boundary layer (height of 2 to 3 km) temperature-difference tendencies are rather homogeneous (Harrison, 1990). For this reason, in order to predict future scenarios the difference in temperature was calculated without requiring altitude adjustments to the GCM temperature data. The future scenarios of air temperature were assessed by adding the GCM temperature difference obtained from the GCM to the present-condition station data. Longitudinally averaged GFDL predictions on air temperature differences are shown in *Figure 9.7*.

9.4.4 Hydrothermal models

To study the impact of climate change on the hydrothermal dynamics of lakes, three models were considered. Two of them are well-known water quality models: Water Quality of River-Reservoir Systems (WQRRS), based on a US Army Corps of Engineers model (HEC, 1978), and the Institute for Water and Environmental Problems model (IWEP), a model from the Siberian Branch of the Russian Academy of Sciences (Zinoviev *et al.*, 1990). For our purposes, the IWEP hydrothermal components were used. A third model was applied solely for generating hypothetical lakes. As already noted, it is a simplified version of the MIT wind-mixing model (Octavio *et al.*, 1977; see appendix). For a complete description of hydrothermal modeling see Chapter 4.

The WQRRS and IWEP models are vertically one-dimensional models of deep stratified reservoirs that assume full mixing in horizontal layers. Both models have been successfully applied to simulate hydrothermal and water quality dynamics of reservoirs. Internal transport of heat and mass occurs only in the vertical direction as a result of a material balance among layers, which is described by the advection-diffusion equation of the form

$$V\frac{\partial T}{\partial t} = -\Delta z Q_z \frac{\partial T}{\partial z} + \Delta z A_z D_c \frac{\partial^2 T}{\partial z^2} + Q_i T_i - Q_o T \pm V\frac{H}{\rho_w c_p} \quad , \qquad (9.1)$$

where T is the water temperature; V is the volume of the fluid element; t is the time coordinate; z is the space coordinate; Q_z is vertical advection; A_z is element surface area normal to the direction of flow; D_c is effective diffusion coefficient; Q_i is lateral inflow; T_i is inflow thermal energy; Q_o is lateral outflow; H is internal distribution of heat source within the water column; c_p is the heat capacity of water under constant pressure; and ρ_w is water density.

Similarities and Differences Between the Two Models

The models are similar in terms of heat source and allocation of inflow. The main difference between the two models stems from the formulation of the diffusion coefficient (representing both molecular and turbulent diffusion mechanisms). In the case of WQRRS, the user can select one of two methods to calculate effective diffusion: the stability method or the wind method. Another difference between the two models is that WQRRS does not simulate surface ice covers or overturn. Also, the IWEP model uses the method proposed by Markofsky and Harleman (1971) and assumes a Gaussian outflow profile. The WQRRS model has two options for the determination of outflow location: the Debler-Craya withdrawal allocation method, or the waterways experiment station method.

9.5 Determination of Hypothetical Lakes

9.5.1 Methodology

Thermal characterization of lakes and reservoirs depends on the atmospheric heat influx and the water body's hydraulics, which in turn depend on morphology. In a reservoir, residence time is a critical determinant of stratification characteristics. Reservoirs with short residence time can be considered mixed water bodies. Others with long residence time can exhibit horizontal stratification. In the case of lakes, negligible inflows and outflows make the thermal household of this system independent of throughflows. To avoid the problem posed by throughflows, it is subsequently assumed that water bodies behave as lakes.

In lakes, morphology is a predominant factor characterizing stratification. A shallow lake is defined as one where wind energy input prohibits the formation of stable thermal stratification. The characterization of a shallow lake depends on the combination of fetch and depth. A lake with large area and moderate depth can be considered shallow, because the fetch is large enough for wind to maintain mixed conditions. In contrast, a deep lake can present isothermal characteristics in a cold season and gradually develop stratification with warming, such that in the warmest periods there will always be two distinct layers: the epilimnion, with well-mixed conditions; and the hypolimnion, which presents isothermal conditions but is practically thermally isolated from the epilimnion by the metalimnion. It is important to stress that the increase of epilimnion depth depends primarily on solar heating and wind entrainment.

The methodology for generating hypothetical shallow and deep lakes is based on the stratification process (see *Figure 9.8*). It is assumed that the reservoir is initially fully mixed with temperature T_H, and that there is an atmospheric source of heat. As a consequence of heating, a thin surface layer will form with a temperature (T_E) that is higher than T_H. Wind action on the surface of the lake will lead to a deepening of this thin surface layer and to the onset of stratification. This layer develops with depth h to create the epilimnion. Further wind mixing under a continuous source of heat causes a deepening of the epilimnion, i.e., the warmer top layer; if this layer with temperature T_E reaches the bottom, stratification is broken and isothermal conditions exist once again.

A lake can be regarded as being shallow if stratification can break during a relatively short period of time. To develop a method for identifying

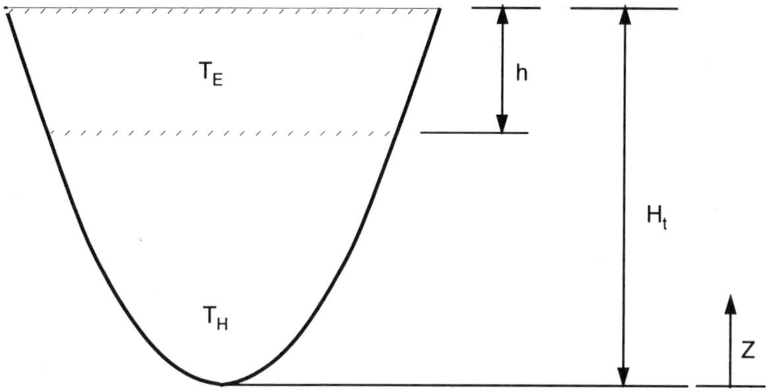

Figure 9.8. Layout of a hypothetical lake morphology and definition of its two layers. In this figure h is the developing depth of the epilimnion; H_t is the total depth of the lake; z is the vertical axis; T_E is the epilimnion temperature; and T_H is the hypolimnion temperature.

shallow lakes from a set of hypothetical lakes situated in different geographical locations and having different morphometry, the ideas described above were applied in the context of the MIT approach (Octavio *et al.*, 1977). This approach calculates the entrainment of hypolimnetic water into the upper mixed layer on the basis of comparing the rate of change in potential energy of lakes. The formulation suggested in this study is a simplified form of the MIT wind-mixing approach, which assumes two distinct lake layers of temperatures T_E and T_H (see *Figure 9.8*). The derivation of the governing equations of the two-layer lake stratification model can be found in the appendix.

9.5.2 Generation of hypothetical lakes

To determine whether a lake is deep or shallow using the approach presented above, an indicator that defines the lake type is required. Because the thermocline in a shallow lake moves downward relatively quickly due to wind-driven mixing, an indicator that defines the mixing stage of a lake can be the percentage of the epilimnion depth to the lake's total depth, which we will call the mixing ratio (h/H_t).

Under this definition a shallow lake reaches a 100% mixing ratio in a short period of time, say a month or less. In contrast, a deep lake can maintain a mixing ratio close to 0%. Moreover, stratification is more likely to occur during the warmest month within a year; therefore, if a lake becomes fully mixed during the warmest month, it is most likely shallow all year round.

With these definitions, determination of hypothetical lakes was performed only for the Northern Hemisphere, under the assumption that Southern Hemisphere characteristics would be equivalent for longitudinally averaged monthly conditions. The month of July was considered the warmest month in the Northern Hemisphere. Because wind is the most important force governing mixing within the epilimnion and movement of the thermocline, it is important to test sensitivities of percentage of mixing due to wind. For this reason, simulations were performed for normal winds (longitudinally historical monthly averages; see Section 9.4), and for a 50% increase and a 50% decrease of normal wind speed.

The surface area of the hypothetical lakes was set at 100 km^2. Linear dependency between the lake depth and the area of horizontal cross section was assumed. Five values for lake depth were selected: 10, 20, 50, 75, and 150 m. Latitudes were systematically changed in five-degree increments. The climate data used at each latitude were those derived from Telegadas and London (1954) (see Section 9.4.1). Initial conditions for the simulation were $h(t = 0) = 0.01$ m, $T_H = 8$°C, and $T_E = 8.05$°C.

It should be noted that the initial temperature of the epilimnion, T_E, is only slightly higher than that of the rest of the lake, T_H. The initial temperature in this case is nothing more than an arbitrary number needed to initiate the warming process.

Figures 9.9 to *9.11* show graphs of the percentage of mixing versus latitude for different lake depths. A comparison of the three figures clearly shows the role of wind, as well as the effect of latitude. The higher the wind speed, the higher the percentage of mixing for a given lake depth. Also, at high and low latitudes percentage of mixing is high for all depths, which is characteristic of polar and equatorial regions. In *Figure 9.9*, isothermal behavior is apparent above 70° latitude, depending on depth. There is a tendency toward higher stratification around 20° and 70° latitude. From an analysis of *Figures 9.9* to *9.11*, lakes with depths of 150 m and 75 m were defined as deep lakes, because for any wind change the mixing ratio is close to 0%. Lakes with depths of 20 m, 10 m, and even 5 m can be considered intermediate-depth lakes. Shallow lakes are characterized by depths less than 5 m. For further analysis, water depths of 75 m, 20 m, and 10 m were selected to analyze deep, intermediate-depth, and shallow lakes, respectively.

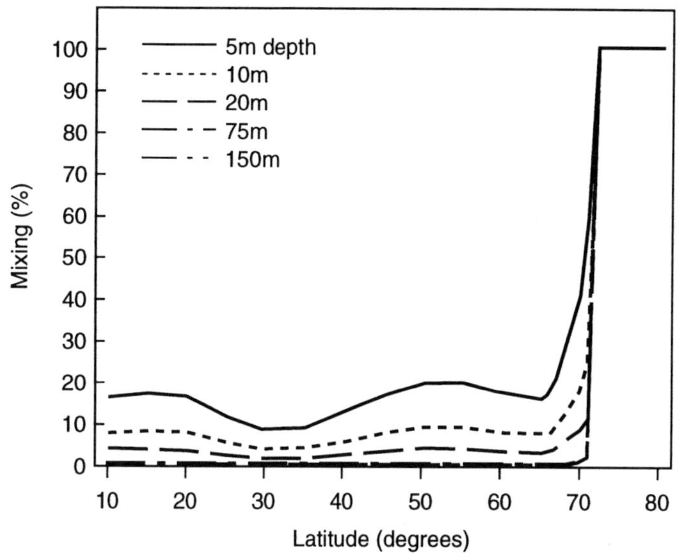

Figure 9.9. Percentage of mixing versus latitude for normal winds.

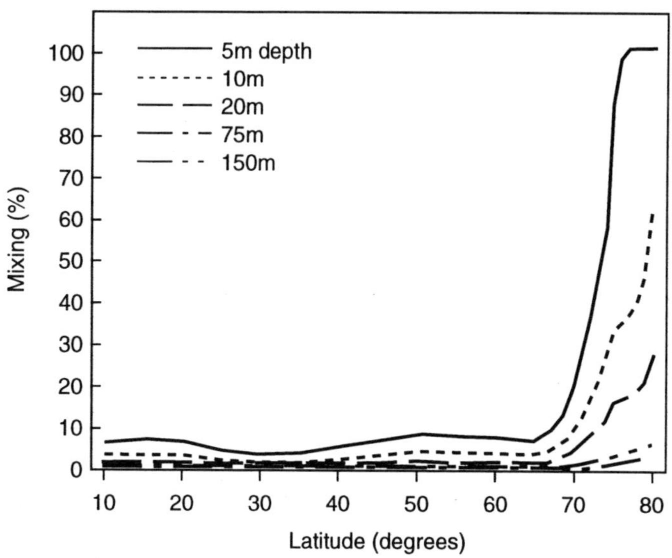

Figure 9.10. Percentage of mixing versus latitude for a 50% decrease in normal winds.

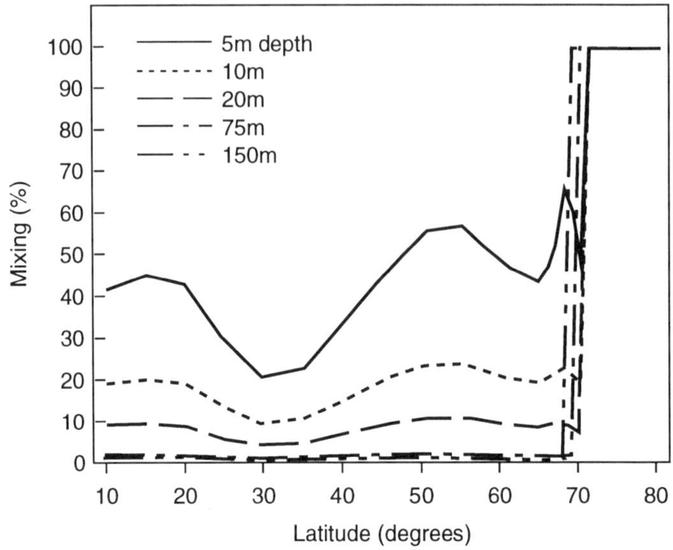

Figure 9.11. Percentage of mixing versus latitude for a 50% increase in normal winds.

9.6 Identification of Sensitive Regions

9.6.1 Criteria for defining overturn

The definition of overturn is somewhat subjective and can take several forms. Accordingly, four criteria for defining overturn were initially examined:

1. Overturn occurs when more than half of the reservoir exhibits a positive density gradient, a phenomenon that induces convective instability.
2. Overturn takes place when the potential energy of the water column prior to mixing exceeds that of the water column after mixing, i.e., the density profile in the water column is such that the potential energy can be used as a source for turbulent energy generation:

$$P_{bov} - P_{mix} > 0 \quad , \tag{9.2}$$

   where P_{bov} is potential energy before overturn and P_{mix} is potential energy after mixing.
3. Overturn is assumed when bottom density is less than top density.
4. Overturn occurs when bottom and top density gradients are positive.

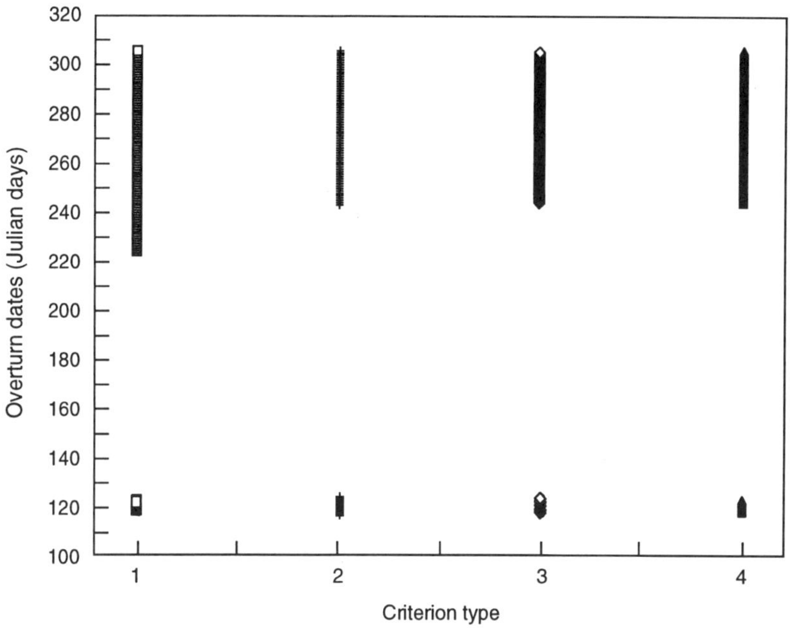

Figure 9.12. Comparison of overturn criteria in an intermediate-depth lake (20 m deep).

Performance of these criteria was tested with two hypothetical lakes, one a deep lake (75 m) and one an intermediate-depth lake (20 m), in a representative middle latitude (55°). As input, longitudinally and monthly averaged climate data were used (see Section 9.4).

Comparisons of the overturn criteria for the two lakes are shown in *Figures 9.12* and *9.13*, which indicate the periods of complete mixing (by Julian calendar days). The figures show that the criteria are similar. Criterion one indicated a somewhat longer period of complete mixing (isothermal conditions) in the late autumn period than did the other three criteria.

All criteria indicated that intermediate-depth lakes are dimictic (i.e., two overturns per year), with the longest overturn period occurring in autumn. Because criterion three proved to be the most robust, sensitive, simple, and physically sound (overturn occurs when the bottom density is less than that at the top), it was accepted for the rest of our study. This criterion is not only indicative of mixing, but also indicates persistence of isothermal conditions.

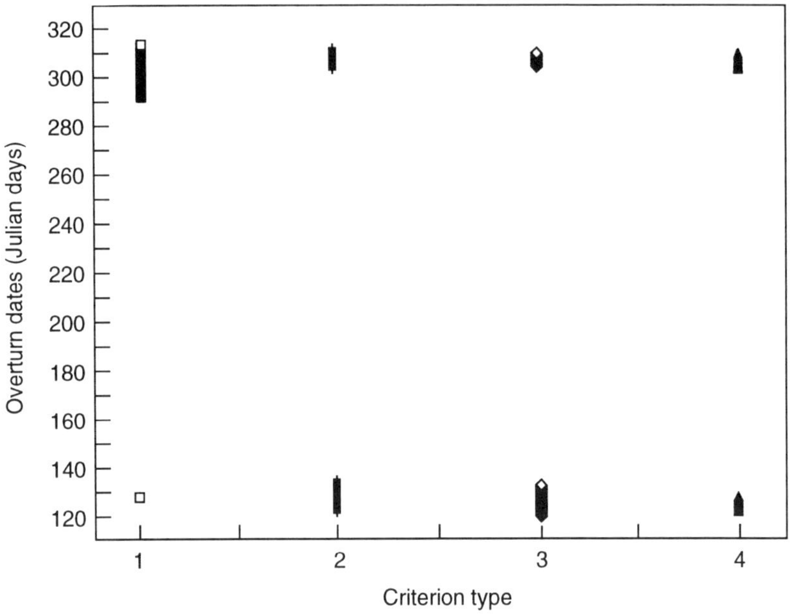

Figure 9.13. Comparison of overturn criteria in a deep lake (75 m deep).

9.6.2 Sensitivity analysis: Definition of sensitive belts of the globe

For the purpose of sensitivity studies, a ±5°C alteration was imposed on the longitudinally and monthly averaged historical air temperature (Section 9.4.1), because this range characterizes the typical average temperature change induced by the greenhouse effect according to most GCM predictions. The procedure was the same as was used before, but now the IWEP model was used to identify those regions of the globe that are sensitive to climate change.

Simulations were performed for three successive year-long periods. The results of the last year are considered to characterize "settled" quasi-stable behavior that is not dependent on the initial conditions. The results of this last modeling year are further analyzed and presented in *Figures 9.14* to *9.26*. The general overview of the occurrence of the overturn events during the last modeling year is shown in *Figures 9.14*, *9.19*, and *9.23*. If at a certain latitude a lake displays overturn conditions, the corresponding day is marked on the plot with a circle. The overview plots combine three cases (one base-case scenario and two sensitivity runs) that differ in the size of

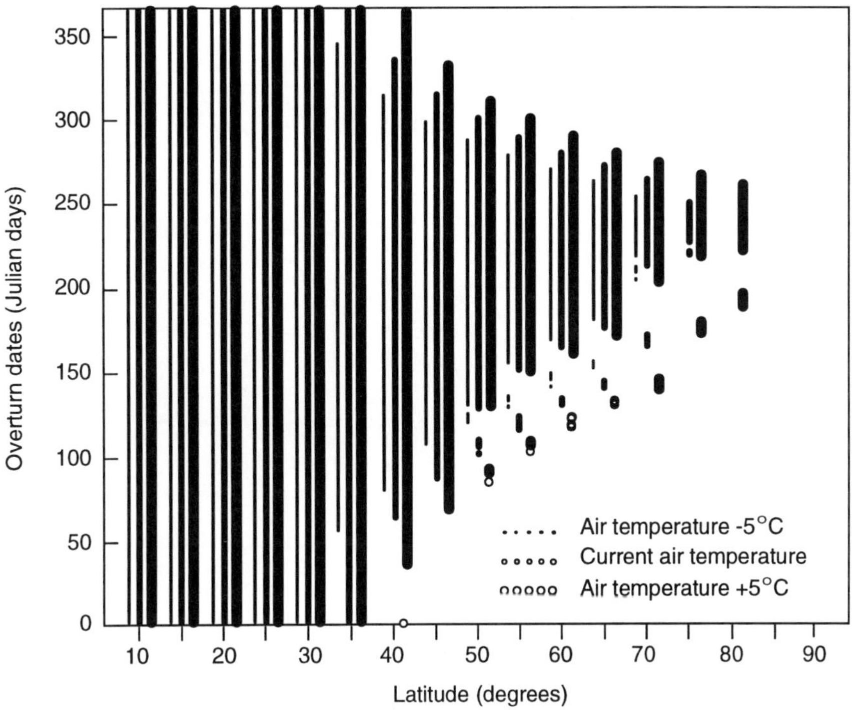

Figure 9.14. Overturn results for a 10-m-deep lake under a ±5°C air temperature change.

the circles. This superimposition allows instant comparison of the scenarios, although it does somewhat complicate the appearance of the plots. The remaining plots are easier to read: *Figures 9.15*, *9.20*, and *9.24* display duration of ice-cover persistence with respect to latitude, and the rest of the plots show duration of the overturn conditions in the lake. Careful examination of the sensitivity results indicates the following:

- The transition latitudes, where lake classification changes from monomictic to dimictic (at around 40° N/S), and from dimictic to monomictic (at around 70° N/S), are well simulated.
- For latitudes higher than 80° N/S, the existence of permanent ice cover (or temperatures close to 0°C) does not induce overturn.
- Stratification in shallow and intermediate-depth lakes was slightly sensitive to changes in air temperature in subtropical regions. On the other hand, stratification of deep lakes was significantly sensitive to air temperature changes in this same region. In temperate and polar regions,

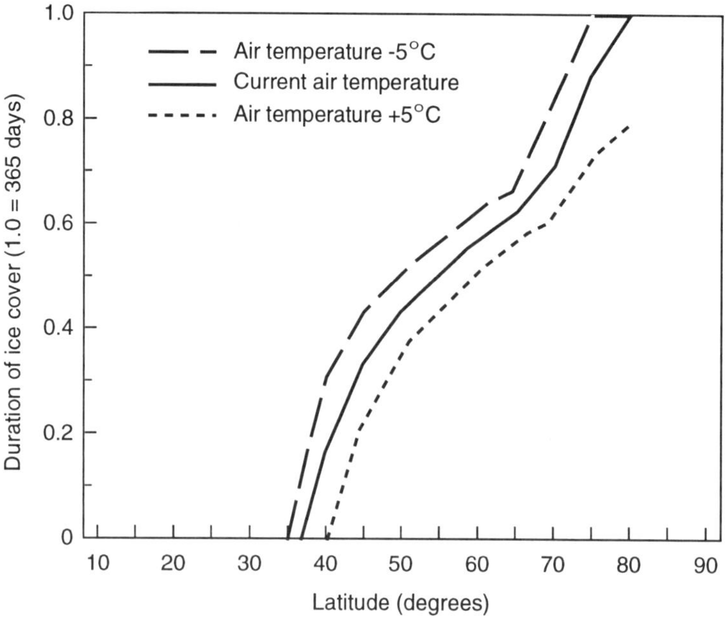

Figure 9.15. Duration of ice cover, 10-m-deep lake.

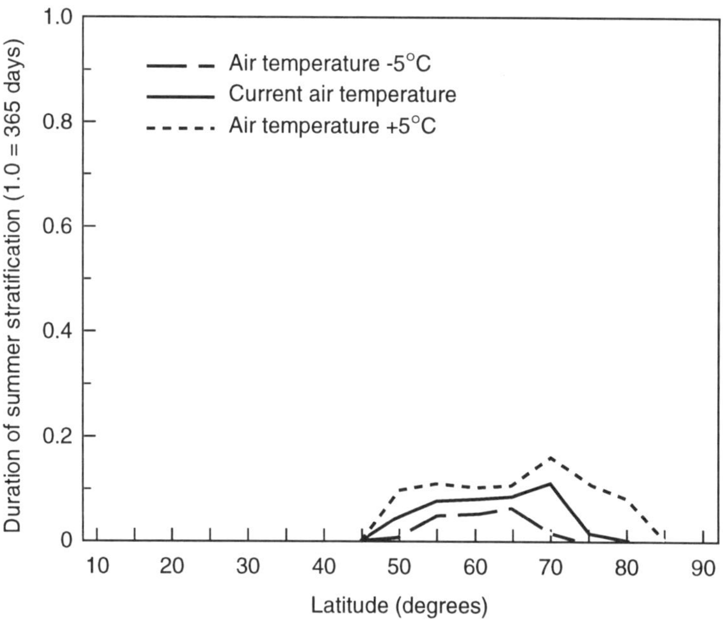

Figure 9.16. Duration of summer stratification, 10-m-deep lake.

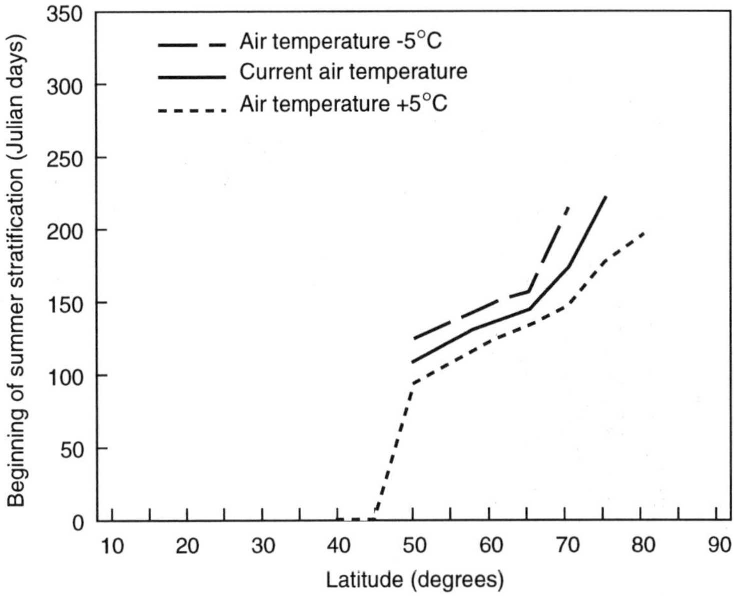

Figure 9.17. Beginning of summer stratification, 10-m-deep lake.

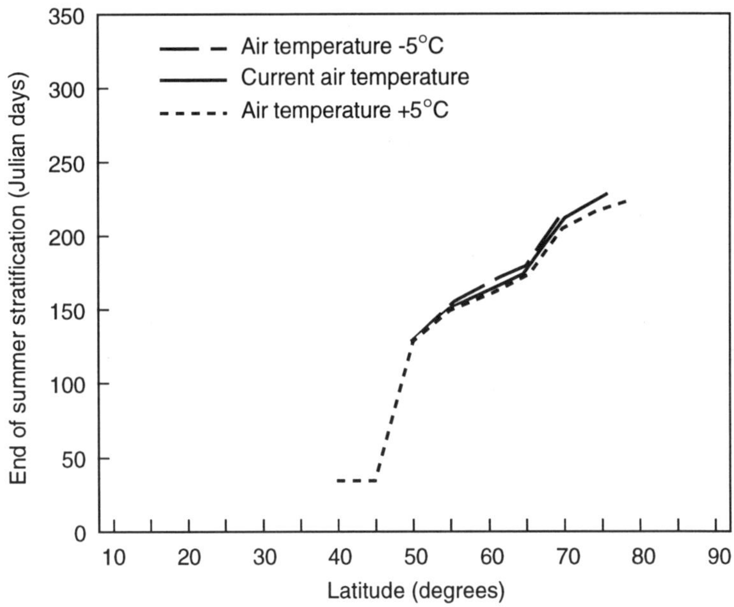

Figure 9.18. End of summer stratification, 10-m-deep lake.

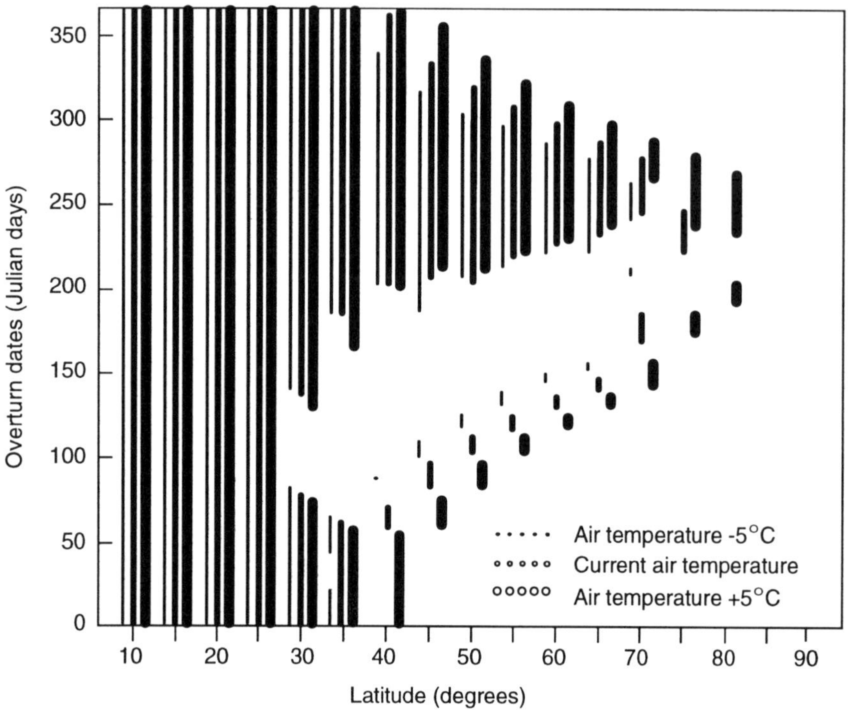

Figure 9.19. Overturn results for a 20-m-deep lake under a ±5°C air temperature change.

the sensitivity of lake stratification to changes in air temperature was considerable for all lake depths.

- As was expected, in all cases during warming seasons turnovers occur earlier in the year. During cooling seasons turnovers are delayed.
- Duration of stratification in subtropical intermediate-depth and shallow lakes is minimally sensitive to climate change; it is considerably more sensitive in deep lakes. In contrast, the sensitivity is equally significant for all lake depths in temperate to polar regions.
- Duration of freezing period is longer in the range of latitudes 70° to 80° N/S (subpolar and polar zones) than in the 50° to 70° N/S latitude range (temperate zone), regardless of lake depth. In the subtropical zones, ice-cover duration is longer in the deeper lakes than in the shallower ones.
- In the 10-m-deep lake, as shown in *Figure 9.14*, the lower latitudes (0° to 30° N/S) develop isothermal conditions for most of the year,

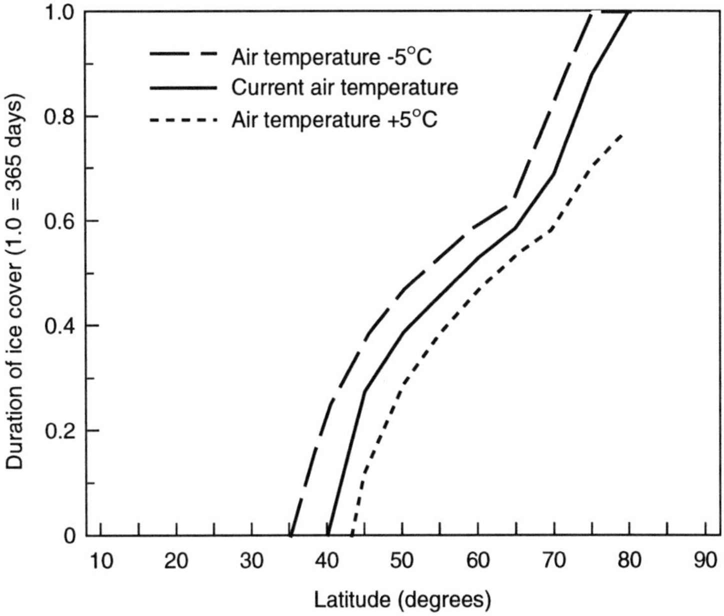

Figure 9.20. Duration of ice cover, 20-m-deep lake.

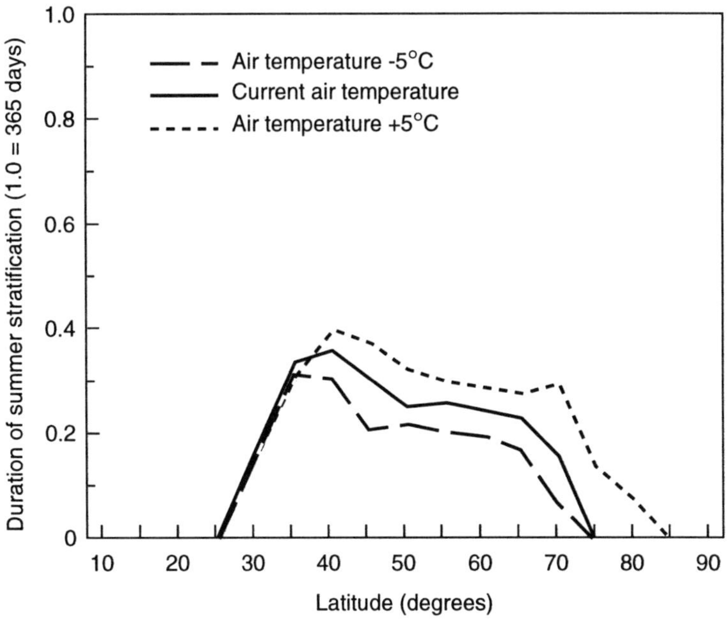

Figure 9.21. Duration of summer stratification, 20-m-deep lake.

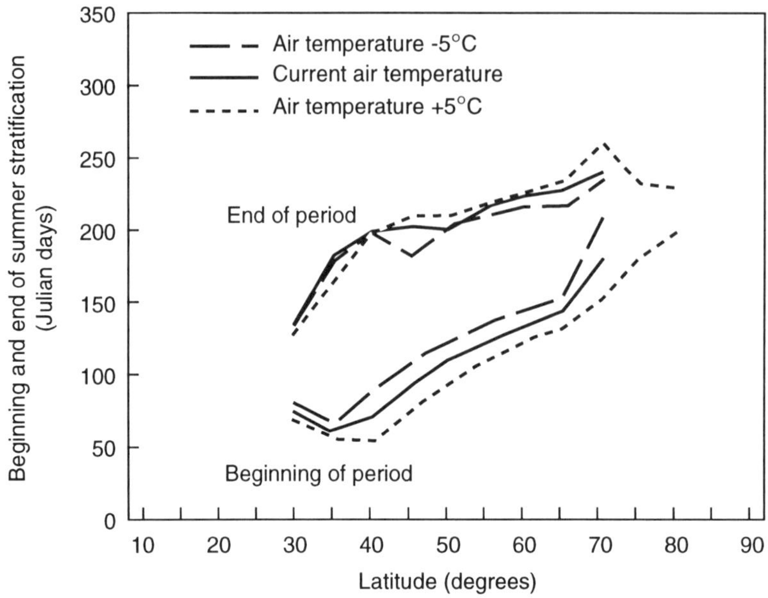

Figure 9.22. Beginning and end of summer stratification, 20-m-deep lake.

most likely because there is not enough heating or cooling to induce stratification except during short periods in the summer. The duration of turnover occurrence can change by 10 to 60 days in middle latitudes (40° to 70° N/S latitude range). At 40° N/S latitude, for example, the turnover period increases by as much as 60 days with a 5°C increase in air temperature. At this same latitude the turnover period decreases by 40 days in the case of 5°C cooling. As shown in *Figure 9.15*, the surfaces can be ice covered from 37° N/S latitude and poleward under normal conditions. The same figure shows that cooling would permit ice cover to begin at lower latitudes; warming would produce the opposite effect. A shift in ice formation toward the lower latitudes is evidenced through a reduced overturn period at 35° N/S latitude. At 75° N/S latitude, turnover is entirely suppressed by cooling. In contrast, at 80° N/S latitude dimictic conditions are induced under warming, whereas under normal conditions the system is amictic. *Figure 9.16* shows that the duration of summer stratification is altered in high latitude ranges, reducing it under cooling and increasing it under warming. *Figures 9.17* and *9.18* verify that at the middle latitudes there is the potential for stratification to shift toward the winter season under warming.

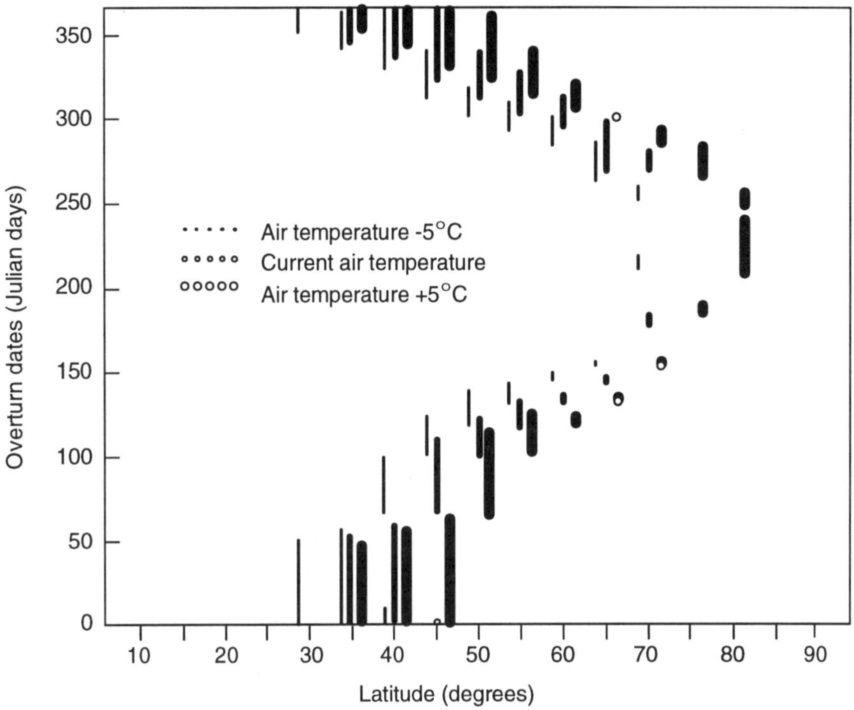

Figure 9.23. Overturn results for a 75-m-deep lake under a ±5°C air temperature change.

- In the 20-m-deep lake (see *Figure 9.19*), changes are more pronounced than are those of the 10-m-deep lake. The winter turnover is altered at 35° N/S latitude under cooling, changing conditions from monomictic to dimictic. This phenomenon can be partly explained by *Figure 9.20*, where it can be seen that under cooling ice-covered surfaces can appear at lower latitudes. Changes in turnover duration can fluctuate from 4 to 40 days in middle latitudes (30° to 70° N/S latitude range). On the other hand, warming generally increases the duration of overturn periods. At 75° N/S latitude, although cooling suppresses overturn resulting in an amictic lake, warming produces an extra overturn period converting the amictic lake to a dimictic lake. *Figure 9.21* shows that the duration of summer stratification generally increases with warming and decreases with cooling. The latter is verified by *Figure 9.22*, which shows the times within the year that summer stratification begins and

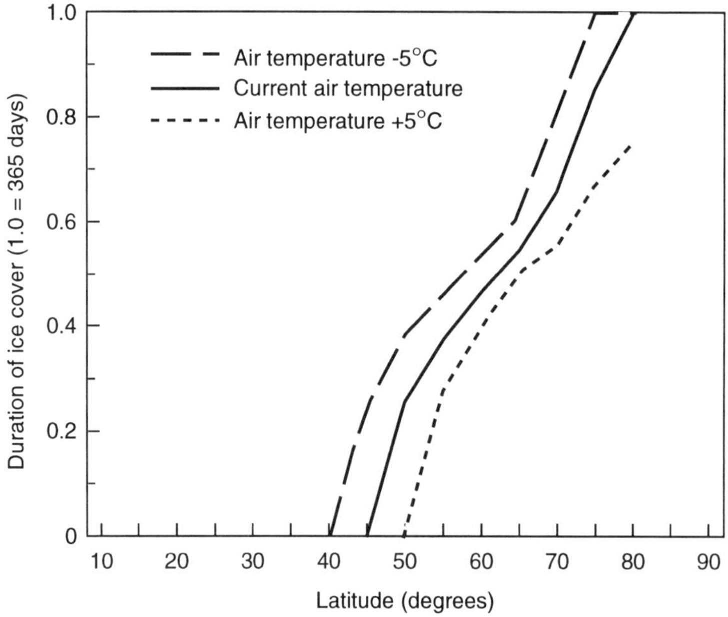

Figure 9.24. Duration of ice cover, 75-m-deep lake.

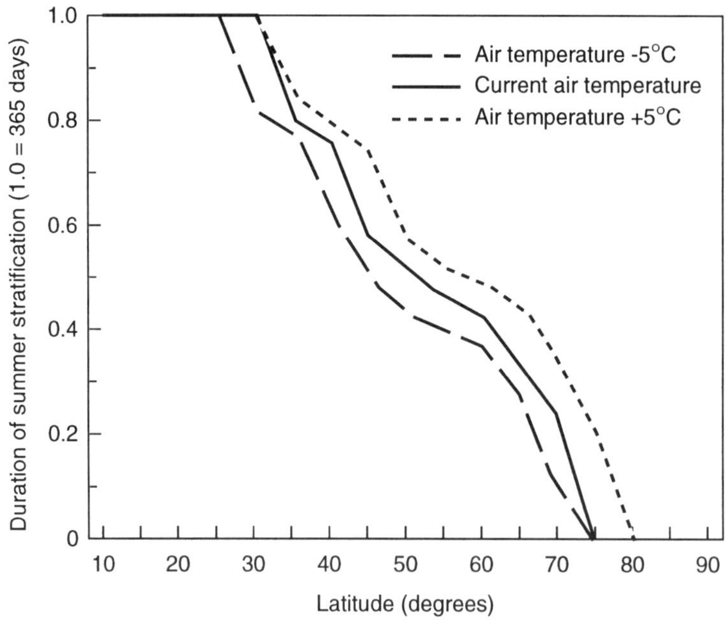

Figure 9.25. Duration of summer stratification, 75-m-deep lake.

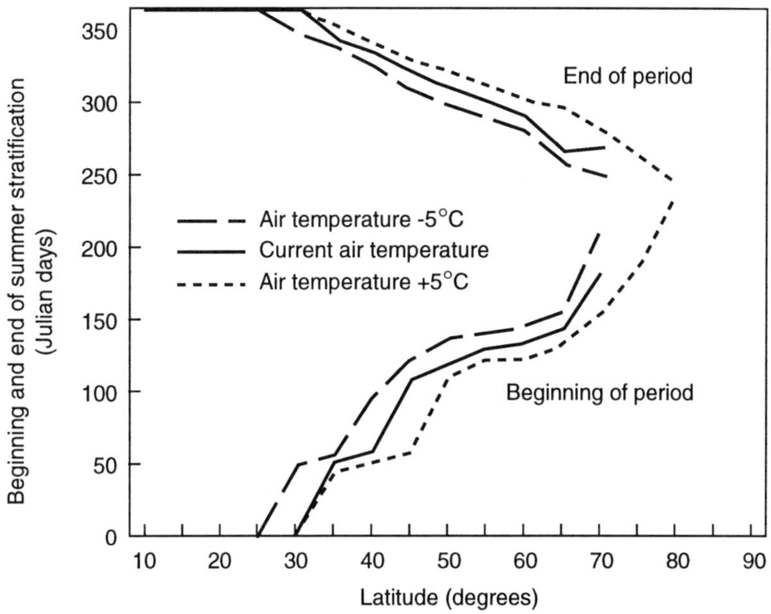

Figure 9.26. Beginning and end of summer stratification, 75-m-deep lake.

ends. As in the case of the 10-m-deep lake in the low latitudes, stratification periods occur earlier in the year, toward the spring season.

- Results for the 75-m-deep lake are shown in *Figure 9.23*. Low latitudes show stratified conditions all year round, due to the combination of the lake depth, which seems to sustain a metalimnion, and the initial water temperature condition (4°C), which is perhaps too cold for warmer surface waters to reach the bottom. Changes begin to occur at 30° N/S latitude, where under cooling amictic lakes have the potential to become monomictic. At 40° N/S latitude dimictic conditions could occur under cooling. In the middle latitudes, the duration of overturn can change by up to 10 days. What seems to be more important is the shift in the times at which overturn occurs. As in the previous cases, at the latitude range of 75° to 80° N/S, warming has the potential to produce dimictic conditions from what are otherwise amictic conditions. *Figure 9.24* consistently shows that ice-cover duration decreases under warming and increases under cooling. Formation of ice-covered surfaces shifts southward under cooling and northward under warming. *Figures 9.25* and *9.26* show a consistent pattern of longer stratification periods under warming; the reverse phenomenon occurs under cooling.

The results suggest that the sensitivity of lakes to changes in air temperature greatly increases in geographic transition regions where lakes change stratification patterns from warm monomictic to dimictic (about 40° N/S latitude), or from dimictic to cold monomictic (about 70° N/S latitude). This phenomenon is expected due to the nonlinear relationship between air and water, which is known to be highly sensitive near transition points. Changes in overturn periods of as much as 10 days, which were common for changes of ±5°C in air temperature, can be particularly important in eutrophic lakes. A 10-day increase in overturn duration could enhance algal growth through longer exposure of the algal cells to light and the availability of nutrients brought to the surface during the overturn. On the other hand, lengthening stratification periods would extend periods of anoxic bottom conditions that would enhance nutrient releases from decaying material, which in the next overturn occurrence would be recycled through the lake, thereby enhancing eutrophication.

As a result of the above findings, we have chosen to define the transition regions as sensitive belts of the globe with latitudes ranging from 30° to 45° N/S and 65° to 80° N/S. These results will be evaluated mainly as indicators of potential changes in the stratification pattern of lakes. To provide more accurate assessments, improvements still need to be made in the criterion defining the occurrence of overturn by improving its mathematical representation. The initial conditions should be revised, especially for deep lakes at low latitudes, specifying more realistic initial bottom temperatures.

9.7 Simulation of the Behavior of Real Lakes

9.7.1 Real lakes selected for simulation

It is difficult to identify real lakes with the required geomorphological information located precisely within the bounds we have established. Thus, we have relaxed the limits somewhat to include a representative set of real lakes. *Table 9.1* presents the selected lakes along with the basic geomorphological characteristics (depth, surface area, and volume). We refer to the selected water bodies as lakes, because they were treated as having negligible inflow.

9.7.2 Simulation results

Simulations were performed to assess changes in the stratification patterns of real lakes due to climate change. The availability of high-quality data

Table 9.1. Lakes located at sensitive latitudes.

Name	Latitude (degrees)	Approx. longitude (degrees)	Depth[a] (m)	Surface area (km^2)	Volume (km^3)
Shasta Lake (California, USA)	42 N	122 W	184 (SH)	120	5.5
Great Bear Lake (Northwest Terr., Can.)	66 N	120 W	149 (mean)	31,153	2,236.0
Lake Seneca (New York, USA)	43 N	77 W	198 (max)	175	15.5
Ezequiel Ramos Reservoir (Neuquén Prov., Arg.)	40 S	66 W	60 (SH)	816	20.2
Lake Geneva (Switzerland and France)	46 N	6 E	310 (max)	584	88.9
Lake Maggiore (Italy and Switzerland)	46 N	8 E	370 (max)	212	37.5
Changshou-Hu Reservoir (Sichuan Prov., PRC)	30 N	107 E	34 (SH)	60	0.7
Lake Ladoga (Russia)	60 N	31 E	230 (max)	18,135	908.0
Lake Biwa-Ko (Shiga Pref., Japan)	35 N	136 E	104 (max)	674	27.5

[a]For lakes the depth is the mean or maximum depth. For reservoirs the depth is the structural height (SH) or mean depth.
Source: Data Book of World Lakes Environment (UNEP, 1987–1990).

collected by the International Lake Environment Committee, ILEC, (UNEP, 1987–1990) made it possible to perform consistent simulations.

Lake morphometry was approximated with a power function dependent on surface area A_o and depth $z : A = A_o z^k$. The power k was calibrated to match the lake morphometry data (volume, maximal depth, and the surface area).

For the base-case scenarios, i.e., existing climate conditions, simulations were performed on a monthly basis using averaged historical data. The data sets used were those of the climatological station nearest to the lakes (see Section 9.4.2).

For analyzing the impact of global climate change, a $2\times CO_2$ scenario, the corresponding GFDL model results were used (see Section 9.4.3). As noted earlier, only temperature changes were considered; this is considered to be a conservative approach because all other climatic (e.g., humidity, wind) and hydrologic factors are kept equal to present conditions. The GFDL temperature differences were interpolated linearly to the station location and added to observed data (see Section 9.4.3). This is found to be a better practice than using GCM temperature results directly (see, for example, Lettenmaier and Gan, 1990). All other meteorological variables (relative humidity, wind speed, and cloudiness) were kept unchanged. One reason for this is the key role that air temperature plays in the heat balance of the lake. Furthermore, there is no clear translation procedure that could lead from the GCM grid data to changes in local meteorology necessary to calculate the water-body heat household.

Simulations were started on June 1, using as an arbitrary initial condition an isothermal profile of 4°C. As mentioned previously, simulations for each lake were performed for three consecutive years. Because the same meteorological conditions were periodically applied each year, quasi-steady-state heat balance was reached, and results of the last-year simulations were used for the analysis. The effect of the initial conditions on the periodically stabilized solution was not significant.

Impacts of climate change were measured through changes in the occurrence of overturn as defined in Section 9.6, changes in the number of days with ice cover, and changes in the duration of stratified periods. *Figures 9.27* through *9.35* represent the overview of the seasonal evolution of lake condition defined by overturn events and ice cover, occurrences of which are marked with the corresponding symbols. The air temperature displacement with respect to the base case is displayed on the vertical axis. This plot layout permits comparison of the lake behavior patterns in the cases of the base scenario and the scenario derived from the GCM results. At the same time one can get an idea about the extent of the changes in regional climate that are causing these behavior alterations. Results of the nine case studies follow.

Case Study 1: Shasta Lake

Shasta Lake on the Sacramento River is a major source of water supply to the state of California. Under present conditions the water delivered from Shasta Lake for most beneficial uses (e.g., municipal, agricultural, recreational, etc.) downstream is of high quality. The most significant quality problem at present is temperature, which if high enough can have a detrimental effect on cold-water habitat. Recent studies have demonstrated that stream temperatures in the Sacramento River are controlled by the availability of cold water in upstream reservoirs, e.g., Shasta Lake, and by the management of these reservoirs (Orlob *et al.*, 1990). Ongoing research on the impact of global climate change on the water quality of Shasta Lake

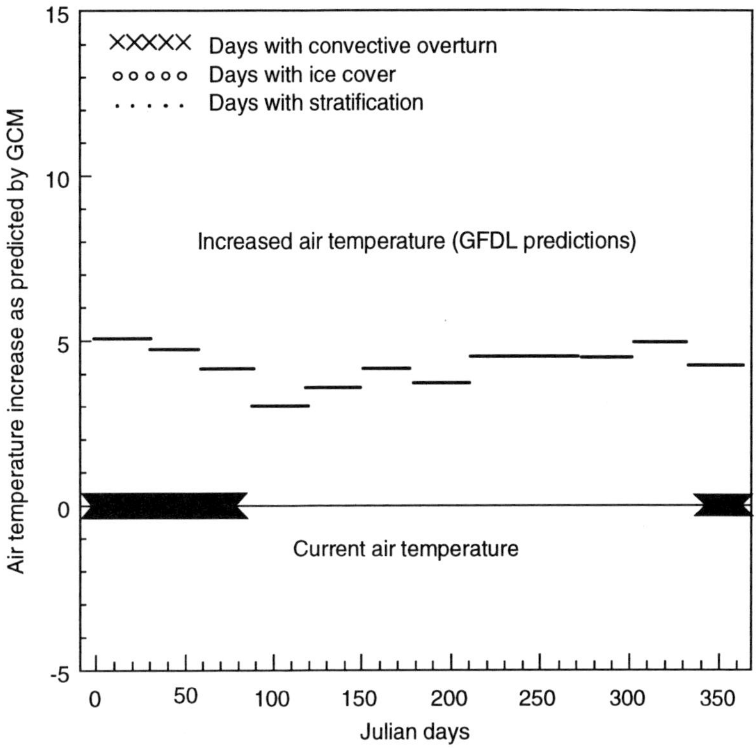

Figure 9.27. Shasta Reservoir, California, USA.

makes this case study an interesting one to analyze, despite its short residence time (half a year, on average).

Base-case simulation results indicate that the lake has an overturn period from December to March (see *Figure 9.27*). Observations show that the reservoir is isothermal from January to March. Thus, Shasta Lake can be classified as warm monomictic. The discrepancies between simulation results and observations can be partly explained by the fact that this study used a meteorologic station located in a colder watershed about 1° latitude north of Shasta Lake.

Simulation results show that under changed climate conditions the isothermal period disappears; thus, the reservoir remains stratified all year round. These results agree with recent investigations, which have shown that under global climate change the isothermal conditions are limited to the month of January and to part of February (Meyer and Orlob, 1992), and water temperature increases year round. The increased temperature

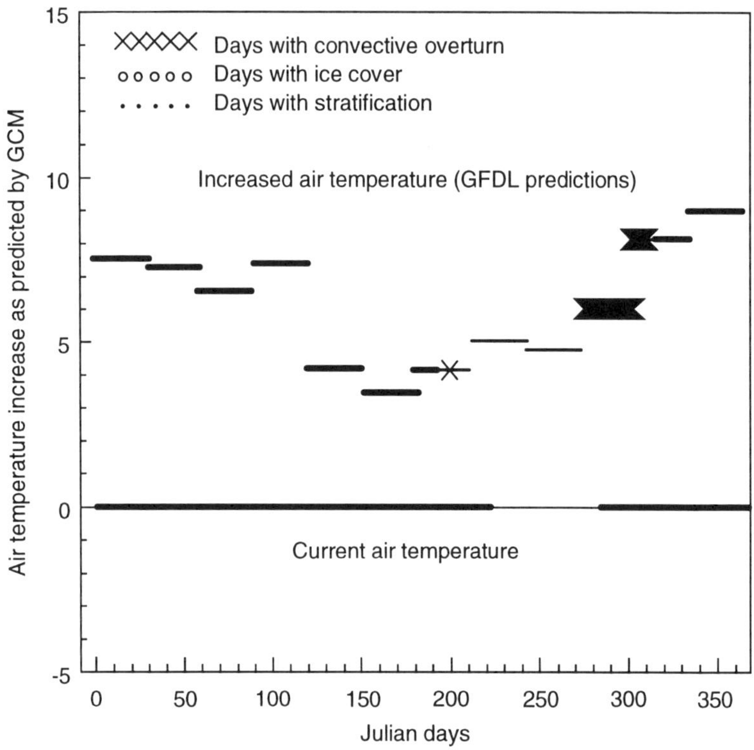

Figure 9.28. Great Bear Lake, Northwest Territories, Canada.

poses a threat to the survival of cold-water habitat already under stress. Moreover, the consequences of high temperatures, such as lower dissolved oxygen concentration at saturation and changes in the reaction kinetics of nutrients and biota, are likely to enhance eutrophication.

Case Study 2: Great Bear Lake

Great Bear Lake in the Northwest Territories is the largest lake in Canada. The residence time of this lake is 124 years. Therefore, neglecting throughflow in the lake is a justified assumption. It should be noted that the meteorological data used for simulation of present conditions was that of the Aklawik meteorological station; although the station at Port Radium is closer to the lake, the Aklawik data were selected because of their completeness.

Figure 9.28 shows the simulation results for this lake. Simulations of present conditions were in general accordance with observed data.

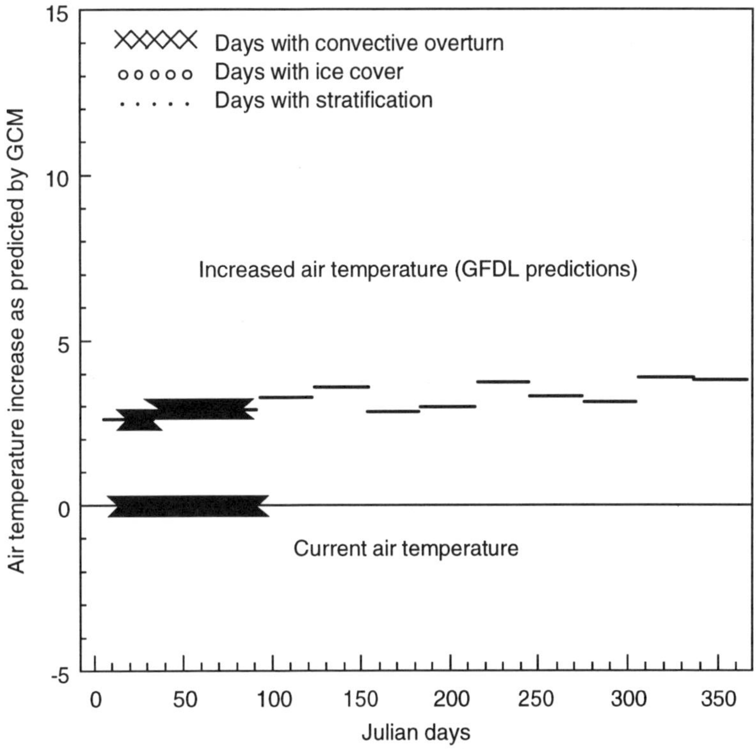

Figure 9.29. Lake Seneca, New York, USA.

According to observations, the lake is ice covered from November to July (UNEP, 1987–1990); our study resulted in an ice-cover period from September to July. This discrepancy can partly be explained by the fact that the Aklawik station, located to the north of Great Bear Lake, has lower average annual air temperatures (–9.2°C) than does the Port Radium station (–7.2°C). Under present climate conditions the simulation indicates that overturn periods do not occur. According to existing observations, the lake experiences complete circulation in only one year out of three. It is therefore classified as cold monomictic.

Under changed climate conditions, simulations indicate a reduction in the period of ice cover from 82% of the year to only 65% of the year. As a result of the predicted air temperature increase, which can reach as much as 9°C (see *Figure 9.28*), strong convective mixing occurs in the spring and autumn. The latter condition implies that the lake can potentially become dimictic, with hydrophysical features typical of temperate lakes.

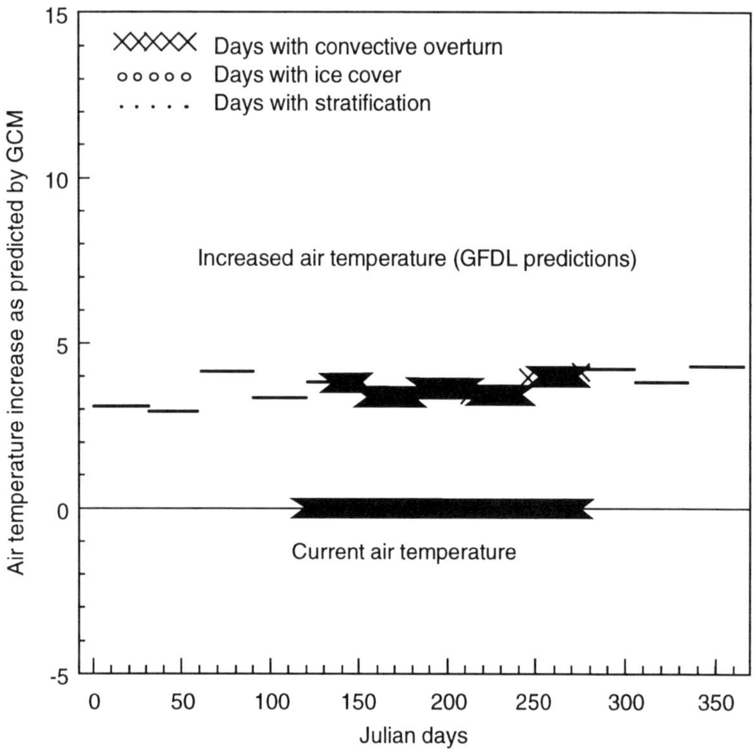

Figure 9.30. Ezequiel Ramos Reservoir, Neuquén, Argentina.

Case Study 3: Lake Seneca

Lake Seneca, New York, USA, is the largest of a group of long and narrow lakes in western New York State known as the Finger Lakes. Observations indicate that the lake has a residence time of 18 years, long enough to justify disregarding throughflows. In an average year the lake is not ice covered; present stratification patterns give the lake a monomictic classification. Results of simulations reflect these observations (see *Figure 9.29*). Global warming simulations did not reveal substantial changes; thus, the lake will remain monomictic, without ice cover.

Case Study 4: Ezequiel Ramos Reservoir

Ezequiel Ramos Reservoir in Neuquén, Argentina, is a multipurpose impoundment used for water supply, transportation, and recreation, and by fisheries. The lake is subject to convective overturns in winter, giving

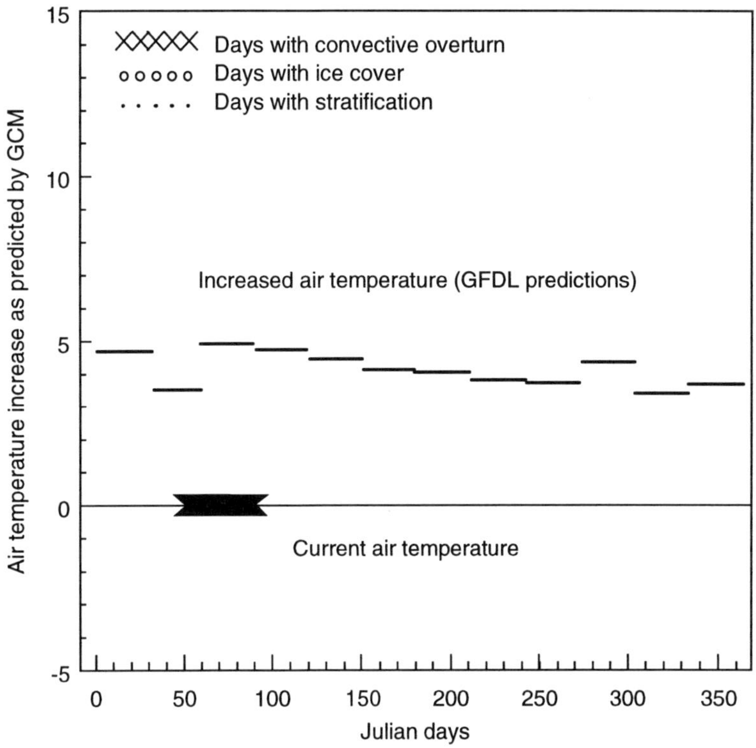

Figure 9.31. Lake Geneva, Switzerland and France.

it a warm monomictic classification This overturn period maintains high dissolved oxygen concentrations at all depths. Present-case simulation results reflect the monomictic character of the lake (*Figure 9.30*). The climate change scenario revealed no change in the stratification pattern; thus, present circulation conditions are expected to hold in the future.

Case Study 5: Lake Geneva

Lake Geneva is a major source of water supply for populations in Switzerland and France. Eutrophication has recently been noticed, resulting in low dissolved oxygen concentrations, particularly in deep lake layers, and in anoxic bottom conditions during most part of the year (Lachavanne, 1980). The residence time of the lake is 12 years.

In accordance with existing observations, present-case simulations indicate that the lake is classified as warm monomictic, with a turnover in the spring (see *Figure 9.31*). Our simulations show that the assumed climate

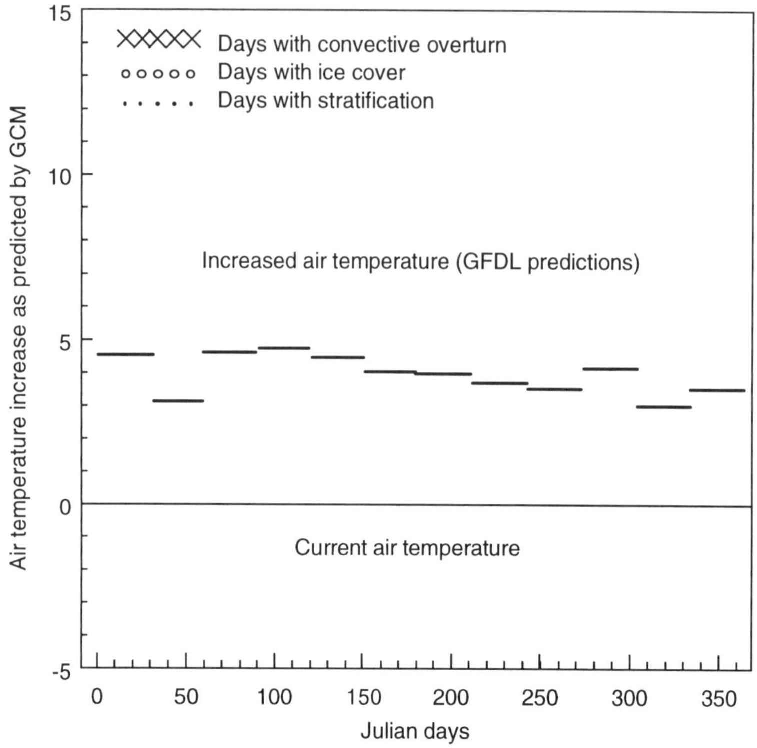

Figure 9.32. Lake Maggiore, Italy and Switzerland.

change will suppress the spring overturn, which will give the lake a new classification of amictic. This result implies that under changed climate, enhancement of anoxic bottom conditions and further eutrophication of the lake are likely.

Case Study 6: Lake Maggiore

Lake Maggiore, located in Italy and Switzerland, is classified as warm oligomictic. Our simulations show that overturns do not occur under present conditions (see *Figure 9.32*), in accordance with limnological observations. Simulations of global climate change do not indicate changes in lake stratification pattern.

Case Study 7: Changshou-Hu Reservoir

Changshou-Hu Reservoir in the Sichaun Province of the People's Republic of China has multiple uses: for transportation, hydroelectric power, and

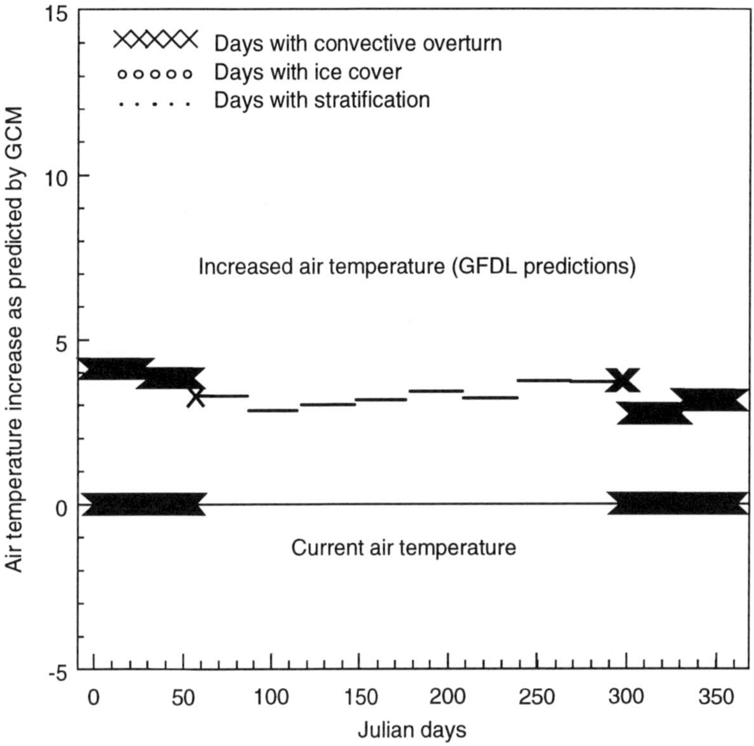

Figure 9.33. Changshou-Hu Reservoir, People's Republic of China.

water supply, and by fisheries. As was accurately reproduced in our simulations, this lake is classified as warm monomictic, subject to winter convective overturns from December to February (see *Figure 9.33*). Under future climate conditions, simulation results reveal no potential changes in the lake stratification pattern.

Case Study 8: Lake Ladoga

Lake Ladoga in northern Russia is Europe's largest freshwater body. It has a residence time of 12.3 years. The current lake stratification pattern gives the lake a dimictic classification. The lake is frozen from February to May (see *Figure 9.34*). Overall, these characteristics are well reflected by our computations, albeit simulations suggest ice cover begins in January, a month earlier than occurs in reality. This discrepancy can largely be explained by the fact that the Petrozavodsk meteorological station used for simulations is located in a colder area and has a lower annual average

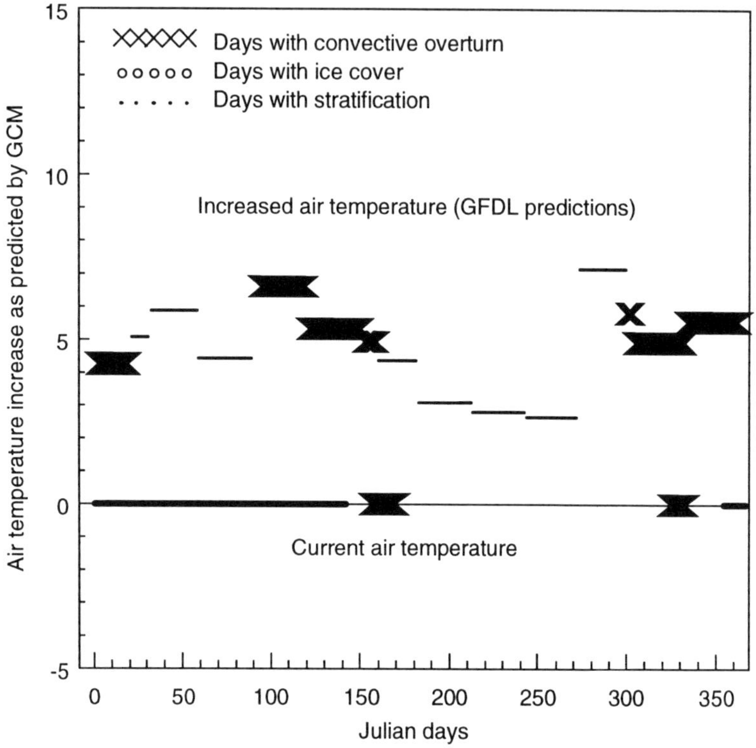

Figure 9.34. Lake Ladoga, Northern Russia.

temperature (2.4°C) than does the Sucho Island station (3.2°C), which is located closer to the lake but has insufficient data.

The simulation results under changed conditions, which reflect an increase in air temperature of up to 7°C, reveal several features. First, there may be an absence of ice cover. Second, the lake will remain dimictic. Intensive cooling during winter will lead to inverse stratification during March and April, and again in late autumn and winter. Third, current summer stratification conditions seem to be preserved. If the simulated future changes take place, environmental conditions will inevitably be affected. The absence of ice cover in winter could stimulate fog formation, producing a subsequent increase in moisture content in the air. The lake's biota, which currently has a number of unique species, might also feel the impact of the absence of ice cover. Convective mixing could have more significant consequences due to the longer duration of overturn periods.

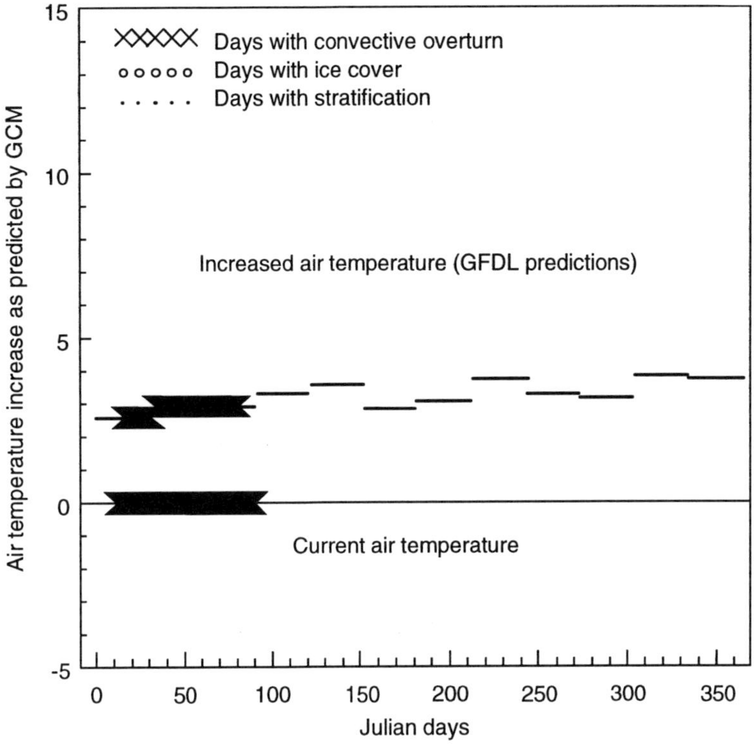

Figure 9.35. Lake Biwa-Ko, Shiga Prefecture, Japan.

Case Study 9: Lake Biwa-Ko

Lake Biwa-Ko, Shiga Prefecture, is Japan's largest lake; it supplies water to 13 million residents in the Osaka/Kyoto/Kobe megalopolis. Thus, water quality of the lake is a serious concern. The lake is formed by two distinct water bodies, called the Northern Lake, with a mean depth of 3.5 m and a residence time of 0.04 years, and the Southern Lake, with a mean depth of 104 m and a residence time of 5.5 years. The lake shows a distinct trend toward eutrophication.

Because its long residence time permits throughflow to be neglected the Southern Lake is the only one considered for simulations. Currently the lake is classified as warm monomictic, as our simulations reproduce (see *Figure 9.35*). Convective mixing occurs during January, February, and March, and stratified conditions prevail from May to December. Simulations under climate change revealed no significant change in the stratification pattern.

The results of these case studies indicate that there is a potential for changes in lake classification due to global warming. The impacts on each lake studied were different, depending on its geomorphologic characteristics and its location. For this reason, five of the nine lakes experienced no significant change in stratification patterns under changed climate conditions. For example, the stratification patterns of the polar lakes responded differently under global warming. For the Great Bear Lake, a very large, cold monomictic lake, the possible increases in air temperatures (up to 9°C) reduced ice cover enough to induce circulation, causing the lake to be reclassified as dimictic. In Lake Ladoga, a large water body that is smaller than Great Bear Lake and is dimictic, temperature increases up to 7°C made ice cover disappear completely; however, intensive cooling in winter led to longer turnover periods, thus maintaining the dimictic classification.

In these transition regions, stratification patterns are the lake characteristics most affected by global climate change. However, the lake's response to global climate change can manifest in characteristics other than mixing classification. Responses may also include impacts on characteristics such as ice formation or overall water temperature changes. In general, the warmer climatic conditions predicted by the GFDL model led to higher-latitude lakes being reclassified into categories representative of lower latitudes. For example, Great Bear Lake, currently a subpolar, cold monomictic lake, could potentially acquire the characteristics of a dimictic temperate lake. Lake Geneva, currently a warm monomictic lake, could become oligomictic. Lake Shasta, currently a warm monomictic lake, could become amictic.

It should be noted that even if no change in stratification pattern occurs, as in the cases of Lake Ladoga, Ezequiel Ramos Reservoir, Lake Maggiore, Changshou-Hu Reservoir, and Lake Biwa-Ko, water temperature changes could be significant enough to generate changes in water quality, biota, and the surrounding environment. The absence of ice cover in Lake Ladoga, for example, would have an immediate impact in local climate because of the consequent enhancement of fog formation.

These results should be interpreted only as a first step in identifying potential impacts of global climate change on the stratification patterns of lakes, and consequently on the water quality of lakes. The major restrictions of this study are the limitations of the overturn criterion used, as explained in Section 9.7.3; the use of only one GCM data set and the uncertainties involved in GCM outcome; and the use of the averaged meteorological data from the stations, which means the variability of local climate was not considered. The effect of climate change on overturn periods and stratification should be further analyzed, using models that include climatic variables other than air temperature and include realistic hydrologic boundary conditions. For water quality considerations, possible alterations should be investigated through monitoring, experiments, and a formalized modeling effort that would account for changes in the ecological processes that govern water quality.

9.8 Summary and Conclusions

The following conclusions can be drawn from the first part of this study, in which hypothetical lakes were simulated using longitudinally and monthly averaged meteorological data for a given latitude.

- The results suggest that the sensitivity of lake stratification to changes in air temperature is greater in the transition zones. The two transition zones are the subtropic zones (about 30° to 45° N/S latitude), where lakes can change from warm monomictic to dimictic, and the subpolar zones (about 65° to 80° N/S latitude), where lakes can change from dimictic to cold monomictic.
- Stratification in shallow and intermediate-depth lakes was slightly sensitive to changes in air temperature in subtropical regions. However, stratification in deep lakes was significantly sensitive to changes in air temperature in this same region. In temperate and polar regions the sensitivity of lake stratification to changes in air temperature was considerable for all lake depths.
- Turnover characteristics of subtopic and subpolar lakes are significantly affected by warmer atmospheric temperatures. In these lakes, turnover was found to begin earlier in the year than under existing conditions. Also, the duration of well-mixed conditions was longer. On the other hand, cooler atmospheric temperatures led to a delay in the onset of lake turnover and to a reduction in the duration of well-mixed conditions.
- The duration of ice cover is most sensitive to changes in air temperature in subpolar and polar lakes, regardless of lake depth. In subtropical regions, ice formation in lakes is less sensitive to atmospheric temperature changes, but it tends to be more sensitive to lake depth. A deep lake is more prone to ice-cover formation than is a shallow lake.
- Changes in the duration of overturn periods of as much as 10 days were common for 5°C changes in air temperature. These changes can be important, particularly in eutrophic lakes that depend on mixing events to offset negative impacts of stratification, e.g., anoxic conditions in the hypolimnion.

- Sensitivity analysis showed that the effect of changing climate is equivalent to a corresponding change in geographic location, approximately 1° latitude per $1^\circ C$ of air temperature.

In the second part of the study, potential impacts of climate change were assessed for real lakes. Present conditions were simulated with historical climate data; $2\times CO_2$ conditions were simulated with GFDL climate model results. Analysis of conditions in a climate change scenario led to the following conclusions:

- In the warmer lower latitudes, periods of stratification are likely to be enhanced. This case is typical of subtropical zones, where a warm monomictic lake can potentially become oligomictic, as in the case of Lake Geneva or Shasta Lake.
- In the colder higher latitudes, the frequency of overturn is likely to increase. There is a potential for subpolar lakes to change from being cold monomictic to being dimictic, as in the case of Great Bear Lake.
- Ice formation in subpolar regions can be reduced or totally suppressed, as in the case of Lake Ladoga.
- Of the nine lakes selected for study, four were subject to changes in stratification patterns and duration of ice cover. Even if no change in stratification pattern occurs, as in the case of Lake Ladoga, Ezequiel Ramos Reservoir, Lake Maggiore, Changshou-Hu Reservoir, or Lake Biwa-Ko, water temperature can change enough to induce alterations in water quality and lake biota, or to effect changes in the local climate, as in the case of Lake Ladoga.

Our study provides a first step in the systematic evaluation of the impacts of global climate change on hydrothermal dynamics of lakes. The study incorporates many simplifying assumptions about both the physical system and the computer modeling. Future research can be pursued in the following areas:

- Investigation of the effects of additional climatic variables, such as relative humidity or cloud cover, by using GCM predictions or by testing sensitivities.
- Use of other GCMs for comparison purposes and uncertainty assessment.
- Use of more realistic representations of lake morphology, lake inflow and outflow, and *in situ* meteorological variables.
- Modeling of a comprehensive range of water quality constituents with an appropriate model.

References

Harrison, S.P., 1990, An Introduction to General Circulation Modelling Experiments with Raised CO_2, WP-90-027, International Institute for Applied Systems Analysis, Laxenburg, Austria.

Hondzo, M., and Stefan, H.G., 1991, Three case studies of lake temperature and stratification response to warmer climate, *Water Resources Research*, **27**(8):1837–1846.

HEC (Hydrologic Engineering Center), 1978, WQRRS: Water Quality For River Reservoir Systems, Computer program description, US Army Corps of Engineers, HEC, Davis, CA, USA.

Kalkstein, S.L., 1991, *Global Comparisons of Selected GCM Control Runs and Observed Climate Data*, Environmental Protection Agency, Policy, Planning and Evaluation (PM-221), 21P-2002, Washington, DC, USA.

Lachavanne, J.B., 1980, Les manifestations de l'eutrophisation des eaux dans un grand lac profond, Le Leman (Suisse), *Schweizer Zeitschrift für Hydrologie*, **42**(2):127–154.

Leemans, R., and Cramer, W.P., 1991, *The IIASA Database for Mean Monthly Temperature, Precipitation, and Cloudiness on a Global Scale*, RR-91-18, International Institute for Applied Systems Analysis, Laxenburg, Austria.

Lettenmaier, D.P., and Gan, T.Y., 1990, Hydrologic sensitivities of the Sacramento–San Joaquin River Basin, California, to global warming, *Water Resources Research*, **26**(1):69–86.

Manabe, S., and Stouffer, R.J., 1980, Sensitivity of a climate model to an increase of CO_2 concentration in the atmosphere, *Journal of Geophysical Research*, **85**(C10):5529–5554.

Markofsky, M., and Harleman, D.R.F., 1971, *A Predictive Model for Thermal Stratification and Water Quality in Reservoirs*, Ralph M. Parsons Laboratory for Water Resources and Hydrodynamics, Report No. 134, Massachusetts Institute of Technology, Department of Civil Engineering, 71-4, Boston, MA, USA.

Meyer, G.K., and Orlob, G.T., 1992, Global climate change effects on water quality: Saving a threatened resource – In search of solutions, in M. Karamouz, ed., *Proceedings, Water Forum '92*, 2–6 August, American Society of Civil Engineers, Baltimore, MD, USA.

Müller, M.J., 1982, *Selected Climatic Data for a Global Set of Standard Stations for Vegetation Science*, Dr. W. Junk Publishers, The Hague, The Netherlands.

Octavio H., K.A., Jirka, G.H., and Harleman, D.R.F., 1977, Vertical Transport Mechanisms in Lakes and Reservoirs, Massachusetts Institute of Technology, Department of Civil Engineering, Report No. 227, August, Boston, MA, USA.

Orlob, G.T., 1983, *Mathematical Modeling of Water Quality*, John Wiley & Sons, New York, NY, USA.

Orlob, G.T., Meyer, G.K., DeGeorge, J., and Christensen, C.L., 1990, Impacts of global warming on water quality in river-reservoir systems, *Proceedings of an International Symposium on Global Warming*, Leningrad, USSR, June.

Schwoerbel, J., 1987, *Handbook of Limnology*, Ellis Horwood Limited, John Wiley & Sons, Halsted Press, New York, NY, USA.

Telegadas, K., and London, J., 1954, *A Physical Model of the Northern Hemisphere Troposphere for Winter and Summer*, New York University College of Engineering Research Division, Department of Meteorology, New York, NY, USA.

UK Meteorological Office, 1966, *Tables of Temperature, Relative Humidity and Precipitation for the World, Part V, Asia*, Her Majesty's Stationery Office, London, UK.

UK Meteorological Office, 1972, *Tables of Temperature, Relative Humidity, Precipitation and Sunshine for the World, Part III, Europe and the Azores*, Her Majesty's Stationery Office, London, UK.

UK Meteorological Office, 1973, *Tables of Temperature, Relative Humidity and Precipitation for the World, Part VI, Australasia and Pacific Ocean*, Her Majesty's Stationery Office, London, UK.

UK Meteorological Office, 1978, *Tables of Temperature, Relative Humidity and Precipitation for the World, Part II, Central and South America, the West Indies and Bermuda*, Her Majesty's Stationery Office, London, UK.

UK Meteorological Office, 1980, *Tables of Temperature, Relative Humidity, Precipitation and Sunshine for the World, Part I, North America and Greenland (including Hawaii and Bermuda)*, Her Majesty's Stationery Office, London, UK.

UK Meteorological Office, 1983, *Tables of Temperature, Relative Humidity, Precipitation and Sunshine for the World, Part IV, Africa, the Atlantic Ocean South of 35° N and the Indian Ocean*, Her Majesty's Stationery Office, London, UK.

UNEP (United Nations Environment Programme), 1987–1990, *Data Book of World Lake Environments (A Survey of the State of the World Lakes)*, Vols. 1 to 4, International Lake Environment Committee, Otsu, Japan.

Walter, H., and Lieth, H., 1960–1967, *Klimadiagramm-Weltatlas*, Gustav Fischer Verlag, Stuttgart, Germany.

Weather Bureau, 1959, *World Weather Records 1941–1950*, US Department of Commerce, Washington, DC, USA.

Wetherald, R.T., and Manabe, S., 1988, Cloud Feedback Processes in a General Circulation Model, *Journal of Atmospheric Sciences*, **45**(8 April):1397–1415.

Wetzel, R.G., 1983, *Limnology*, 2nd ed., Saunders College Publishing.

Zinoviev, A.T., Kuzmin, A.A., and Masliev, I.E., 1990, Mathematical modelling of thermal stratification of deep reservoirs, application of computers in hydrotechniques and protection of water resources, Verna, BAN [in Russian].

Appendix

A simplified wind-mixing model for lakes to calculate entrainment of hypolimnetic water into the upper mixed layer

Let us consider a lake with total depth H, depth of a fully mixed epilimnion h with temperature T_E, and a hypolimnion with temperature T_H, as shown in *Figure 9.8.*

The thermal energy exchange equation for the hypolimnion can be written as follows:

$$\frac{d}{dt}\left(\int_0^{H_t-h} c_H \rho_H T_H A(z) dz\right) = -q_E c_H \rho_H T_H - \phi_0 \ , \qquad \text{(A9.1)}$$

Rate of change of heat due to entrainment in hypolimnion	Flux of heat from the hypolimnion	Heat exchange with the lake bed

where T_H is hypolimnion temperature, $T_H = T_H(\mathrm{t})$; A is cross-sectional area, $A = A(z)$; ρ_H is density of the hypolimnion, $\rho_H = \rho_H(t)$; t is time; z is a vertical coordinate; H_t is total depth of the lake; h is epilimnion depth; q_E is flow rate of the hypolimnetic water to the epilimnion due to entrainment: $q_E = A(H_t - h)dh/dt$; c_H is the heat capacity of the hypolimnetic water; and ϕ_0 is heat exchange with the lake bed, which will be neglected.

The heat exchange of the epilimnion is as follows:

$$\frac{d}{dt}\left(\int_{H_t-h}^{H_t} c_E \rho_E T_E A(z) dz\right) = -\phi_s \quad + q_E c_p \rho_H T_H \quad - \phi_0 \ , \qquad \text{(A9.2)}$$

Rate of change of heat contents in the epilimnion	Surface heat flux	Influx of heat from the hypolimnion	Heat exchange with the bottom

where T_E is epilimnion temperature, $T_E = T_E(t)$; ρ_E is density of the epilimnion, $\rho_E = \rho_E(t)$; ϕ_s is surface heat flux; c_E is the heat capacity of the epilimnetic water; and other variables are as previously defined.

The following is the mechanical energy budget equation:

$$\frac{dP}{dt} = k' \ , \qquad \text{(A9.3)}$$

where P is modified potential energy, the sum of gravitational and buoyancy forces (Imberger, 1990), and is expressed as follows:

$$\frac{dP}{dt} = \left(\frac{\partial P}{\partial T}\right)_{\phi_s=0} + \left(\frac{\partial P}{\partial t}\right)_{dh/dt=0} \ , \qquad \text{(A9.4)}$$

Potential energy due to gravity	Potential energy due to buoyancy

where T is an arbitrary water temperature variable, and where k' is the rate of change of kinetic energy supply:

$$k' = \rho_E u_*^3 \quad , \tag{A9.5}$$

where u_* is the shear velocity across the water–air interface (wind-related stress).

In the resulting set of three ordinary differential equations [equations (A9.1) to (A9.3)], the unknown variables are T_E, T_H, and h. Integration of equation (A9.1) neglecting the heat exchange with the lake bed reduces the hypolimnion heat exchange balance to the following:

$$c_H \rho_H V_H (H_t - h) \tfrac{dT_H}{dt} = 0 \quad , \tag{A9.6}$$

where V_H is the volume of the hypolimnion.

Therefore, T_H = constant (with the accuracy of neglected heat exchange with lake bed).

Integration of equation (A9.2) gives the heat balance equation for the epilimnion:

$$c_E V_E \tfrac{d}{dt}(T_E \rho_E) + c_p \rho_E T_E \tfrac{dh}{dt} A(H_t - h) = \phi_s + c_p \rho_H T_H A(H_t - h) \tfrac{dh}{dt} \quad . \tag{A9.7}$$

The gravity portion of potential energy is as follows:

$$\left(\tfrac{\partial P}{\partial t}\right)_{\phi_s=0} = g \int_0^{H_t - h} \rho_H A(z) z dz + g \int_{H_t - h}^{H_t} \rho_E A(z) z dz \quad . \tag{A9.8}$$

This expression can be further simplified if we consider the following points:

- T_H is a constant that implies that ρ_H is also a constant.
- A lake morphometry function $Z_E(h)$ is defined as follows:

$$Z_E(h) = \tfrac{1}{V_E} \left[\int_{H-h}^{H} Azdz - (H - h)V_E \right] = \tfrac{1}{V_E} \int_0^h tA(t - H + h)dt \quad . \tag{A9.9}$$

  where V_E is the volume of the hypolimnion.
- Adiabatic conditions are preserved during mixing, implying that one can set $\partial\rho/\partial T = 0$.

Thus, the gravity portion of potential energy [equation (A9.8)] becomes:

$$\left(\tfrac{\partial P}{\partial t}\right)_{\phi_s=0} = Z_E(h)(\rho_H - \rho_E) g A(H_t - h) \tfrac{dh}{dt} \quad . \tag{A9.10}$$

To calculate buoyancy forces, consider an infinitesimal layer at the surface, δh, experiencing a change in density $\delta\rho$. Instantaneous mixing of this layer then changes the epilimnion density to ρ_E, and under these circumstances the infinitesimal potential energy change due to buoyancy can be calculated as follows:

$$\left(-\frac{\delta P}{g}\right)_{\frac{dh}{dt}=0} = \int_{H_t-h}^{H_t-\delta h} \rho_E Azdz + \int_{H-\delta d}^{H_t} (\rho_E + \delta\rho)Azdz - \int_{H_t-h}^{H_t} \rho_E Azdz \ . \tag{A9.11}$$

Equation (A9.11) can be simplified on the basis of following assumptions:

- For a small δh, the area A can be considered constant ($A = A_s$).
- With $\rho = M/V$, where M is mass and V is volume, the change in density is expressed as follows:

$$\rho_H - \rho_E = \delta\rho \frac{\delta h A_s}{V_E} \ . \tag{A9.12}$$

- Changes in temperature due to surface heating can be approximated as

$$\phi_s \delta t A_s = \delta V c_E \rho \delta T \ , \tag{A9.13}$$

and

$$\delta T = \delta\rho \frac{1}{\left(\frac{d\rho}{dt}\right)} \ . \tag{A9.14}$$

After rearrangements, the rate of change of potential energy due to buoyant forces can be expressed as follows:

$$(\delta P)_{\frac{dh}{dt}=0} = -g(h - z_E(h))A_s\left(\frac{\phi_s \delta t}{c_E \rho}\right)\left(\frac{dp}{dT}\right) \ . \tag{A9.15}$$

Finally, setting the rate of change of kinetic energy equal to the total rate of change of potential energy yields the necessary differential equation for $h(t)$ (cf. equation A9.5):

$$k' = -g(h - z_E(h))\frac{\phi_s A_s}{c_E \rho_E}\left(\frac{d\rho}{dT}\right)|_{T=T_E} + g(\rho_H - \rho_E)\frac{dh}{dt} Z_E(h)A(H_t - h) \ . \tag{A9.16}$$

Water density (ρ) as a function of water temperature was calculated using the Gephard formula (Henderson-Sellers, 1984):

$$\rho = \rho_o(1 - k_1|T - T_o|k_2) \ , \tag{A9.17}$$

where T_o is the reference temperature (=4.029°C); ρ_o is the reference density (=999.973 kg/m^3); k_1 is a constant (=9.207 10^{-6}); k_2 is a constant (=1.895).

A numerical solution of the resulting set of ordinary differential equations was obtained using the fourth-order Runge-Kutta method.

Chapter 10

Effects of Climate Change on Water Quality in the Central Valley of California

Gabriela K. Meyer, Gerald T. Orlob, and Christian Jokiel

10.1 Introduction

The Central Valley of California is one of the most productive agricultural areas of the world, due to a moderate Mediterranean climate, excellent soils, good drainage, and abundant supplies of water. However, the principal sources of water for irrigation of semiarid growing areas are in the northern and eastern mountain areas, whereas the greatest demand occurs in the southern and western coastal regions. Moreover, runoff from precipitation – which is greatest in northern and western regions – is dominated by snowmelt and reaches peak levels well after the beginning of the growing season. About half of the natural runoff from the Cascade Mountains and the Sierra Nevada occurs during the late spring and early summer months – April, May, and June. Until the advent of major flood-control reservoirs, this was a period of intense flooding.

To control floods and to conserve water for agriculture and other uses, the state and federal water agencies and local agricultural and municipal water districts have designed and constructed one of the most extensive systems of dams, reservoirs, canals, pipelines, power stations, and pumping plants ever conceived. Principal features of the State Water Project (SWP) and the federal Central Valley Project (CVP) are reservoirs with

Z. Kaczmarek et al. (eds.),
Water Resources Management in the Face of Climatic/Hydrologic Uncertainties 274–299.

total storage capacities of about 14.97×10^9 m^3, roughly 45% of the total annual runoff of the Central Valley (which is about 43% of the total for the state). The Shasta-Trinity complex in the Sacramento Valley plays a key role in regulation of the Central Valley's water supply. The Shasta-Trinity system comprises the following: two major reservoirs, Shasta, with a capacity of 5.615×10^9 m^3 and Trinity (Clair Engle Lake) with a capacity of 3.020×10^9 m^3; three small regulating afterbays, Keswick, Lewiston, and Whiskeytown; three power stations; and interconnecting channels, pipelines, and penstocks. Water stored and conveyed by this system is derived from the Sacramento River basin, with a drainage area of approximately 1.7×10^6 km^2 at Shasta Dam, and from diversions from the Trinity basin, with a drainage area of about 188,500 km^2.

The hydrologic basin has both topographic and climatic extremes. Elevations range between about 200 m at Keswick Dam and about 4,300 m at the peak of Mount Shasta, a dormant volcano in the Cascade Mountains. Much of the basin is in the rain shadow of the Cascade Mountains, so precipitation varies widely over the region of the Shasta-Trinity system, from as much as 1,900 mm per year in the coastal drainage area of the Trinity River, to as little as 300 mm per year in the northeastern portion of the Shasta drainage area. Average annual precipitation for the entire drainage area of the system is about 1,140 mm, of which approximately 40% (roughly 450 mm) is delivered as net runoff to the Sacramento River below Keswick Dam. Consumptive use above this location (evapotranspiration from forested and grassland areas and evaporation from water surfaces) has an average maximum potential of about 400 mm per year.

Annual runoff follows the seasonal cycle of snowmelt from higher mountain areas, with a peak inflow to Shasta Lake occurring in the month of June. Minimum rates of runoff usually occur in the winter, from November to January, when precipitation is in the form of snow at the higher elevations. Historical hydrologic records indicate runoffs ranging from a minimum of 41% of average annual runoff to a maximum of about 45%. The second-driest year on record, 1977, produced a runoff of about 580 mm over the drainage area.

10.2 Water Quality Problems

Under current conditions, water delivered through the Shasta-Trinity system is of high quality for most beneficial uses (agriculture, municipal water supply, recreation, etc.) downstream. It is generally low in total dissolved solids, pH is neutral to slightly acidic, and water released through power outlets usually contains sufficient dissolved oxygen (DO) to support aquatic life. Nutrients are generally below concentrations likely to induce eutrophication. A major quality problem at the present time – and the one most likely to become worsened by climate change – is water temperature.

Historically the Sacramento River and its tributaries have been the habitat for anadromous salmon species, especially the chinook salmon, *Oncorhynchus tshawytscha*. The life cycle of this migratory fish requires that it spawn in fresh water in the gravel beds of streams where waters are cold and well oxygenated. Laboratory and field investigations (SDFG, 1968) have demonstrated that water temperatures greater than about 13°C will interrupt the normal cycle of development from the fertilized eggs to the larval stage, resulting in increased rates of mortality as temperatures rise above this critical threshold.

Since the advent of the California water projects, notably the Shasta-Trinity system, there has been a well-documented decline in the number of upstream-migrating salmon reaching the spawning areas below project facilities. The decrease in habitat as a result of limited upstream access has been a contributing factor in the decline of salmon runs, although this has been partially offset by planting of hatchery-reared fingerlings. The decline has been especially notable in the autumn runs of chinook salmon, which are most likely to be exposed to the warmer waters occurring in the late summer–early autumn period. Also, the problem has become more serious in recent years due to a prolonged drought affecting both the quantity and quality of water available from upstream storage.

Field investigations of stream and reservoir temperature (SDFG, 1968) and preliminary modeling studies have demonstrated that stream temperatures in the Sacramento River spawning areas are controlled by the availability of cold water in upstream reservoirs, e.g., Shasta Lake, and by the manner of regulation and release of water from storage. Also, model studies (Meyer and Orlob, 1992) suggest that climate changes, such as might be expected to accompany a doubling of atmospheric carbon dioxide (CO_2), could exacerbate the problem of maintaining an acceptable habitat for salmon. It is expected that, not only would the area of habitat at risk be enlarged, but the duration of the extremely critical temperatures in spawning areas might be greatly lengthened.

The critical element in this situation is the upstream reservoir, i.e., Shasta Lake, and its response to climate change. This case study explores the consequences of climate changes imposed on the surface drainage and how they might influence the quality of water released downstream. The emphasis will be on temperature, but the implications for other quality characteristics and processes will be assessed.

10.3 Climate Change Scenarios

In this study, climate change scenarios are simulated using mathematical models that are sensitive to global climate changes manifested in new meteorologic and hydrologic conditions due to a doubling of atmospheric CO_2. In particular, such a model has to be capable of simulating the heat exchange mechanisms at water surfaces through which meteorological changes are reflected. The Hydrologic Engineering Center (HEC, 1978) Water Quality for River-Reservoir Systems (WQRRS) model is implemented with surface heat exchange routines derived by the engineering laboratory of the Tennessee Valley Authority (TVA, 1972); thus, it is used in this study to simulate temperature and other water quality/ecologic variables of concern in Shasta Lake and part of the Sacramento River (e.g., DO, nutrients, algae, etc.).

10.4 Shasta Lake Water Quality

10.4.1 Model structure

In the WQRRS model, the reservoir is represented conceptually as a series of one-dimensional horizontal slices. Each horizontal slice or layered volume element is represented by an area, a thickness, and a volume. Within each element the water is assumed to be fully mixed, making the model one-dimensional in the vertical sense. External inflows and withdrawals are represented as sources or sinks within each layer and are instantaneously dispersed and homogeneously mixed throughout each element, from the headwaters of the impoundment to the dam.

10.4.2 Meteorological boundary conditions

In this study, a base-case ($1\times CO_2$) scenario has been developed using historical meteorological data obtained from observations at Redding, California, near Shasta Lake, as well as historical hydrologic, water quality, and reservoir-operation data from the US Bureau of Reclamation, which operates the Shasta-Trinity system.

Global climate change ($2\times CO_2$) scenarios are generated according to predictions by general circulation models (GCMs) of hydrologic and meteorologic boundary conditions interpolated (in altitude and geographic

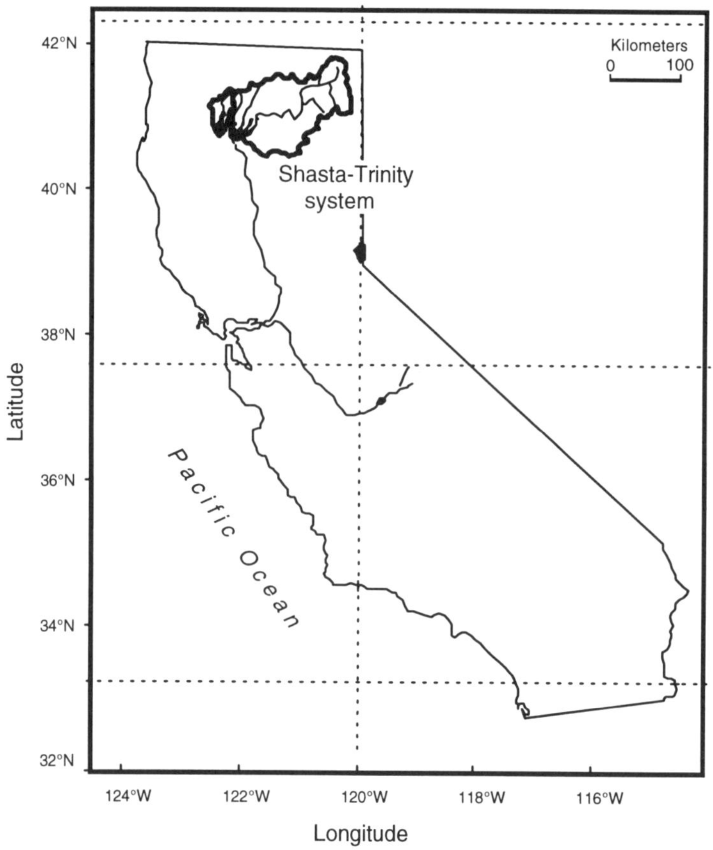

Figure 10.1. California in relation to GFDL general circulation model grid.

location) to the study site. In particular, this study uses the results derived from the GCM of the Geophysical Fluid Dynamics Laboratory (GFDL) for the grid space covering the southwestern USA (*Figure 10.1*) as a guide for estimation of changes in primary variables. Ideally, results from GCMs could be translated to regional scales to produce a set of boundary conditions that could then be used in hydrologic and water quality models to predict ecological impacts under changed climate conditions. However, as noted in Chapter 4, such a capability in the form of mesoscale atmospheric models is not yet equal to this task. Alternatively, for this case study it was necessary to develop a simple method for adjusting key GCM outputs to conform to the specific historical boundary conditions of the Shasta-Trinity site and to facilitate estimation of corresponding boundary conditions for a climate change scenario. The method developed for this case study is outlined briefly as follows:

1. Mean monthly predictions of air temperature, wind velocity, precipitation, runoff, and mixing ratio (water vapor mass to dry air mass) were obtained for each GFDL grid intersection, as shown in *Figure 10.1*, for $1\times CO_2$ and $2\times CO_2$ scenarios at dynamic equilibrium.
2. The air temperatures for each of these data sets were adjusted for differences in altitude between the grid location and the Redding station by applying an adiabatic lapse rate to the elevation change.
3. Meteorologic quantities at the Redding station were estimated by inverse distance interpolation of the data from the surrounding GFDL grid intersections.
4. Boundary conditions for the $1\times CO_2$ scenario were taken as historical monthly means for the key variables at the Redding station for the period from 1986 to 1991.
5. Boundary conditions for the $2\times CO_2$ scenario were estimated by adjusting historical values ($1\times CO_2$) in accord with changes predicted by the GCM as follows:
   - Air temperature and dew point:* add the incremental temperature change indicated by the GFDL output.
   - Wind velocity, precipitation, and runoff: multiply historical values by the ratio of GFDL output.
   - Cloud cover: multiply historical values by the ratio of GFDL precipitation estimates.

The calculated increments and factors used to produce the $2\times CO_2$ boundary conditions for the model are summarized in *Table 10.1*. These adjustments were made to the historical data for the Redding station, producing the key boundary conditions for simulations of water quality changes in Shasta Lake. These are summarized in *Table 10.2*.

10.4.3 Hydrologic boundary conditions

In producing climate change scenarios we are faced with several difficulties. The first is that GCMs produce average cell point values at the global scale and, therefore, are unable to accurately represent regional or local

*Dew point derived from mixing ratio estimates by GFDL, i.e., water vapor mass:dry air mass.

Table 10.1. Differences and ratios used to produce a $2\times CO_2$ scenario.

Month	Air temperature (°C) $\theta_{a2} - \theta_{a1}$	Dew point[a] (°C) $\theta_{d2} - \theta_{d1}$	Wind u_2/u_1	Precipitation[b] P_2/P_1
January	3.83	4.79	1.40	1.16
February	3.92	4.64	1.77	1.33
March	4.15	4.12	1.41	1.40
April	4.50	3.32	0.75	0.86
May	4.84	2.50	0.72	0.52
June	5.10	1.85	1.14	0.79
July	5.21	1.57	0.94	0.60
August	5.12	1.72	0.70	0.66
September	4.87	2.27	0.47	0.46
October	4.54	3.05	1.34	1.42
November	4.19	3.89	0.82	1.25
December	3.93	4.52	0.91	0.68

[a]Derived from mixing ratios given by GFDL.
[b]Used to adjust historical cloud cover to $2\times CO_2$ coordinates.

Table 10.2. Key boundary conditions: $1\times CO_2$ and $2\times CO_2$ scenarios.

	Base case, $1\times CO_2$				Climate change, $2\times CO_2$			
Month	Air temp. (°C)	Dew point (°C)	Wind (m/s)	Sky-cover fraction	Air temp. (°C)	Dew point (°C)	Wind (m/s)	Sky-cover fraction
January	8.06	0.13	2.86	0.66	11.89	4.92	4.00	0.77
February	10.67	1.21	3.22	0.55	14.59	5.85	5.70	0.73
March	12.44	3.51	3.62	0.60	16.59	7.63	5.10	0.84
April	15.67	5.36	3.40	0.62	20.17	8.68	2.55	0.53
May	20.39	6.62	3.75	0.49	25.23	9.12	2.70	0.25
June	24.94	8.93	3.67	0.37	30.04	10.78	4.19	0.29
July	28.61	9.78	3.31	0.16	33.82	11.35	3.11	0.10
August	27.33	9.21	3.04	0.20	32.45	10.93	2.13	0.13
September	24.56	6.84	2.99	0.26	29.43	9.11	1.41	0.12
October	19.00	6.32	2.91	0.36	23.54	9.37	3.90	0.51
November	12.11	4.63	2.91	0.53	16.30	8.52	2.39	0.66
December	8.50	1.26	3.08	0.54	12.43	5.78	2.80	0.37

climate and corresponding hydrologic processes. Second, runoff predictions by GCMs use the so-called bucket method, by which runoff is predicted as a balance between precipitation and evaporation, thus producing an "immediate runoff" occurring mostly during storm periods. A review of the literature suggests that for regional predictions GCMs are best suited to estimation of the percentage of the changes in hydrologic responses (Gleick, 1986; McCabe and Ayers, 1989). This suggestion is strengthened by the contention that, despite the inherent deficiencies of GCMs, "their internal consistency and strong physical basis may provide more accurate estimates of relative changes in the variables they predict" (McCabe and Ayers, 1989). Therefore, it is proposed here that to form $2\times CO_2$ scenarios historic records should be proportioned by the runoff ratios predicted by the GFDL model.

Previous studies have used water balance models to predict runoff changes using GCM predictions of precipitation and temperature changes as input data (Arnell *et al.*, 1990; Bultot *et al.*, 1988; Cohen, 1986; McGabe and Ayers, 1989; Mimikou *et al.*, 1991; Schaake, 1990; Lettenmaier and Gan, 1990; and Gleick, 1986). However, all of these studies dealt with specific conditions in a given water basin. Lettenmaier and Gan's (1990) case study of the McCloud River watershed, a tributary to Shasta Lake, provides some insight into the sensitivity of runoff to global climate changes in this region. The projected hydrograph in the present study is in accord with the simulation produced by Lettenmaier and Gan for the McCloud River basin using the Goddard Institute for Space Studies (GISS) predictions. Moreover, sensitivity analyses of global climate change impacts on hydrologic processes (Nemec and Schaake, 1982; Kaczmarek and Krasuski, 1991; see also Chapter 4 of this monograph) indicate that changes in runoff due to global climate changes range between –75% and 180%, depending on season and local climate conditions. Our results, developed from the GFDL model (see *Table 10.3*), indicate a maximum increase in runoff of 88% in March, no change in the spring and summer months, and up to a 50% decrease in the winter. The inflow hydrographs for the $1\times CO_2$ case and the $2\times CO_2$ case based on this GCM are illustrated in *Figure 10.2*.

10.4.4 Water quality boundary conditions

The critical water quality boundary condition for water quality simulation is the temperature of the inflows to the impoundment. In this case study, it is assumed that this temperature corresponds to the equilibrium temperature of the tributary streams. Following the method outlined in Chapter 4, the equilibrium temperature for the Shasta basin was calculated for each month of the year using historical mean values of the key parameters of the thermal energy exchange (see *Table 10.2*).

Because application of the equilibrium-temperature concept requires knowledge of heat conduction through the soil–water interface, as well as

Table 10.3. The GFDL-predicted runoff ratios used to produce $2\times CO_2$ scenarios.

Month	Runoff R2/R1
January	1.30
February	1.48
March	1.88
April	1.00
May	1.00
June	1.00
July	1.00
August	1.00
September	1.00
October	1.00
November	0.75
December	0.52

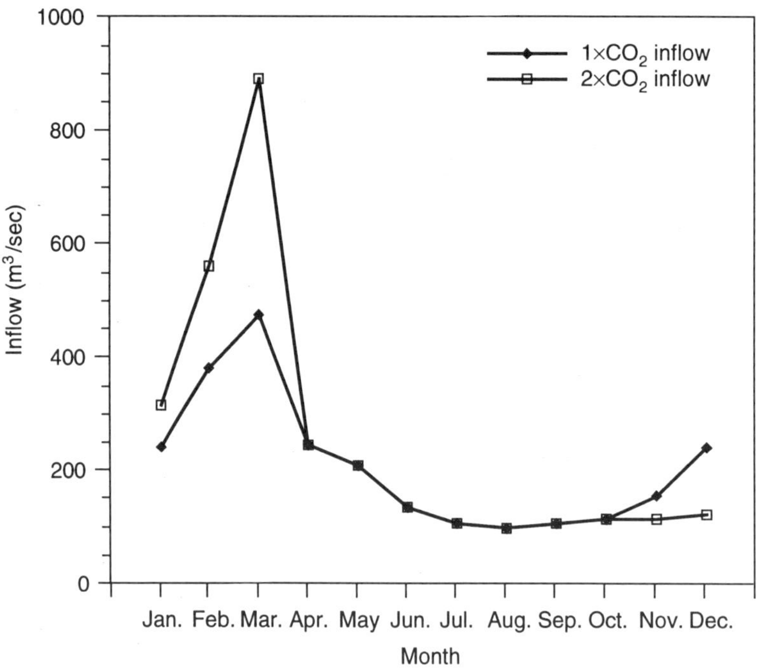

Figure 10.2. Inflow to Shasta Lake: $1\times CO_2$ and $2\times CO_2$ scenarios.

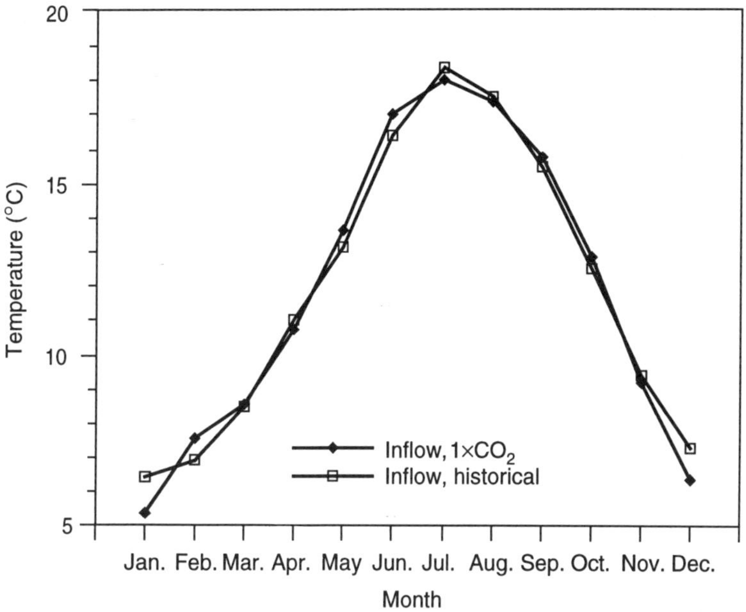

Figure 10.3. Historical inflow temperatures to Shasta Lake and calculated equilibrium inflow temperatures.

exchanges at the air–water interface, it was first necessary to estimate the coefficient of heat conduction, K_g. This was accomplished as follows:

1. The net heat exchange through the air–water interface was calculated in the conventional manner to determine the apparent equilibrium water temperature in the absence of heat conduction from the ground.
2. The flow-weighted average temperature of streams that are tributary to Shasta Lake was then determined and compared with the calculated equilibrium temperature from step one. The differences in temperature were attributed to heat conduction to (or from) the ground through the stream bottom.
3. Assuming an average ground temperature of 10°C at the time of peak equilibrium temperature (July), the heat conduction flux required to achieve a balance corresponding to the historical observation from step two was calculated.
4. Based on the differential temperature between water and ground and the heat flux required for balance, the heat conduction coefficient, K_g, was calculated.

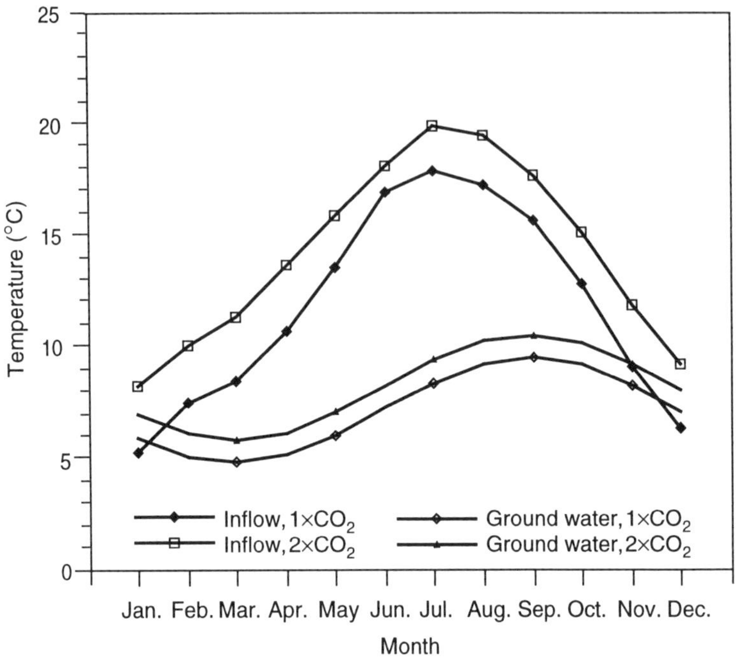

Figure 10.4. Inflow and ground water temperatures: $1\times CO_2$ and $2\times CO_2$ scenarios.

5. Considering a constant value of K_g and the flux required for equilibrium at other times of the year, the historical cycle of ground temperature was estimated. For this case study the cycle was roughly sinusoidal; for convenience, in subsequent analysis the calculated values were fitted to a sine wave with a period of one year and the equilibrium temperature was recalculated using this sine fit.

Comparison of the calculated equilibrium temperatures for the headwaters with those observed under historical meteorologic conditions, as shown in *Figure 10.3*, reveals a good approximation of the annual temperature cycle using the equilibrium method described above.

Application of this same procedure for the $2\times CO_2$ scenario required use of the ground temperature cycle resulting from the $1\times CO_2$ scenario, shown as the lower sinusoidal curve in *Figure 10.4*. This cycle was increased by 1°C to account for the feedback of thermal energy to the ground as a result of atmospheric warming. The assumed ground water temperature cycle and the inflow temperature cycle are shown in *Figure 10.4*.

Table 10.4. Assumed outflows from Shasta Lake.

Month	$1\times CO_2$ scenario (cm/s)	$2\times CO_2$ scenario (cm/s)	Ratio
January	141.58	283.17	2.0
February	283.17	424.75	1.5
March	283.17	424.75	1.5
April	283.17	283.17	1.0
May	293.48	293.48	1.0
June	339.20	339.20	1.0
July	341.93	341.93	1.0
August	307.76	307.76	1.0
September	94.14	94.14	1.0
October	37.35	37.35	1.0
November	52.57	52.57	1.0
December	47.03	47.03	1.0

Dissolved oxygen boundary conditions for all inflows were fixed at saturation levels for the corresponding temperatures. Although nominal concentrations of nutrients and biochemical oxygen-demanding substances were provided through tributary inflows, these were generally below levels likely to induce significant biological responses in the impoundment. The effect of climate change on the biogeochemical cycles of Shasta Lake is the subject of a continuing investigation.

10.4.5 Reservoir operation conditions

The outflow from Shasta Lake is governed by power generation and water supply in most years, with occasional adjustments to accommodate special demands, such as releases of colder water from river outlets below the power intakes at times of salmon spawning. In the $1\times CO_2$ scenario, releases of Shasta Lake were taken to be equal to the schedule adopted in 1987 (see *Figure 10.2*) under the assumption that it was an average water year. As *Table 10.4* shows, high outflows were adopted during the first quarter of the year for the $2\times CO_2$ scenario to prevent spill from the reservoir.

10.5 Simulation Results, Shasta Lake

10.5.1 Reservoir temperature

Having set meteorologic, hydrologic, and water quality boundary conditions of the $1\times CO_2$ and $2\times CO_2$ scenarios, the WQRRS model was employed to simulate water quality changes over typical annual cycles. As noted in Chapter 4, the primary driving variables that determine water quality characteristics of a stratified impoundment are the hydrologic flux and temperature. Thermal stratification and the temperature of outflows are largely dictated by the thermal energy influx to the impoundment, i.e., the product of the boundary inflow and its temperature, and heat exchanges at the air–water interface. *Figures 10.5(a)* and *10.5(b)* illustrate the annual cycles of thermal energy distribution in Shasta Lake for the two scenarios.

The resulting patterns of isotherms in the depth-time field represented in the figures are typical of observed behavior of stratified impoundments in northern temperate climate zones where ice cover does not occur in the winter (see Orlob, 1984). The water body is monomictic, that is, it experiences only one convective mixing cycle annually, usually initiated in late autumn and extending into the winter. As shown in the figures, both thermal histories for the cases simulated begin with near isothermal conditions throughout the full depth of the reservoir in January. Heat energy supplied by inflow and exchanged at the water surface results in gradual warming in the spring and early summer months, mostly near the surface. A thermocline forms in the spring, and the body of warm water in the upper part of the reservoir deepens, with the maximum temperature at the surface occurring in late summer. Cooling begins in the autumn, with convective mixing extending progressively deeper as winter approaches.

A comparison of the two scenarios reveals some distinctive differences that can be attributed to climate change. First, under $2\times CO_2$ conditions temperatures are generally elevated throughout the annual cycle in all parts of the reservoir. Near the bottom, temperatures are about 1.5°C higher during the entire year. At the surface, temperatures are higher at all times, but this is particularly noticeable during the period from September through December, when temperature increases attributable to climate change are as much as 4°C. Next, the volume of warm water, say above 13.3°C (the upper threshold for downstream temperature control to benefit salmon propagation), is generally increased by climate change, while the volume of cold water is decreased. Finally, the period during which water colder than 13.3°C can be released at the fixed outlet elevation is substantially reduced.

10.5.2 Outflow temperature

Outflow temperature histories for the two cases are shown in *Figure 10.6*. In neither case is it possible to meet the temperature target of 13.3°C in the summer or autumn months. The effect of climate change is indicated by

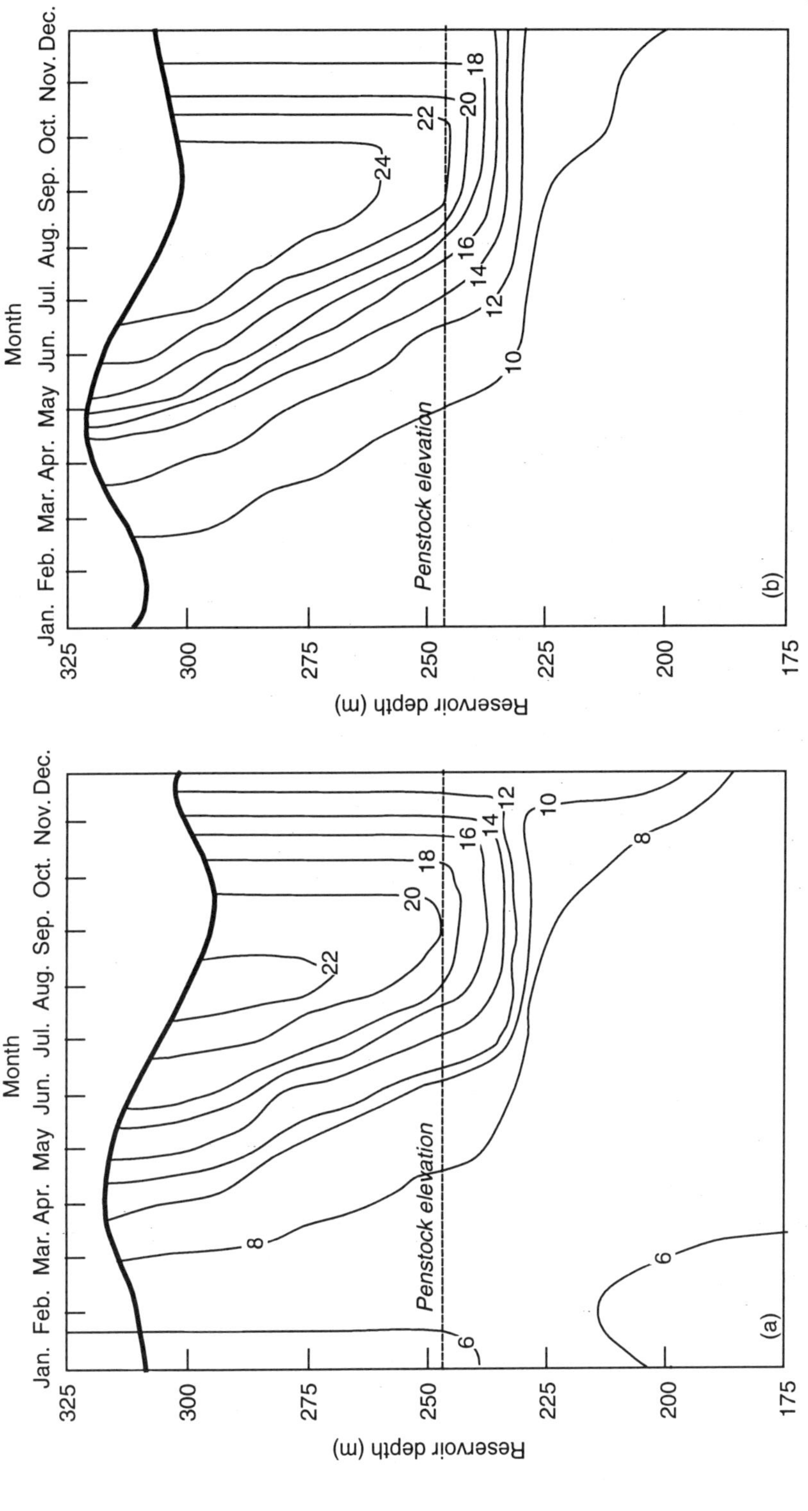

Figure 10.5. Annual cycle of thermal energy distribution in Shasta Lake: (a) $1 \times CO_2$ scenario; (b) $2 \times CO_2$ scenario.

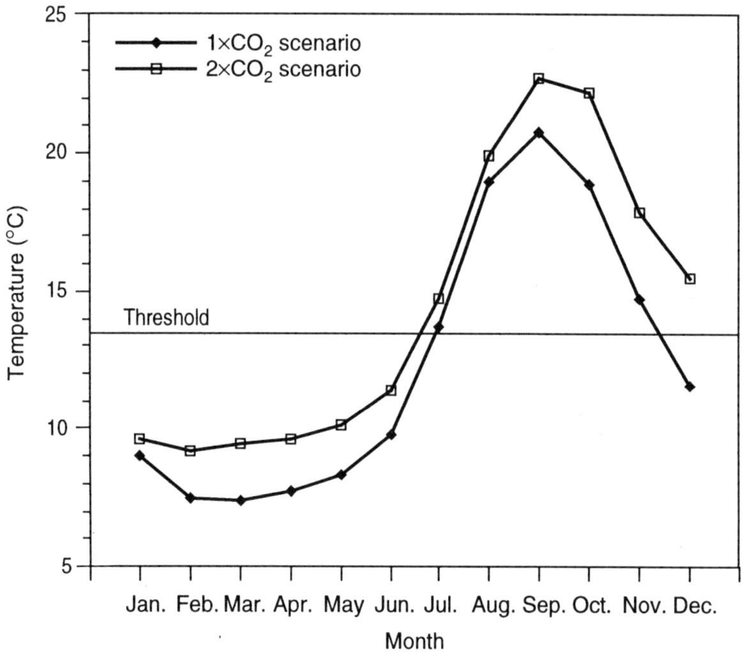

Figure 10.6. Outflow temperature through the penstock.

outlet temperatures that are 2° to 4°C higher during these months, with the larger increases occurring in the later part of the year. The period during which temperatures at the outlet elevation would be higher than the target was increased from about five months for the $1\times CO_2$ scenario to about seven months for the $2\times CO_2$ scenario.

The consequences of these climate-induced changes in the thermal regime of Shasta Lake will be more apparent downstream, where water temperatures may be further increased by heat exchange at the surface of the river. Water temperatures in the river are, of course, strongly influenced by releases from the reservoir, which define the headwater conditions for the stream system.

10.5.3 Dissolved oxygen in Shasta Lake

From a limnological standpoint, Shasta Lake may be classified as oligotrophic; that is, nutrient input is comparatively low, primary production is also low, and oxygen resources are usually sufficient to maintain a healthy habitat for higher forms of aquatic life. The distribution of DO within the

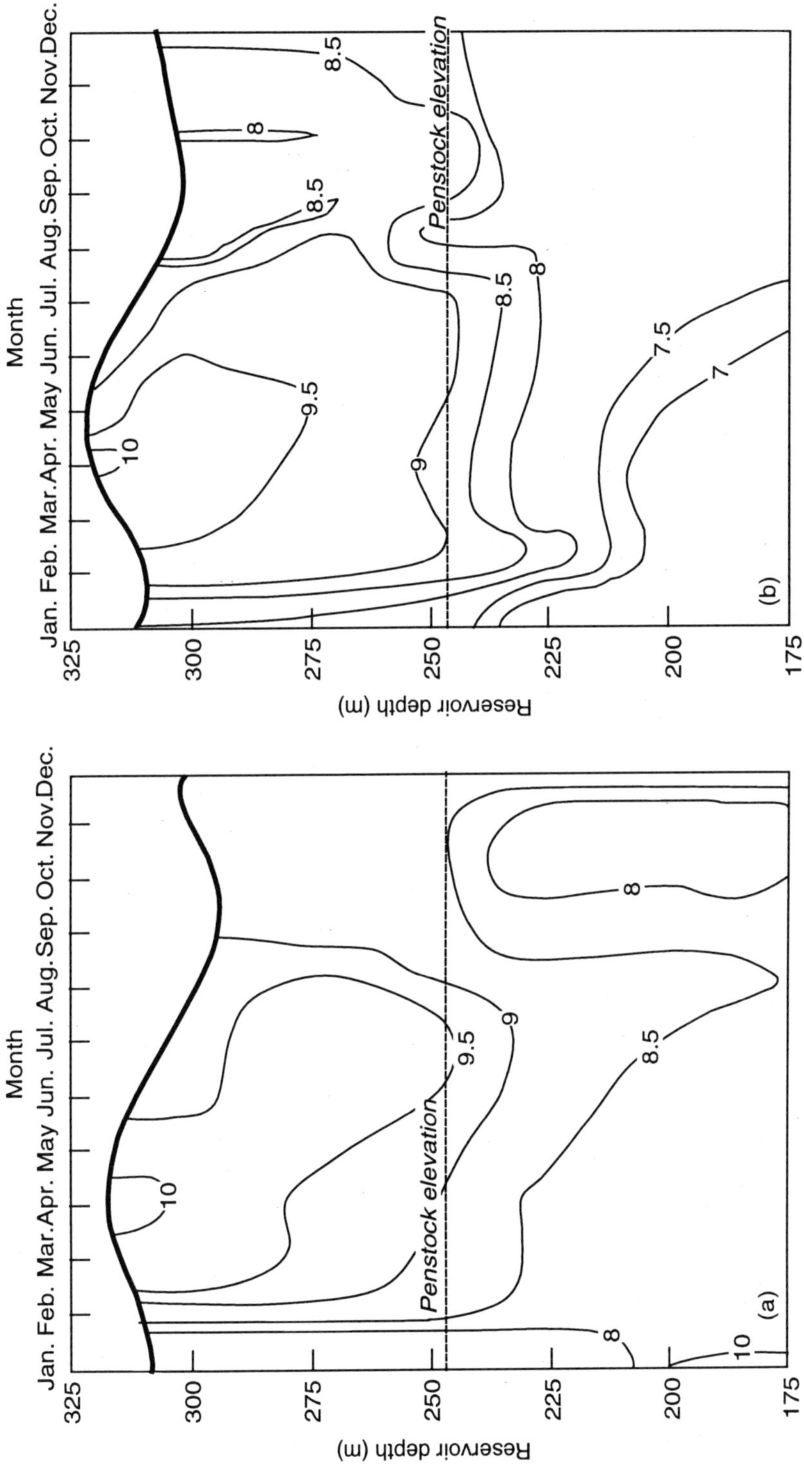

Figure 10.7. Annual cycle of oxygen distribution in Shasta Lake: (a) $1\times CO_2$ scenario; (b) $2\times CO_2$ scenario.

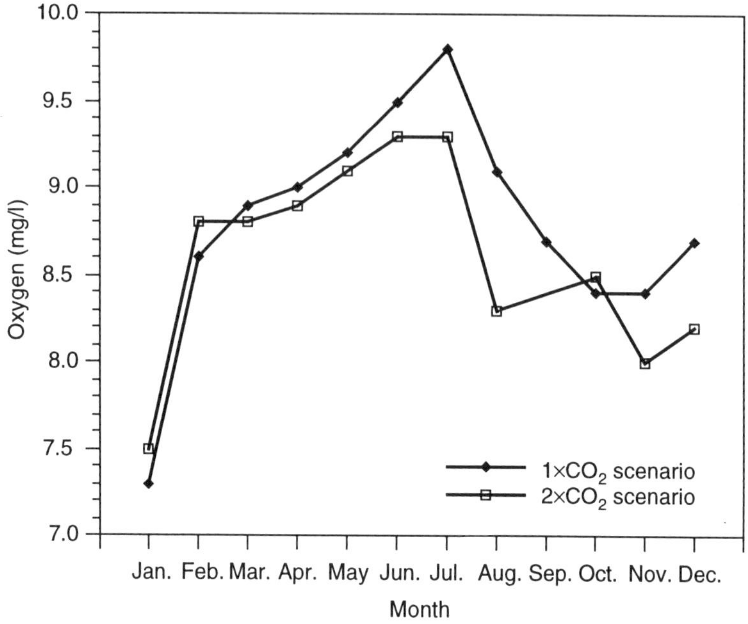

Figure 10.8. Oxygen content in outflow: $1\times CO_2$ and $2\times CO_2$ scenarios.

reservoir in its present condition is controlled primarily by concentrations in the inflow, which are usually close to saturation. However, because oxygen saturation is temperature dependent, decreasing as temperatures increase, it is reasonable to expect significant differences in the distribution of DO in a stratified reservoir due to increases in water temperature alone. As noted previously, biogeochemical activity, although modeled in these simulations, was maintained at low levels, hence the effects on DO (e.g., by photosynthesis) were generally minimal.

The WQRRS simulations previously provided a means of assessing the potential effects of climate change on DO in Shasta Lake. The DO distributions for the $1\times CO_2$ and $2\times CO_2$ scenarios are represented in *Figures 10.7(a)* and *10.7(b)*, in the same depth-versus-time format used to characterize temperature changes. Isopleths of DO concentration indicate spatial and temporal variations over the annual cycle. It should be noted that the DO concentration for September for the $2\times CO_2$ case is interpolated between August and October values.

10.5.4 Outflow dissolved oxygen

Figure 10.8 shows the oxygen content that characterizes the outflow. It is noted that oxygen concentrations in the outflow from Shasta Lake are generally below saturation at the corresponding temperature. For example, the highest concentration occurs in July in the $1\times CO_2$ scenario, about 9.8 mg/l, corresponding to 93% saturation at a temperature of about 13.5°C at this same time. In the other months of the year the percentages of saturation were notably lower.

10.6 Sacramento River Water Quality

10.6.1 Model structure

To simulate water quality in the Sacramento River, the HEC (1978) WQRRS model has been applied. This model is divided into two separate modules: the stream hydraulics package (SHP) and the stream quality module (WQRRSQ). SHP, a stand-alone program, creates a data file that provides hydrodynamic information to the stream quality module.

The WQRRS water quality/ecological model is one-dimensional and capable of simulating both steady-state and dynamic conditions. A linear network of volume elements represents the stream system depicted in *Figure 10.9*. The 75-km-long reach of the Sacramento River from Redding to Red Bluff is divided in the model into 20 volume elements, with provisions for including five tributaries: Clear Creek, Cow Creek, Bear Creek, Cottonwood Creek, and Battle Creek. Releases from Shasta Lake provide headwater inflows and define upstream boundary conditions.

It should be noted that the small afterbay below Shasta Lake, Keswick Reservoir, is assumed to have little impact on water quality, temperature, or hydraulic conditions in the Sacramento River. For this reason it was not explicitly included in the present investigation of the incremental effects of climate change, although it will become an integral part of the river-reservoir system in future studies.

10.6.2 Meteorological boundary conditions

The meteorological boundary conditions for the Sacramento River were assumed to be identical to those for Shasta Lake, as described previously and summarized in *Tables 10.1* and *10.2*.

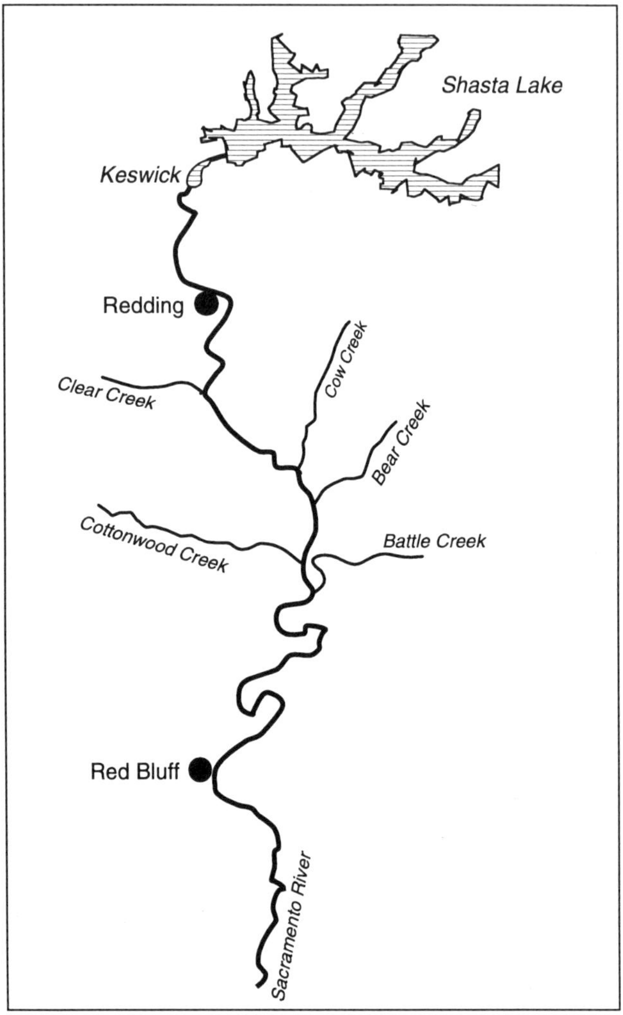

Figure 10.9. The Sacramento River system.

10.6.3 Hydrologic boundary conditions

As in the water quality case, inflows to the system vary depending on the type of boundary as follows:

- Headwaters: This flow is identical to releases from Shasta Lake, as simulated before, for either a $1 \times CO_2$ or $2 \times CO_2$ scenario (see *Table 10.4*).

- Clear Creek: For the $1\times CO_2$ scenario, the inflows are taken to be equal to gauged flows from 1987, which is assumed to be a typical, unextraordinary hydrologic year. Because Clear Creek is a regulated tributary, the runoff ratio used for a $2\times CO_2$ scenario is assumed to be equal to that of Shasta Lake (see *Table 10.4*).
- Other nonregulated tributaries (Cow Creek, Bear Creek, Cottonwood Creek, Battle Creek): For the $1\times CO_2$ scenario these inflows were taken as equal to gauged flows from 1987. For the $2\times CO_2$ change case, monthly values of inflow were proportioned by the runoff ratio derived by the GFDL general circulation model for the $1\times CO_2$ and $2\times CO_2$ scenarios, using the values from *Table 10.3* as the inflows to Shasta Lake.

10.6.4 Water quality boundary conditions

Sacramento River water quality inputs are subdivided into headwater inflows, inflows from Clear Creek (regulated by releases at Whiskeytown Reservoir upstream), and other nonregulated tributaries. Temperatures for each type of inflow were fixed as follows:

- Headwaters: Inflow temperatures through this boundary corresponded to the simulated outflow from Shasta Lake for either a $1\times CO_2$ or $2\times CO_2$ scenario.
- Clear Creek: Inflow temperatures through this boundary were considered equal to the outflow temperature simulated for Shasta Lake under the hypothetical assumption that the quality of water released from Whiskeytown reservoir is similar to that of Shasta Lake.
- Other nonregulated tributaries (Cow Creek, Bear Creek, Cottonwood Creek, Battle Creek): Inflow temperatures through these boundaries were considered equal to the equilibrium temperature determined for the inflows to Shasta Lake for both scenarios.

As noted previously for the case study of Shasta Lake, DO boundary conditions taken as input to the Sacramento River model were generally below saturation levels. Nutrients and biochemical oxygen demanding substances were at minimal levels. Dissolved oxygen concentrations in tributary streams were set at or slightly below saturation.

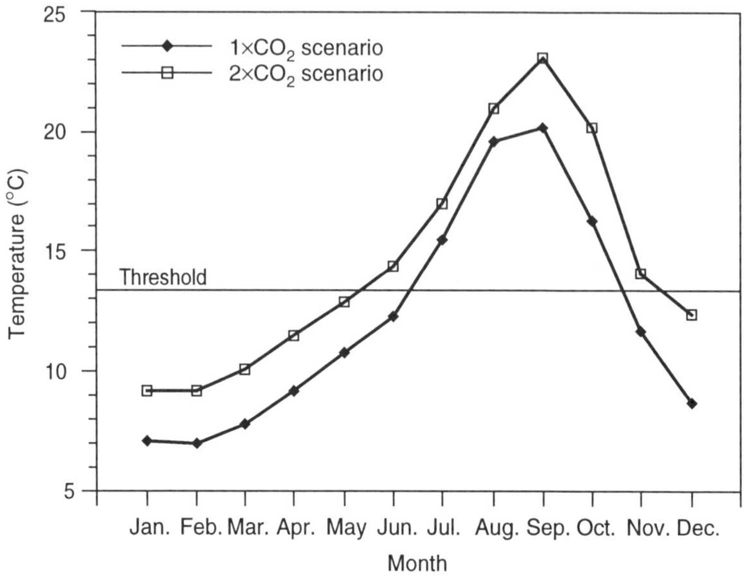

Figure 10.10. Yearly distribution of temperatures at Red Bluff: $1\times CO_2$ and $2\times CO_2$ scenarios.

10.7 Simulation Results, Sacramento River

10.7.1 River temperature

As noted previously, WQRRSQ, the stream module in the WQRRS package, was applied to simulate water quality changes over typical annual cycles. The temperature distribution within the Sacramento River is driven mainly by boundary inflows and their temperatures, and by heat exchange with the river bottom and the atmosphere.

Simulated yearly distributions of the temperatures at Red Bluff for the $1\times CO_2$ scenario and the $2\times CO_2$ scenario, as illustrated in *Figure 10.10*, are typical for a river in the middle latitudes of the Northern Hemisphere. Temperatures are at their lowest levels in January or February. Warming progresses in the spring and summer, reaching peak temperature levels in August and September. Cooling begins in the early autumn, and temperatures drop rapidly as the winter approaches.

Although the annual pattern of temperature variation over the year is similar for both cases, a comparison of the two scenarios reveals some distinct differences that can be attributed to climate change. Increased atmospheric CO_2 results in higher temperatures throughout the year. Differences

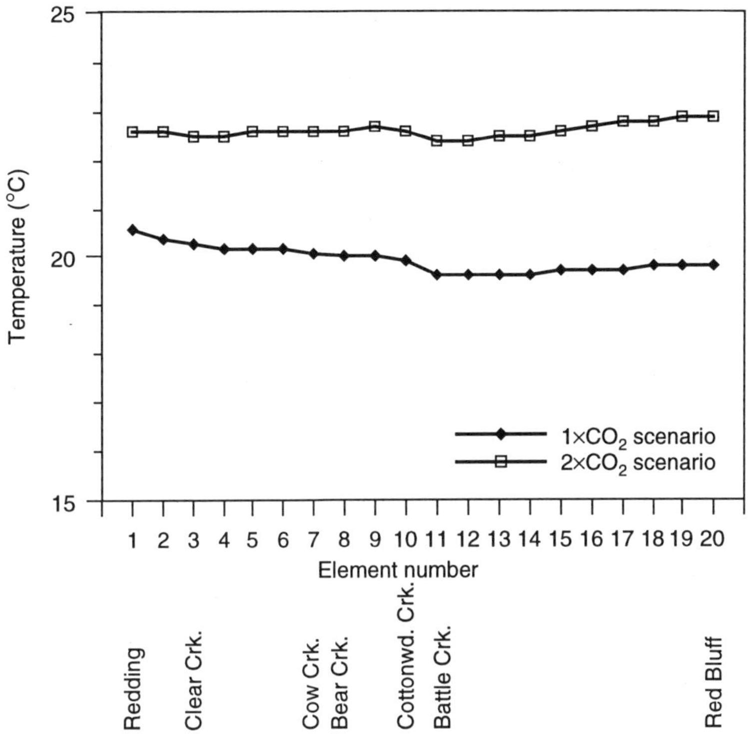

Figure 10.11. Temperature distribution along the Sacramento River for the month of September, Redding to Red Bluff: $1\times CO_2$ and $2\times CO_2$ scenarios.

between the two scenarios vary from about 1.5° to nearly 4.0°C. Peak values occur in September in both cases: about 20.3°C for the $1\times CO_2$ scenario and 23.2°C for the $2\times CO_2$ scenario. Due to the temperature increase attributed to climate change, the period when temperatures are above 13.3°C (the upper limit for successful salmon propagation) is prolonged significantly. The critical period is extended from about five months to about seven months of the year by climate change. Both cases would prove detrimental to the survival rates of salmon runs in the Sacramento River.

The temperature distributions along the Sacramento River from Redding to Red Bluff for the critical month of September are illustrated in *Figure 10.11.* In both cases temperatures are sustained throughout the reach at levels well above the critical threshold. Colder tributary inflows tend to reduce the temperatures from Redding to Battle Creek, but not enough to restore the river to conditions favorable to salmon propagation.

Downstream from Battle Creek, temperatures tend to increase again along the river, apparently due to the tendency of the river to equilibrate by heat exchange at the air–water interface. The differential due to climate change effects appears to increase gradually in the downstream direction, from about 2°C at Redding to more than 3°C at Red Bluff. Under the conditions simulated, the entire reach from Redding to Red Bluff would be unsatisfactory for salmon propagation in September. These results are in general agreement with earlier studies of climate change effects on the Sacramento River (Orlob *et al.*, 1990).

10.7.2 Dissolved oxygen in the Sacramento River

The distribution of DO within the simulated reach of the Sacramento River is strongly influenced by the DO concentrations of inflows; i.e., the releases from Shasta Lake and the five tributary inflows. As shown earlier, releases from Shasta Lake are simulated to be below saturation throughout most of the year. Saturation conditions generally prevail for the five tributary inflows. Consequently, it is reasonable to anticipate that near-saturation conditions would occur at most times and locations along the Sacramento River. Dissolved oxygen distributions for both scenarios are represented in *Figure 10.12*. Because oxygen saturation is temperature dependent, the decrease in DO concentration indicated for the $2\times CO_2$ scenario seems reasonable. It was noted that changes in DO are inversely proportional to temperature changes, i.e., an increase in temperature due to climate change is accompanied by a corresponding decrease in DO concentration, due for the most part to the effect of temperature on saturation.

10.7.3 Ecological impacts

The primary ecological impact of climate change in this case study is on the survival of anadromous salmon species in the Sacramento River below the Shasta-Trinity system of reservoirs. According to these results under the assumptions specified, as well as to the results of earlier studies, river temperatures below Shasta Lake and as far downstream as Red Bluff are unsuitable for salmon propagation. Temperatures are likely to be above 13.3°C for much of the year, especially during the late summer and early autumn, the period of an important salmon run. Although this study deals only with a representative average condition, the target temperature is exceeded to such an extent in this case, even for present conditions, that it

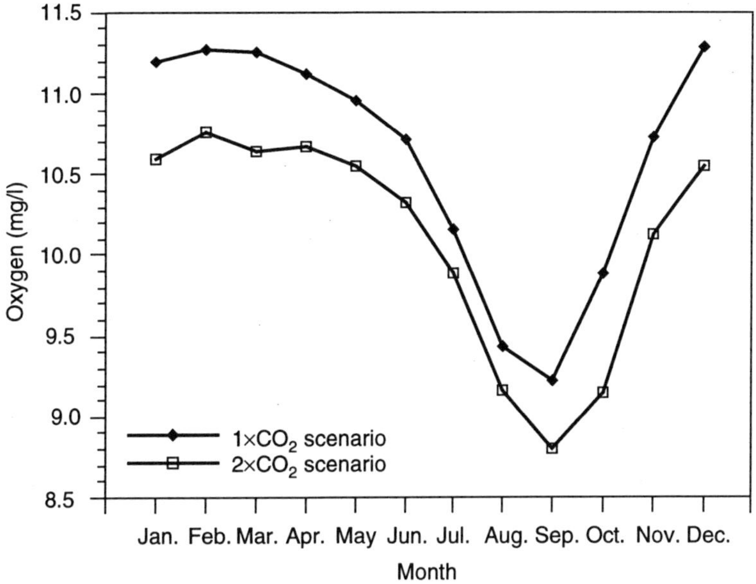

Figure 10.12. Yearly distribution of oxygen content at Red Bluff: $1\times CO_2$ and $2\times CO_2$ scenarios.

is likely that the target would be exceeded in most years. The potential of temperature increases due to global climate change amplifies the risk of permanent environmental damage.

Dissolved oxygen conditions in the reservoir and below lake Shasta currently are not near critical levels and do not appear to be seriously affected by the climate change conditions considered in this study. However, the potential exists for significant changes in the accretion of nutrients and organic substances from the drainage basins tributary to the reservoir and river due to changes in climate. These could add to the demand on the oxygen resources of the system, possibly reducing concentrations to levels that could stress existing ecosystems. Such a possibility is the subject of continuing investigation of the effects of climate change on the Shasta-Trinity system.

10.8 Conclusions

The WQRRS model was used to evaluate the potential water quality changes that might be anticipated in the Shasta-Trinity system due to a doubling of CO_2 in the atmosphere. Boundary conditions for model application were developed from the mean monthly output of the GFDL general circulation model for a typical annual cycle. Present meteorologic, hydrologic, and water quality conditions were taken as corresponding to recent observations ($1\times CO_2$); the $2\times CO_2$ conditions were developed by adjusting the $1\times CO_2$ scenario by incremental differences or ratios determined by the GFDL model for $1\times CO_2$ and $2\times CO_2$ scenarios.

The primary impacts of climate change were changes in water temperature and hydrologic fluxes to Shasta Lake and the Sacramento River. Inflow temperatures increased, in some cases by nearly 4°C, and the hydrodynamic regime in the reservoir was significantly altered, both in the distribution of temperatures throughout the depth and in the volumes of cold and warm water retained in the region accessible at the level of the fixed outlet. These conditions, defining the boundary for simulation of water quality in the river downstream, were shown to have a severe impact on the potential for propagation of salmon species. River water temperatures, already at levels detrimental to spawning fish, would be further increased and the period of critical conditions would be further extended under conditions of climate change.

Dissolved oxygen conditions in the reservoir and river would also be modified by climate change, primarily due to the decrease in saturation concentration associated with increases in temperature. Although oxygen conditions are currently favorable for indigenous aquatic species, the potential exists for additional depletion of oxygen resources if climate changes are accompanied by increases in accretions of nutrients and organics to the aquatic environment. This a subject for continuing investigation.

A potential for alleviating adverse temperature conditions exists in modification of the intake system for the hydropower plant in Shasta Dam. At present the intake is at a fixed elevation, precluding access to cold water stored at lower elevations unless river outlets that bypass the power plant are opened. A modified system is currently being designed. However, the long-range success of a more versatile outlet structure that could conserve cold water for release during critical periods of salmon migration and spawning depends on cold water being available in necessary quantity. These studies indicate that under conditions of climate change the volume of available cold water is likely to be significantly reduced, perhaps to the point that it will not be sufficient to meet downstream temperature control targets. This potentiality is also a subject for future investigation.

References

Arnell, N.W., Brown, R.P., and Reynard, N.S., 1990, Impact of Climate Variability and Change on River Flow Regimes in the UK, Report No. 107, Institute of Hydrology, Wallingford, UK.

Bultot, F., Coppens, A., Dupriez, G.L., Gellens, D., and Meulenberghs, F., 1988, Repercussions of a CO_2 doubling on the water cycle and on water balance: A case study for Belgium, *Journal of Hydrology*, **99**:319–347.

Cohen, S.J., 1986, Impacts of CO_2-induced climate change on water resources in the Great Lakes Basin, *Climatic Change*, **8**(2):135–153.

Gleick, P.H., 1986, Methods for evaluating the regional hydrologic effects of global climate changes, *Journal of Hydrology*, **88**:97–116.

HEC (Hydrologic Engineering Center), 1978, WQRRS: Water Quality for River-Reservoir Systems, Computer Program Description, US Army Corps of Engineers, Davis, CA, USA.

Kaczmarek, Z., and Krasuski, D., 1991, Sensitivity of Water Balance to Climate Change and Variability, WP-91-047, International Institute for Applied Systems Analysis, Laxenburg, Austria.

Lettenmaier, D.P., and Gan, T.Y., 1990, Hydrologic sensitivities of the Sacramento–San Joaquin River Basin, California, to global warming, *Water Resources Research*, **26**(1):69–86.

McCabe, G.J., Jr., and Ayers, M.A., 1989, Hydrologic effects of climatic change on the Delaware River Basin, *Water Resources Bulletin*, **25**(6):1231–1242.

Meyer, G.K., and Orlob, G.T., 1992, Global climate change effects on water quality, in M. Karamovs, ed., *Water Resources Planning & Management: Saving a Threatened Resource, in Search of Solutions*, American Society of Civil Engineers, Baltimore, MD, USA.

Mimikou, M., Kouvopoulos, Y., Cavias, G., and Vayainos, N., 1991, Regional hydrological effects of climate change, *Journal of Hydrology*, **123**:119–146.

Nemec, J., and Schaake, J.C., 1982, Sensitivity of water resources systems to climate variations, *Hydrological Sciences Journal*, **27**(3):327–343.

Orlob, G.T., 1984, Water quality modeling, streams, lakes, and reservoirs, in G.T. Orlob, ed., *Mathematical Modeling of Water Quality*, Vol. 12 of *IIASA State of the Art Series*, Wiley Interscience, London, UK.

Orlob, G.T., Meyer, G.K., DeGeorge, J., and Christensen, C.L., 1990, Impacts of global warming on water quality in river-reservoir systems, *Journal of Hydrologic Science and Technology*, **5**(1–4):10–19.

Schaake, J.C., 1990, From climate to flow, in P.E. Waggoner, ed., *Climate Change and U.S. Water Resources*, John Wiley & Sons, New York, NY, USA.

SDFG (State of California Department of Fish and Game), 1968, Recommendations on Thermal Objectives for Water Quality Control, Policies on the Interstate Waters of California, Water Project No. 7, Sacramento, CA, USA.

TVA (Tennessee Valley Authority), 1972, Heat and Mass Transfer Between a Water Surface and the Atmosphere, Engineering Laboratory Report No. 14, Norris, TN, USA.

Chapter 11

Impact of Climate Change on the Lake Kariba Hydropower Scheme

Kazimierz A. Salewicz

11.1 Introduction

The Zambezi River, situated south of the equator between 12° and 20° latitude (see *Figure 11.1*), is the largest African river flowing into the Indian Ocean. It is approximately 2,500 km from its source in the Central African plateau to its outlet (Balon and Coche, 1974; Balek, 1977). It consists of three sections (Balon and Coche, 1974): *upper*, from its sources to the Victoria Falls; *middle*, from the Victoria Falls to the Cahora Bassa rapids; and *lower*, from Cahora Bassa to the Indian Ocean. The river's total catchment is about 1,300,000 km^2; it is shared by eight countries, for some of which it constitutes the main water resource. The total population of the river basin is about 20 million (UNEP, 1987).

Although the catchment has great potential for development, the main water uses so far have been limited to three large hydroelectric schemes: Kariba and Cahora Bassa on the Zambezi itself, and Kafue Gorge – Itezhi-Tezhi on the Kafue River, one of the Zambezi's main tributaries. The Kariba and Kafue Gorge hydropower schemes provide more than 70% of the electricity to the interconnected energy systems of Zambia and Zimbabwe (Williams, 1984; ZESA, 1986); therefore, their efficient operation is very important for these two countries.[1] This importance will increase in the future, because historically economic growth has been impossible without at least an equal increase in the electric energy production and supply.

Z. Kaczmarek et al. (eds.),
Water Resources Management in the Face of Climatic/Hydrologic Uncertainties 300–321.

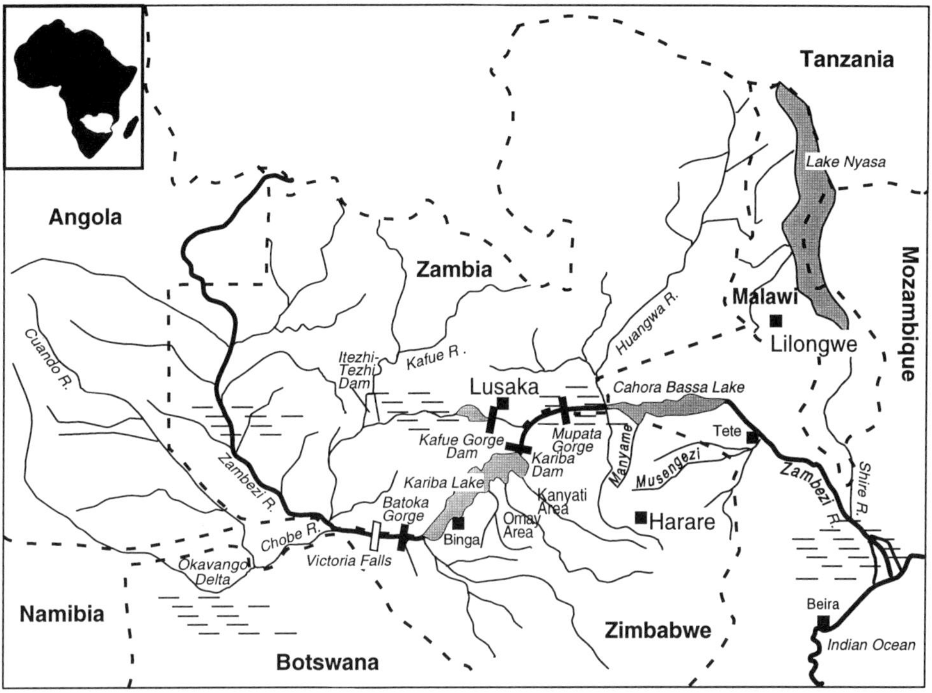

Figure 11.1. The Zambezi River basin.

Due to the difficult economic situations of all the countries in this region and the lack of financial resources for investments, one cannot expect significant growth of energy produced by new thermal plants. Thus, hydropower generation at existing schemes probably will remain the main source of energy for Zimbabwe and Zambia in the coming years.

In this chapter the operation of the Lake Kariba hydropower scheme is analyzed. The chapter follows the optimization analysis of the operating policies conducted for this scheme in the actual climatic and hydrologic conditions (Gandolfi and Salewicz, 1990, 1991). Analysis of the results characterizing performance of the Lake Kariba scheme under current hydrological conditions raises questions about the future: What might be the impact of climate change on the hydropower generation potential of this scheme? Will energy output increase or decrease? Will currently applied reservoir operating policies remain relevant to a new hydrological regime?

These and similar questions cannot yet be fully answered; but even at the present state of knowledge about climate change and its impact on hydrology, the results that can be obtained are instructive and worthy of consideration.

11.2 Hydrologic and Climatic Conditions in the Zambezi River Basin

Geographically, the Zambezi River basin is located south of the equator between 12° and 20° latitude. In this region the cool and dry season takes place between May and September. The headwaters, situated in the northern part of the basin, belong to the tropical summer rainfall zone. As one moves downstream, the climate becomes progressively more arid due to a prolongation of the dry season. Thus, the upper and the northern part of the middle Zambezi regions can be classified as warm temperate regions with dry winters; the Ocavango Delta and the southern part of the middle Zambezi belong to the arid steppe with a mean annual temperature of more than 18°C. Finally, the western portion of the lower Zambezi belongs to the arid climatic zone, and the eastern portion at the outlet to the Indian Ocean has an equatorial climate with dry winters.

The upper stretch of the Zambezi River extends from the river's source in a marshy bog near the Kalene Hills, about 1,500 m above sea level, to the Victoria Falls, near the town of Livingstone. The Victoria Falls, 1,650 m wide with a mean annual discharge of 1,237 m^3/sec, form a 98 m high barrier between the upper and the middle Zambezi. The upper stretch is the most natural portion of the Zambezi River due to the absence of impoundments and the lack of significant human activity. Rainfall in the drainage basin varies from 1,400 mm/year in the headwater zone to 700 mm/year farther south. Average yearly temperatures reach 20° to 22°C, with a range of 6° to 8°C. The rainy season begins in November and lasts until April; the dry season is from May to July. The catchment area upstream of Kariba Gorge is approximately 664,000 km^2 (Santa Clara, 1988). The Zambezi River was gauged at Livingstone near Victoria Falls, approximately 400 km upstream of Kariba Gorge, as early as 1905. At the Livingstone gauging station, the flow starts to increase in December, peaks in April or May (3,500 m^3/sec), and then decreases to about 350 m^3/sec in October. However, because of the erratic rainfall pattern in the region, the flow of the Zambezi varies greatly from year to year.

Balon and Coche (1974) have estimated that it normally takes approximately three to four weeks for the flood peak to move from the sources of the river to Lake Kariba. Additional inflow to Lake Kariba comes from the lower catchment, located between Victoria Falls and the Kariba Dam. Almost 60% of the 9×10^9 m^3 of average annual flow from the lower catchment is concentrated between January and March (Santa Clara, 1988).

Since the creation of the Kariba Dam, the natural situation of the middle stretch of the river has changed drastically. Before the creation of the impoundment the river was defined as a "sand bank" river, like the Limpopo River (Jackson, 1961). In its natural stage, the middle Zambezi had two high flood periods every year: the first occurred in February and was less turbulent and carried local runoff; the second came in April and was higher than the first flood. After the creation of the dam, the inundating effects of the floods could be observed upstream of Lake Kariba. The dam has caused an increase in the dry-season flows and a delay and reduction of the peak flood flows during the wet season.

11.3 The Lake Kariba Hydropower Scheme

When the Kariba reservoir filled to capacity in 1962, it was the largest man-made lake in the world; currently it is the fourth largest. At maximum retention level the lake covers an area of more than 5,600 km^2 and has an active storage exceeding 70 km^3. Hydropower generating facilities consist of six 111 MW turbo generators located on the southern bank (Zimbabwe) and four 150 MW units on the northern bank (Zambia). The scheme is operated jointly by Zambia and Zimbabwe and the electricity generated is shared equally by the two countries. Since the completion of the generating facilities in 1977, the scheme has supplied a monthly average of about 600 GWh, with an almost constant distribution throughout the year.

The central management problem of the Kariba operation is how to balance two conflicting objectives: the maximization of the hydropower output and the maintenance of a safe capacity at the beginning of the rainy season to store high flows and to avoid peak discharges through the floodgates.[2] From the operational viewpoint, the main task of Lake Kariba management is to maintain the maximal fixed level of energy production. The second objective is to avoid very high discharges through the floodgates. There are also other objectives related to human activities and to environmental protection of wildlife areas downstream of the reservoir. These are, however, very difficult to quantify and were not considered in the formulation of the optimization problem. Additional studies dealing with these issues should be conducted in the future.

11.4 Optimization Analysis of the Lake Kariba Management Problem

Given the characteristics of the Lake Kariba management problem formulated above, a two-phase study was conducted. The first phase of the study (Gandolfi and Salewicz, 1990) was based on the following assumptions:

- The present firm energy target of 600 GWh/month is a rigid constraint that should not be violated.
- The operation of the reservoir is conducted according to a simple operating rule, shown in *Figure 11.2*, which specifies the value of release (r_t) as a function of the estimated amount of water (y_t) that is available at the beginning of time period t and that will be stored in the reservoir during this period of time. In general, the amount of water (y_t) available at the beginning of time period t can be expressed as the sum of the amount of water currently stored in the reservoir (s_t) plus the forecast inflow to the reservoir during this period of time ($\hat{i}_t$) minus estimated evaporation losses ($\hat{v}_t$):

  $$y_t = s_t + \hat{i}_t - \hat{v}_t \quad . \tag{11.1}$$

  This type of information pattern (i.e., the type of data used) in the decision-making process associated with the reservoir operation requires knowledge about future inflows to the reservoir. Often there are no forecasting services or models, or there is a high level of uncertainty which is common in studies dealing with climate–hydrology interactions. In these cases, the most appropriate form of the information pattern to be applied in derivations of the operating rule for the reservoir is one that directly relates release from the reservoir to actual value of storage:

  $$y_t = s_t \quad . \tag{11.2}$$

  Consequently, this form of the information pattern (free of additional assumptions and sources of uncertainty) has been used in the reported study.

The particular operating rule considered at the first stage of the study is such that for every time period t, until a certain value Y_t is reached, the release decision r_t is equal to the amount of release r_E necessary to generate a given value of the monthly energy production target E. When Y_t is exceeded, r_t increases steadily with coefficient α as a linear function of water available in the reservoir (see *Figure 11.2*) to avoid excessive reservoir

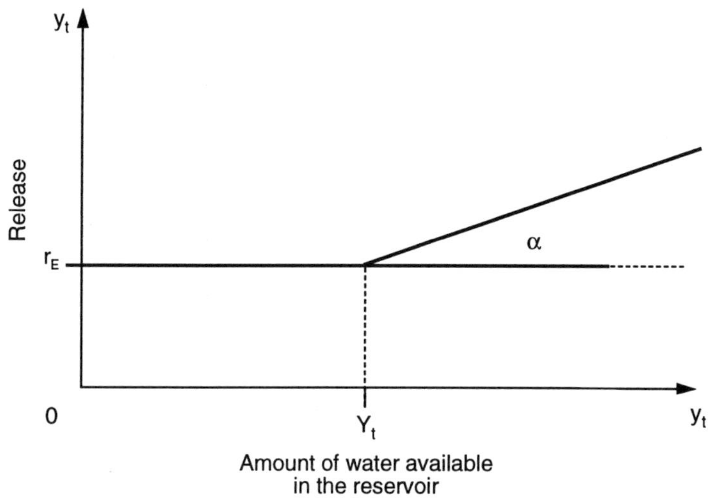

Figure 11.2. Initial form of the operating rule with the value of Y_t dividing storage capacity into two zones: the lower one with the release determined by the value of energy production target, and the upper one with linearly growing release.

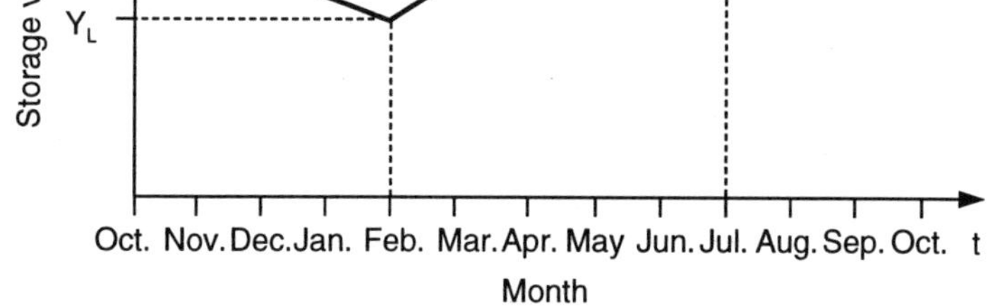

Figure 11.3. Line representing the time-dependent border between storage zones. Variations of the values reflect the desire to prepare free storage in the reservoir before the beginning of the flood period [the lowest value (Y_L) in February], and subsequently to fill the reservoir before the beginning of a dry season [maximum storage value (Y_U) in July].

filling and, consequently, to maintain a flood reserve needed to store peak flows and avoid floodgate discharges. Additionally, the value of Y_t varies according to the risk of occurrence of flood inflows from the lake catchment (*Figure 11.3*), which is determined based on an analysis of inflows to the reservoir and historical operating policy (Gandolfi and Salewicz, 1990). This type of reservoir operating policy means that the active storage is divided into two zones by the line representing values of Y_t. For every time period t and amount of water available in the reservoir y_t, the release r_t is univocally identified by three parameters, α, Y_L, and Y_U, as denoted symbolically by:

$$r_t = r(t, y_t, \alpha, Y_L, Y_U) \ . \tag{11.3}$$

The management problem was thus expressed in terms of the optimization task with the objective being to minimize, over a given period of time and with respect to the parameters of the operating policy, the maximum value F of the floodgates discharge f exceeding release necessary to meet the given energy production target E; that is,

$$\min_{(\alpha, Y_L, Y_U)} [F] \ , \tag{11.4}$$

where

$$F = \max_{0 \leq t \leq T} [f_t] \ , \tag{11.5}$$

and where T is the length of the optimization horizon, and where the floodgates discharge f_t associated with the release higher than the release r_E necessary to meet the energy production target is defined as follows:

$$f_t = \begin{cases} r_t - r_E & \text{if} \quad r_t > r_E \\ 0 & \text{if} \quad r_t \leq r_E \end{cases} \ . \tag{11.6}$$

The formulation of problem (11.4) was completed by the set of constraints involving the reservoir continuity equation and a number of inequality constraints on storage and release necessary to secure feasibility of the solution (Gandolfi and Salewicz, 1990).

The solution of problem (11.4), if feasible, gives the operating rule that produces the lowest value of the peak floodgates discharge while maintaining in any time period energy production of at least 600 GWh/month.

The solution of the optimization problem given above was found by simulating the system operation for a number of different operating rules (for example, tripling α, Y_L, Y_U), using inflow data covering the period from 1950 until 1984. A random-search algorithm (Karnopp, 1963) in the space of (α, Y_L, Y_U) was used to limit a region around the optimum, and then a deterministic method (Himmelblau, 1972) was applied for refinement of the results.

The computations proved that a firm energy production target of 600 GWh/ months could be guaranteed with reasonable values of peak flood discharges.

The operating rule obtained as the result of optimization was then verified for its robustness, and the reservoir operation was simulated using the historical sequence of inflows from the period from 1924 to 1949. In this simulation, the firm energy output of 600 GWh/month could be maintained and the maximum monthly floodgate discharges did not exceed values computed for the first sequence of inflows. The results of simulations demonstrated that, except for a very few critical years, the reservoir storage remains well above its minimum value and huge amounts of water are lost through floodgates discharge. The results suggest that a higher energy output could be achieved if a more flexible operating policy were adopted.

To check this possibility, the following approach was used:

- A modified operating rule was proposed, as shown in *Figure 11.4*. This rule differs from the first rule in that the range of values between 0 and Y_t is divided into two subzones by a new parameter (Y_t'): in the lower subzone, the release is equal to the release r_{E1} necessary to generate $E1 = 600$ GWh/month; in the upper subzone, an increased energy target, $E2 = 700$ GWh/month, was set and the release r_{E2} was varied accordingly. The borders between storage subzones Y_t' and Y_t vary in time according to the same simple law through time-dependent parameter β_t representing the relative degree to which the reservoir is filled. This parameter has been introduced to reflect seasonal variability of inflow and to force specific behavior of the reservoir operating policy: i.e., to prepare enough free storage to catch flood-season inflows and then fill the reservoir before the beginning of the drought period (see also *Figure 11.5*, which shows dependence of β_t on parameter β_L and time):

$$Y_t' = \beta_t \cdot Y_L \quad , \tag{11.7}$$

$$Y_t = \beta_t \cdot Y_U, \quad . \tag{11.8}$$

Consequently, the new operating rule is identified by the four parameters α, β_L, Y_L, and Y_U:

$$r_t = (t, y_t, \alpha, \beta_L, Y_L, Y_U) \quad . \tag{11.9}$$

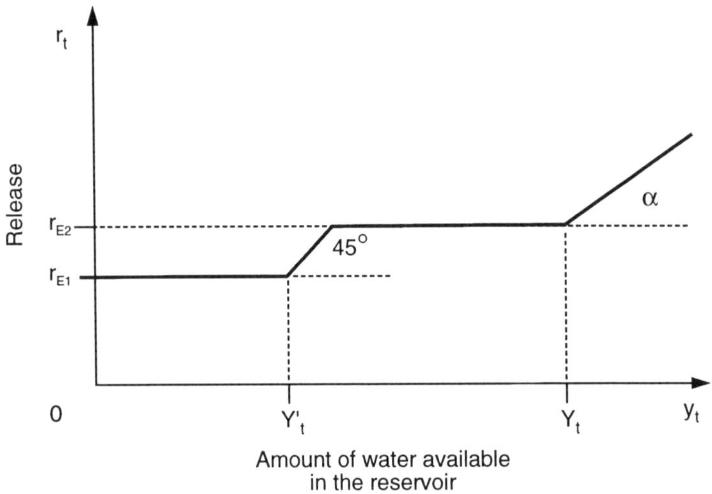

Figure 11.4. Modified (more flexible) version of the operating rule.

- A second objective was introduced to express the desire to achieve the highest possible energy output in the following form:

$$\min_{(\alpha,\beta_L,Y_L,Y)} [N] \ , \tag{11.10}$$

where

$$N = \sum_{t=1}^{T} n_t \ , \tag{11.11}$$

when T is the length of optimization horizon and when n_t is defined as follows:

$$n_t = \begin{cases} 1 & \text{if} \quad e_t \leq \frac{E1+E2}{2} \\ 0 & \text{if} \quad e_t > \frac{E1+E2}{2} \end{cases} \ , \tag{11.12}$$

where e_t denotes desired energy production target equal to the average value of the sum of two components: minimum energy production target ($E1 = 600$ GWh/month) and energy production ($E2 = 700$ GWh/month) reflecting increasing aspiration levels for energy output.

Thus, the new formulation of the management problem was given as follows:

$$\min_{(\alpha,\beta_L,Y_L,Y_U)} [F, N] \ , \tag{11.13}$$

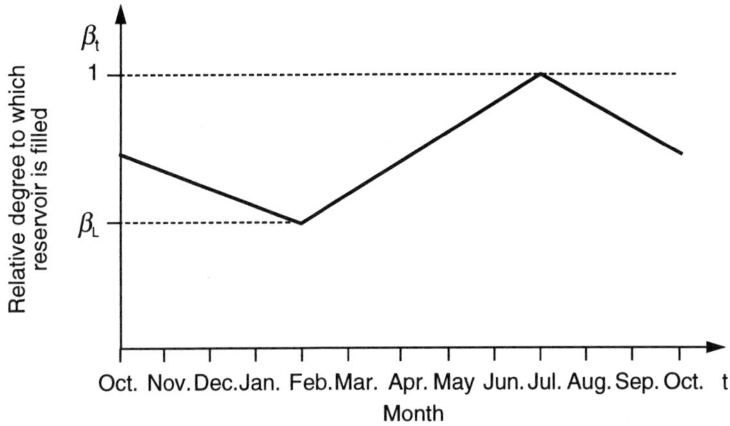

Figure 11.5. Time-dependent values of β_t. Variations of β_t reflect the desire to prepare free storage in the reservoir before the beginning of the flood period (β_L), and subsequently to fill the reservoir before the beginning of the dry season ($\beta_t = 1$).

subject to operating rule (11.9) and to the same set of constraints as in problem (11.4). The set of solutions that were efficient in the Pareto sense was determined using the method of weights and the same optimization algorithm as was used in problem (11.4).

A series of computations led to a Pareto-optimal solution, for which the energy output of 700 GWh/month or more could be achieved for 81.2% of the time over the period of 35 years; for which energy output lower than 650 GWh/month occurred only 8 times (1.9% of the time); and for which the minimum monthly energy production did not fall below 600 GWh/month. At the same time, the maximum release obtained during the period from 1977 to 1984 (7,906 $\times$ 10^9 m^3) was lower than the historically registered maximum release.

These results prove that it is possible to improve the operation of the Kariba hydropower scheme during current climatic and hydrologic conditions.

11.5 Analysis of the Lake Kariba Operation for Changing Climatic Conditions

Given that changes are expected in climatic and hydrological conditions within the life span of this scheme, what might be the impact of the climate change on the hydropower generation output from the Kariba scheme? To what extent will current operating policies be applicable to new climatic and hydrologic conditions?

The answers to these and other related questions require knowledge about the future hydrological regime, which will be influenced by changes in global geophysical processes. Studies on the mechanisms causing changes in the earth's energy balance, and consequently in meteorological processes, are highly advanced at many research centers; however, numerous uncertainties concerning climate variability and variations remain. Global warming is one of the possible symptoms of climate change, but it, itself, is not the problem: changes in spatial and temporal distribution of other weather characteristics, such as rainfall and soil moisture, may take place even before there is any detectable warming.

Climate is commonly understood as *average weather*. This is, however, a simplified definition of a complex, multivariate stochastic process with significant seasonal components. In a catchment (atmosphere–land surface) system, the climate process is transformed into a hydrologic process with runoff, evaporation, and soil moisture as the most important variables. This means that to evaluate possible changes in hydrological characteristics it is necessary to describe the way the catchment system operates and then to assume changes in the climatic input process. Unfortunately, the state of the art of hydrological modeling techniques and the ability to predict future climate are far from being perfect.

Today, the main tools used for climate simulation are general circulation models (GCMs), in which some input information is transformed into a set of output values (such as temperature and precipitation), on a grid of rather coarse resolution. For the present generation of GCMs, the horizontal distance between grid points may be several hundred kilometers; GCMs do not "see" processes that occur on a scale smaller than the grid resolution. A review of climate models can be found in the IPCC Reports (IPCC 1990, 1992), and in a number of papers (see Harrison, 1990). For water resources impact studies, it is necessary to translate the GCM information into a catchment-scale regional climate scenario. For the purpose of the Kariba study, the following procedure was applied to define the climatic input to hydrological models for the two equilibrium conditions equivalent to $1\times CO_2$ and $2\times CO_2$ concentrations of trace gases:

1. Climatological data based on the National Center for Atmospheric Research (NCAR) World Monthly Surface Station Climatology data sets (Spanger and Jenne, 1988) and on records from the Lake Kariba station

were used at first to calculate various climate characteristics: precipitation, air temperature, vapor pressure, and relative sunshine duration.

2. These data were then used to calculate mean monthly values of potential evapotranspiration (PET) by means of the Budyko-Zubenok method (see USSR Committee for the International Hydrological Decade, 1978).
3. Values of the temperature increments and precipitation ratios expressed in the 0.5° latitude × 0.5° latitude grid interpolated by the IIASA Biosphere Project (Leemans, 1990) and based on the outputs from the GCMs developed by the Geophysical Fluid Dynamic Laboratory (GFDL) of the National Oceanographic Administration (1980 version), and the Goddard Institute for Space Studies (GISS) of the National Aeronautics and Space Administration (1985 version) have been added to observed temperature and precipitation data to create $2\times CO_2$ climate scenarios.
4. Relative air humidity and sunshine duration values were assumed to be the same under $1\times CO_2$ and $2\times CO_2$ conditions.
5. Based on data determined in steps (3) and (4), the $2\times CO_2$ PET values were calculated by means of the Budyko-Zubenok method.

The "historic" and $2\times CO_2$ climate characteristics were then used as input data to hydrological models to obtain scenarios of inflow to the Lake Kariba scheme.

Most of the existing hydrological rainfall-runoff models have been constructed to predict flood characteristics due to heavy rainfall or snowmelt processes. Only a few of them can be used for evaluating large-scale catchment processes over the whole year, which is essential for climate impact studies. For the Lake Kariba studies, a stochastic water balance model in a form of the CLIRUN software package developed by Kaczmarek and Krasuski (1991) has been applied to prepare flow scenarios to the reservoir for the two alternative climate scenarios described above. The PC software package CLIRUN (Kaczmarek, 1993) permits the calculation of mean values and standard deviation of the main water balance components, namely runoff, catchment storage, and actual evapotranspiration. Then, combined with the standard simulation technique, CLIRUN can be used to generate time series of monthly or seasonal inflows to the reservoir.

The main characteristics of the CLIRUN are the following:

- The relative storage of the catchment $z = S/S_{max}$ (with $0 < z < 1$) is divided into k equal intervals. For every time period of the length T (e.g., one month) and for every pair of discrete storage values z_i, z_j, where $i, j = 1, 2, \ldots, k$, the catchment water balance differential equation

$$\frac{dz}{dt} = \frac{[P - R_s(z, P) - R_g(z) - E(PET, z)]}{S_{max}} \tag{11.14}$$

is solved to get the value of effective precipitation P. In equation (11.14), R_s and R_g represent surface and subsurface runoff, respectively, and actual evaporation E is a function of PET and relative storage of the catchment.

- Assuming that the effective precipitation P is represented by a discrete stochastic process with known probabilistic properties, the matrix storage theory (see Klemeš, 1981; or Phatarfod, 1989) is applied to calculate the ergodic stochastic matrices of the catchment storage levels for every time period t.
- Finally, mean values and standard deviations of the runoff, catchment storage, and actual evapotranspiration are calculated.

The model parameters, namely S_{max} and parameters describing the runoff function $R_s(z, P)$, have to be calibrated to minimize the mean square deviations between the observed (historical) values of the runoff and those calculated by the model.

The modeling scheme described above has been applied to the Zambezi River catchment upstream of Lake Kariba. The calibration and all other calculations had been performed separately for the following three sub-catchments:

- The upper part of the basin, upstream of Chavuma Falls [catchment area $(A) = 76,000$ km^2].
- The middle part of the basin between Chavuma Falls and Livingstone/ Victoria Falls ($A = 244,000$ km^2).
- The lower Kariba basin between Victoria Falls and the Kariba Dam ($A = 157,000$ km^2).

Runoff of the Chobe River tributary to the lake is relatively small (about 3 mm/year) and was therefore assumed to be constant within the year.

The average yearly values of temperature (T) and precipitation (P), obtained from meteorological records and subsequently calculated for the GFDL and GISS 2×CO_2 scenarios, are given in *Table 11.1*. The mean monthly inflows to Lake Kariba, estimated on the basis of hydrological observations and subsequently calculated by the means of CLIRUN model, are presented in the *Table 11.2*. It can be seen that the change of total inflow

Table 11.1. Climatic scenarios for the Lake Kariba catchment.

	Historical data		$GFDL_{2\times CO_2}$		$GISS_{2\times CO_2}$	
Sub-catchment	T (°C)	P (mm)	T (°C)	P (mm)	T (°C)	P (mm)
Upper basin	21.7	1,290	25.2	914	25.8	1,390
Middle basin	21.0	921	25.1	884	25.4	1,004
Lower basin	21.8	732	25.2	871	25.3	849

Table 11.2. Inflow to the Lake Kariba (m^3/sec).

Month	Q_{obs}	Q_{calc} $1\times CO_2$	Q_{GFDL} $2\times CO_2$	Q_{GISS} $2\times CO_2$
October	347	366	306	364
November	385	386	368	457
December	913	688	902	1,036
January	1,497	1,326	2,015	2,006
February	2,361	2,380	2,798	2,926
March	3,003	3,172	3,549	2,894
April	3,585	3,508	2,957	3,807
May	2,862	3,026	2,121	3,017
June	1,848	2,110	1,147	1,839
July	1,021	1,096	892	1,116
August	633	721	566	662
September	454	488	408	463
Yearly average inflow	1,576	1,606	1,502	1,714

Q_{obs} = estimate based on hydrological observations.
Q_{calc} = estimate based on calculations from CLIRUN model.
Q_{GFDL} = estimate based on calculations from Geophysical Fluid Dynamics Laboratory GCM.
Q_{GISS} = estimate based on calculations from Goddard Institute for Space Studies GCM.

is not very significant, but its seasonal distribution varies substantially for various climate scenarios.

The last step of the hydrological analysis is generating a 300-year series of the monthly inflows to the reservoir by means of the simple Markov-type random generator for inflow parameters based on historic observations and $2\times CO_2$ climate scenarios.

Optimization analysis of the operation of the Lake Kariba hydropower scheme in the face of climate change thus has been based on two basic flow scenarios over the 300-year time interval calculated for climatic parameters obtained from GFDL and GISS models, respectively. To verify the correctness of the modeling procedure, a 300-year flow sequence was also generated for current climatic conditions.

The impact of eventual climate change can only be analyzed in relative terms when current characteristics are used as the reference values for the comparisons. In our study, the results of the Lake Kariba optimization analysis, obtained as the solution of the two-objective optimization problem (11.13) and solved for historical flow data, were used as the reference for evaluating consequences of the climate change. As noted, the objective function of the optimization problem (11.13) has two components: term F used to evaluate the maximum flood discharge, and term N specifying number of months when energy production level drops below desired energy production target E'. Although the optimization analysis has been performed with respect to the two-objective problem, in the analysis and evaluation of optimization results more emphasis has been put on those elements of the solution that deal directly with the energy production estimates. This is because the uncertainty associated with the generation of flood events by the CLIRUN model is much greater than the uncertainty associated with the calculation of the average runoff (and inflow) characteristics, which in fact decide long-term energy production output. Moreover, for the changing climate conditions it is not as important to analyze individual values of the performance function (which depend upon various initial parameters such as initial storage, desired energy production target, weight coefficients, etc.) as it is to take into account indicators that in an aggregate form can characterize the river's hydropower generation potential and the extent to which current operating policies could be applicable to new climatic and hydrologic conditions.

Taking into account these factors, the following quantitative characteristics of the performance of the hydropower scheme have been used as the basis for analysis and comparisons:

- Attainability of the currently guaranteed minimum energy production level $E = 600$ GWh/month.
- Percentage of time when the energy output drops below desired energy production target $E' = 650$ GWh/month.

In the first stage of the study the optimization and consequent simulation of the Lake Kariba operation were conducted using the 300-year flow scenario equivalent to current climatic and hydrological conditions, which was generated by the stochastic catchment model. These computations were performed to check the reliability of the stochastic modeling procedure used to generate flow scenarios, and to determine the differences between the solution obtained for the observed (historical) flow data record and the flows generated based on current climatic characteristics.

Initial optimization and simulation results obtained for the 300-year flow sequence generated by the stochastic catchment model for current climatic conditions were consistent with the results obtained for the 35-year historical flow record. Assuming that the desired energy production target $E' = 650$ GWh/month, the resulting minimum energy production output was 601.2 GWh/month compared with 601.4 GWh/month obtained for historical (observed) flows. At the same time, the energy output dropped below the given energy production target 3% of the time; the value obtained for historical flow data was 1.9%. These results clearly demonstrate that the modeling procedure applied to derive flow sequences from the climatic data is as accurate as the aggregated, long-term characteristics of the real flow record, and therefore can be used to generate sequences of inflow to the reservoir for new climatic conditions.

The study then focused on the combined optimization and simulation analysis of the Lake Kariba scheme for flow scenarios developed on the basis of appropriate outputs from the GISS and GFDL general circulation models. The formulation of the optimization problem and the problem-solving technique were exactly the same as were used for the management problem (11.13).

Several runs of the combined optimization and simulation procedure performed for the flow scenario generated by the GISS model demonstrated that the desired energy production target $E' = 600$ GWh/month in the current climate conditions can be exceeded by approximately 10%, to values in the range of 700 to 710 GWh/month, without violation of the minimum energy production output boundary (600 GWh/month). In *Figures 11.6* and in *11.7*, the set of efficient solutions obtained for the two values of the desired energy production target, $E' = 700$ and 710 GWh/month, respectively, can be seen. With only a small difference in the value of desired energy production target (10 GWh/month, approximately 1.5%), respective differences in the values of peak flood discharge (indicator F on the vertical axis) and the number of months when the energy production drops below E' are few times higher. This indicates that the limit on the desired energy production target has been reached.

Values of the peak flood discharge obtained for the generated flow data (over 3.3×10^{10} to 3.4×10^{10} m^3/month) are much higher than respective historically observed values. This can be attributed to the fact that the observed flow record (about 60 years long) has been too short to produce

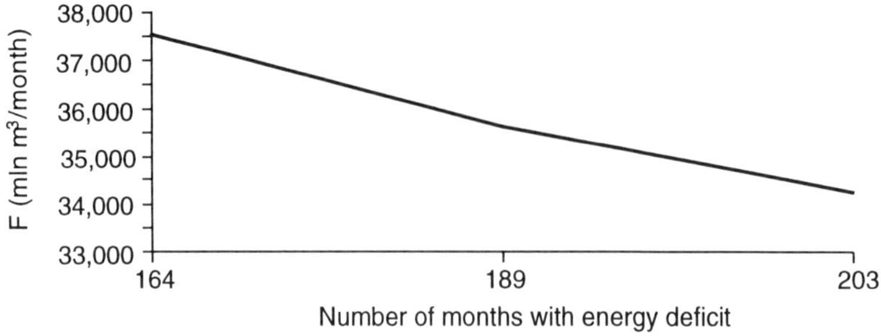

Figure 11.6. The set of efficient solutions obtained for the flow scenario generated by the GISS model for energy production target $E' = 700$ GWh/month.

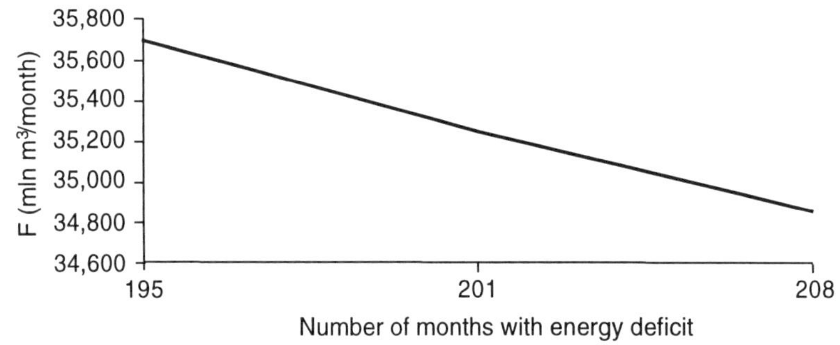

Figure 11.7. The set of efficient solutions obtained for the flow scenario generated by the GISS model for energy production target $E' = 710$ GWh/month.

extremely high floods that may happen once over 200 to 300 years. Taking into account available data, this explanation seems to be more justified than the claim that the climate change may bring an increase in the magnitude or frequency of extreme events.

Analysis of the results obtained from several optimization and simulation experiments performed for the flow scenario developed on the basis of the GISS general circulation model output are summarized in *Table 11.3*. These results suggest that realization of the climate scenario predicted for the Zambezi River basin by the GISS model may cause a 10% increase of the minimum energy production output, from 601.2 GWh/month to 661 GWh/month. This increase is estimated very conservatively because a 100% guarantee of the energy production output is required. An output of 710

Table 11.3. Results for the flow scenario generated from the GISS general circulation model data.

Range of desired energy outputs (GWh/month)	Energy production target E' (GWh/month)	Minimum energy output (GWh/month)	Time with energy output lower than E' (%)
650–750	700	651.0	5.6
660–760	710	661.0	7.2
680–780	730	19.2	3.1
700–800	750	19.2	4.4

Table 11.4. Results for the flow scenario generated from the GFDL general circulation model data.

Range of desired energy outputs (GWh/month)	Energy production target E' (GWh/month)	Minimum energy output (GWh/month)	Time with energy output lower than E' (%)
580–680	630	581.0	11.3
590–690	645	128.8	3.4
600–700	650	98.8	5.9

GWh/month or more can be obtained with a 92.8% guarantee of energy production output. Moreover, one can expect that the energy production can even exceed 750 GWh/month with a 95.4% guarantee of energy production output. In this case, the lowest energy output falls to 19.2 GWh/month; however, this happened for only one month during 300-year time period, thus its probability is lower than 0.001.

The next series of optimization and simulation experiments was performed in an analogous way for flow scenarios generated on the basis of output from the GFDL model. These results are summarized in *Table 11.4*.

Qualitatively speaking, realization of the $2\times CO_2$ scenario predicted by the GFDL model for the Zambezi River basin would cause effects opposite to results obtained for the output from the GISS model. The guaranteed energy production output would drop from 601.2 GWh/month to 581.0 GWh/month, and the value of the second indicator used for comparisons would deteriorate, namely, the percentage of time when the energy production output is lower than energy production target. In current climatic conditions, desired energy production target E' is equal to 650 GWh/month and can be achieved or exceeded with 98.1% confidence; for the climate scenario obtained from GFDL model, the equivalent energy production target is equal to 630 GWh/month and can be achieved with the probability of 88.7%.

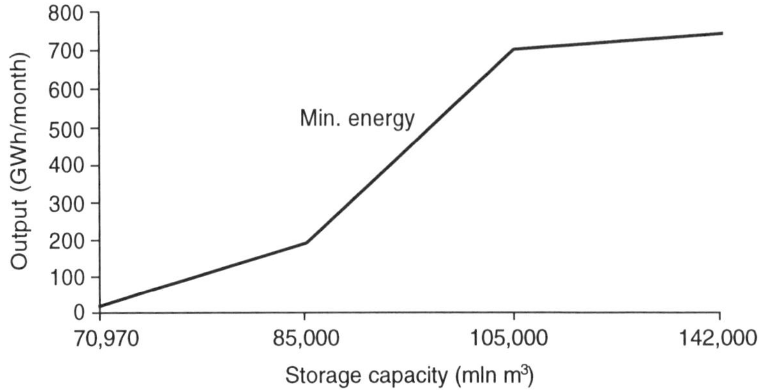

Figure 11.8. The relationship between maximum storage capacity and minimum monthly energy production output obtained for the flow scenario generated by the GISS model.

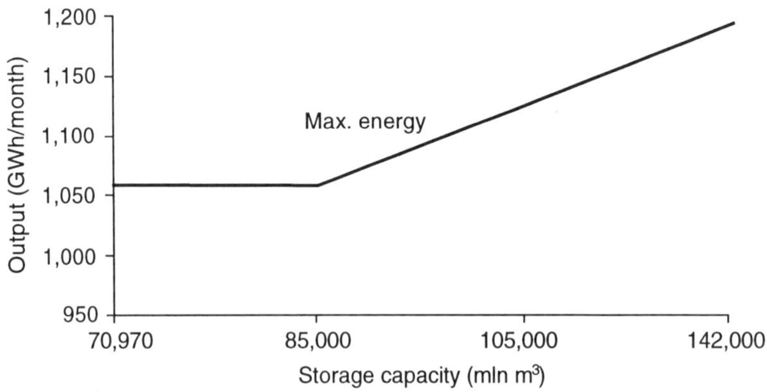

Figure 11.9. The relationship between maximum storage capacity and maximum monthly energy production output obtained for the flow scenario generated by the GISS model.

Results of the analysis obtained for the flow scenarios based on the GISS model demonstrated the possibility of the increase of the peak flood flows. Therefore, as the next element of the study, we have analyzed what the effect would be of increased storage capacity of the Lake Kariba reservoir. The analysis was based on the assumption that storage capacity increases by 20%, 50%, and 100%, respectively, while the capacity of turbines remains unchanged (due to a lack of data characterizing hydraulic and energetic properties of the generating facilities). Summarized results of the analysis are presented in *Figures 11.8* and *11.9*, respectively.

Considering these results with due caution, the current storage capacity of the reservoir may be too small to regulate the flows predicted by the GISS model for the new climate conditions. Analysis of the results obtained so far for the flow scenario calculated based on GFDL model output data also support this thesis.

11.6 Concluding Remarks

Numerical and qualitative results obtained during the course of the study discussed in this chapter shed some light on possible consequences of climate change for the hydropower generation potential of the Lake Kariba scheme located on the Zambezi River. Hydrological scenarios, developed on the basis of climatic data obtained from the GISS and GFDL general circulation models, yield opposite results concerning guaranteed energy production level, minimum energy production level, and percentage of time the energy output is between guaranteed energy production level E and desired energy production target E'. The climate scenario from the GISS model predicts up to a 20% increase of the hydropower generation potential of the scheme; however, for the climate scenario obtained from GFDL model one should expect an approximately 5% deterioration of the performance. Taking into account the amplitude of these differences, it can be concluded that the impact of climate change evaluated through the changes in the hydropower generation potential of the scheme may be less dramatic than impacts predicted for other phenomena.

Careful analysis of the generated flow sequences and the results of the simulation of the operation of the Lake Kariba scheme conducted for both historical and generated flow sequences demonstrate that the flow regime obtained for new climatic conditions is more variable than is the current one: there are more extremely high flows and, simultaneously, dry periods are longer and more severe than are those currently observed. This is one of the reasons why for the higher average values of flows obtained for new climatic conditions the overall increase of the hydropower potential of the Lake Kariba scheme is limited to a maximum of 20%. Consequently, it is reasonable to conclude that the rather stiff operating policies currently applied should be modified in the future to increase their elasticity and secure efficient operation of the scheme in the face of hydrologic conditions vacillating from one extreme to the other. It also seems justified to claim that the actual storage capacity of the reservoir may be inadequate to regulate future flows characterized by higher variability.

Notes

[1] The rest of the energy is generated by a few small hydropower plants and a few midsize thermal plants.

[2] For a more detailed explanation of this aspect, see Gandolfi and Salewicz, 1990.

References

Balek, J., 1977, *Hydrology and Water Resources in Tropical Africa*, No. 8 of *Development in Water Science Series*, Elsevier, Amsterdam, The Netherlands.

Balon, E.K., and Coche, A.G., 1974, Lake Kariba: On man-made tropical ecosystems in central Africa, *Monographiae Biologicae*, **24**.

Gandolfi, C., and Salewicz, K.A., 1990, Multiobjective Operation of Zambezi River Reservoirs, WP-90-031, International Institute for Applied Systems Analysis, Laxenburg, Austria.

Gandolfi, C., and Salewicz, K.A., 1991, Water resources management in the Zambezi valley: Analysis of the Lake Kariba operation, in F.H.M. Van De Ven, D. Gutknecht, D.P. Loucks, and K.A. Salewicz, eds., *Hydrology for the Management of Large River Basins*, International Association of Hydrological Sciences (IAHS) Publication No. 201, IAHS Press, Wallingford, UK.

Harrison, S.P., 1990, An Introduction to General Circulation Modelling Experiments with Raised CO_2, WP-90-27, International Institute for Applied Systems Analysis, Laxenburg, Austria.

Himmelblau, D.L., 1972, *Applied Nonlinear Programming*, McGraw Hill, New York, USA.

IPCC (Intergovernmental Panel on Climate Change), 1990, *Climate Change: The IPCC Scientific Assessment*, J.T. Houghton, G.J. Jenkins, and J.J. Ephraums, eds., Cambridge University Press, Cambridge, UK.

IPCC (Intergovernmental Panel on Climate Change), 1992, *The Supplementary Report to the IPCC Scientific Assessment*, J.T. Houghton, B.A. Callander, and S.K. Varney, eds., Cambridge University Press, Cambridge, UK.

Jackson, P.B.N, 1961, Ichthyology: The fish of the Middle Zambezi, *Kariba Studies*, Manchester University Press, Manchester, UK.

Kaczmarek, Z., 1993, Water balance model for climate impact analysis, *Acta Geophysica Polonica*, **XLI**(4):423–437.

Kaczmarek, Z., and Krasuski, D., 1991, Sensitivity of Water Balance to Climate Change and Variability, WP-91-047, International Institute for Applied Systems Analysis, Laxenburg, Austria.

Karnopp, D.C., 1963, Random search techniques for optimization problems, *Automatica*, **1**(1):111–122.

Klemeš, V., 1981, Applied stochastic theory of storage in evolution, *Advances in Hydrosciences*, **12**:79–141.

Leemans, R., 1990, personal communication.
Phatarfod, R.M., 1989, Riverflow and reservoir storage models, *Mathematical and Computer Modeling*, **12**(9):1057–1077.
Santa Clara, J.M.A., 1988, The Hydrological Operation of the Kariba Hydroelectric Scheme: Past, Present and Future, Paper presented at 16th Congress on Big Dams, San Francisco, CA, USA.
Spanger, W.M.L., and Jenne, R.L., 1988, World Monthly Surface Station Climatology and Associated Datasets, National Center for Atmospheric Research, Scientific Computing Division, Boulder, CO, USA.
UNEP, 1987, *Agreement on the Action Plan for the Environmentally Sound Management of the Common Zambezi River System – Final Act*, Harare, Zimbabwe.
USSR Committee for the International Hydrological Decade, 1978, *World Water Balance and Water Resources of the Earth*, UNESCO Press, Paris, France.
Williams, G.J., 1984, Problems of a shared resource: Hydropower on the Middle Zambezi, *Swansea Geographer*, **21**:17–26.
ZESA (Zimbabwe Electricity Supply Authority), 1986, *Annual Report and Accounts*, Harare, Zimbabwe.

Chapter 12

Climate Change and Water Resources Management Strategy: The Terek River Basin*

Vladimir M. Shnaydman and Jeffrey D. Niemann

12.1 Introduction

The investigation of water resources planning strategies must include scenarios of climate changes and development of water-related sectors. Climate scenarios can be used to estimate the changes in meteorological and hydrological indices, and in volumes and schedules of irrigation demands. Current strategies can then be evaluated under different hydrological and water demand scenarios. The reliability of decisions can be increased by using expert analysis and by considering the system's performance during average years, as well as during extraordinary events.

12.2 The Case Study Area

This study considers the Terek River water resources system (WRS) in North Caucasus, Russia. North Caucasus is situated on the vast plains between the Black Sea and the Caspian Sea to the north of the Caucasus Mountains. Fertile soils, level topography, and mild temperatures favor

*A detailed description of the technique used in this study was published in Shnaydman (1993).

Z. Kaczmarek et al. (eds.),
Water Resources Management in the Face of Climatic/Hydrologic Uncertainties 322–338.

large-scale agricultural development in this region. The agricultural sector's main constraint is the limited amount of natural precipitation. On the western side of this region, annual precipitation amounts to about 500 mm; on the eastern side, precipitation totals only about 200 mm per year. The climate conditions across the region vary from being humid to extremely dry. Droughts repeat regularly, hindering agricultural stability, and crop yields vary greatly from year to year due to rainfall variations. Despite this variability, however, the area ranks as Russia's most agriculturally productive region.

The Terek River receives its water from moderate rain, heavy showers, glacial waters, and ground water. The annual runoff of the Terek River basin above the Mozdok site varies from 8.2 km^3 in an average year to 7.2 km^3 in a year of 95% probability. The entire river basin is 623 km long, with an area of 35.4×10^3 km^2. The river's annual runoff is 350 m^3/s, with 10.1 km^3 of runoff in a year of 75% reliability.

The region's largest water user is the agricultural sector. In fact, irrigation accounts for approximately 60% of the 6.3 km^3 of water use. This water is generally used for the production of wheat, vegetables, and fodder. The region's fisheries require 3.6 km^3 of the water annually for fish ponds and for sturgeon spawning in the middle and lower river reaches.

From April to July, there are conflicts over water use for irrigation and for fisheries in the river basin. The total water resources deficit reaches more than 1.0 km^3 in dry years. However, reservoirs can be used to solve the Terek River water problems. Terek streamflow regulation makes it possible to double the area of land under irrigation. Because the main irrigated plots are situated in the zones with insufficient or unstable precipitation, irrigation demand varies greatly, both intra-annually and interannually. There is a limited amount of historical data for hydrological and meteorological indices.

There are several major reasons for choosing this area for a case study. First, the WRS is expected to be highly vulnerable to climate change, because the climate has large variations of rainfall and drought frequency. Crop yields are also rather sensitive to climate variation, as was shown by the results of joint studies implemented by the International Institute for Applied Systems Analysis (IIASA) and the United Nations Environment Programme (UNEP) (Nikonov *et al.*, 1987) in North Caucasus (Priazhinskaya, 1992). A map of the Terek River basin is shown in *Figure 12.1.* A linear diagram of the Terek River basin is shown in *Figure 12.2.* This diagram shows the seven sites and two reservoirs used in the model.

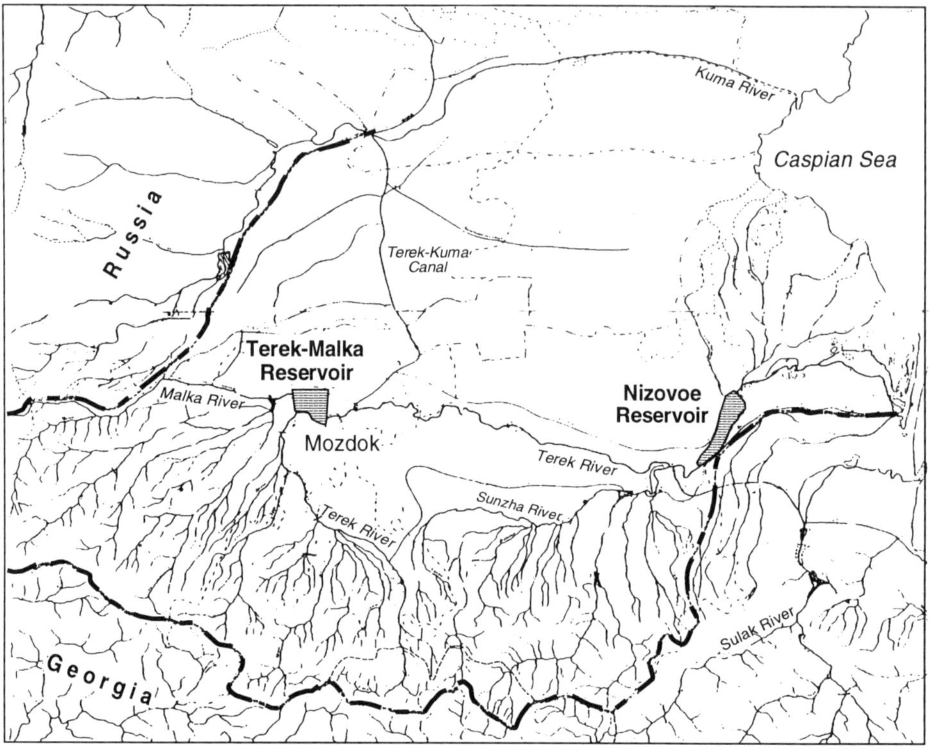

Figure 12.1. The Terek River basin.

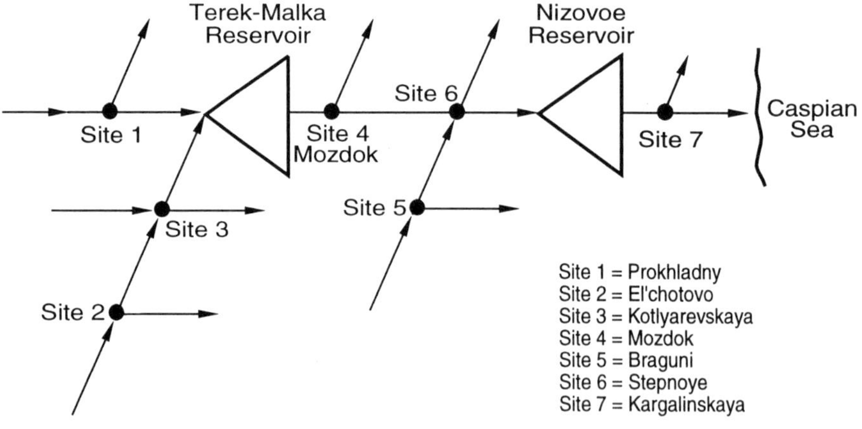

Figure 12.2. Linear diagram of the Terek River basin. Arrows indicate users of irrigated water.

12.3 Climate Change Scenarios

A number of problems prohibit the use of a single climate change scenario in modeling WRSs under climate change. First, the temporal and spatial resolutions of climate scenarios are too coarse. Second, the commonly used set of climatic parameters is insufficient for reliable determination of the climate-versus-runoff function and the climate-versus-water demand function. Third, estimates of the impact of climate change on precipitation, evaporation, and river runoff cannot be very accurate. Finally, the degree of aggregation of climate scenario data and the degree of detail required for WRS simulation cannot be matched. Therefore, the problem is one of decision making under uncertainty. To a certain extent, the problem's formulation and solution depend on geographic and climatic conditions, as well as on available data.

As an experiment, two scenarios of global warming were considered. The first scenario is a decreased-precipitation hypothesis corresponding to the year 2010. The other scenario is an increased-precipitation hypothesis corresponding to the year 2030. In the first case, the climatic variations will result in a decrease of runoff and increase of irrigation demand. In the second case, the climatic variations will result in an increase of runoff and decrease of irrigation demand.

12.4 Modeling Technique

Various techniques have been developed for estimating WRS sensitivity to climate change. The first technique (Klemeš, 1985), based on a deterministic model of river runoff, is most effective for local catchments with uniform conditions for runoff formation. The second technique (TVA, 1983) is based on the selection of fragments of hydrological sequences and the establishment of a connection between these sequences and the sequences of climatic indices. The second method is the most efficient for river basins with long historical sequences of data. The third method (Schwartz, 1977) investigates the sensitivity of a WRS (including a reservoir) assuming that climate change influences the statistical characteristics of the runoff. Therefore, simulation of the hydrological sequences is performed by using different hypotheses concerning the expectations and variances of the sequences. However, this last method does not relate runoffs to climate scenarios.

Unfortunately, an analysis of these various approaches demonstrates that none will work for the assessment of the Terek River WRS. For this

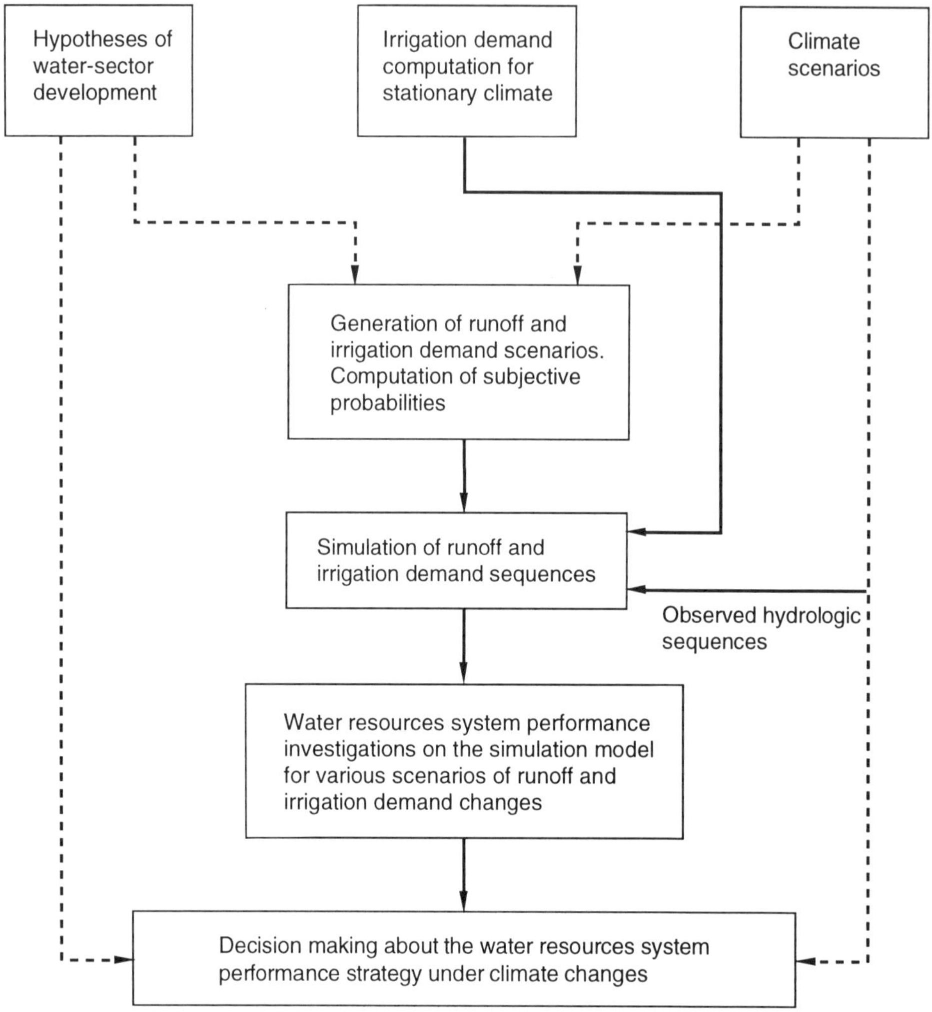

Figure 12.3. A diagram of the solution of an irrigation WRS performance strategy under climate change.

reason, another technique oriented toward irrigation, WRS simulation, and climate change was suggested (see *Figure 12.3*) (Shnaydman, 1989). This method includes a set of interrelated problems: generation of river runoff and irrigation demand scenarios and estimation of subjective probabilities; simulation of the irrigation demand schedule; joint modeling of river runoff and irrigation demand sequences; and basin-wide WRS simulation, including computation of performance indices and evaluation of the operation strategy under climate change.

12.4.1 Generation of river runoff and water demands

Hydrological and climatological investigations allow experts to suggest a set of river runoff and irrigation demand sequences for a simulated river basin. Such scenarios are based both on model investigations for various catchments and on geographical analysis of similar regions. The scenarios differ in their temporal levels, power output hypotheses, CO_2 concentrations, etc. Every climate scenario also contains a set of possible sequences of runoffs and water demands.

For each scenario, one can define a set of transformation functions that relate the sequences to the climate scenario. Because of the considerable uncertainty in the climate scenarios and the transformation functions, the fuzzy-sets theory is applicable (Zadeh, 1965). This method requires subjective probabilities reflecting expert confidence that a sequence can match a scenario.

Two approaches can be used to estimate the subjective probabilities. The first approach is based on expert opinion (Larichev, 1979; Fiering and Matalas, 1990). The second approach is based on computation of the probabilities. For the latter method, the range of annual or seasonal runoffs is divided into a certain number of intervals. The historical or simulated runoffs are then distributed into these intervals. The frequency of occurrence within an interval can be viewed as the frequency of occurrence (i.e., probability) of a scenario. This method also applies to irrigation demands.

12.4.2 Simulation of the irrigation demand schedule

Unfortunately, the irrigation demand scenarios cannot be based on data because such data are not available. This deficiency is mainly because of variations of crop planting due to the economic and agrotechnical aspects of land cultivation, worsening soil fertility, and variations of water use technologies. Therefore, irrigation demand scenarios must be developed from individual crop water demands and computation of irrigation rates and schedules.

To determine the irrigation demand for river basin sites, the irrigated areas are divided into subareas with uniform meteorological, economic, and social conditions. Every irrigated plot's crop production indices (crop types, crop yields, irrigated areas, etc.) and climatic indices (temperature, precipitation, radiation balance, etc.) are determined. If the irrigation rates and schedules have been computed for every crop, the rates and schedules can be determined for the aggregated WRS sites.

A technique suitable for the determination of a crop's irrigation water demand has two requirements: first, a wide range of climatic parameters must be taken into account; second, the initial information must be available. The Budagovsky method (Budagovsky and Dzhogan, 1980) meets these requirements. The technique is based on solving the heat exchange and moisture exchange equations for both the soil and the plants. The model consists of several units that compute radiation balance, total evaporation, current and critical moisture content, irrigation rates and schedules, and irrigation demand for every site.

Irrigation rates and schedules are computed from a set of equations determining the dynamics of moisture content. Watering is assigned in an interval of the vegetation period for an irrigated crop if the moisture content reaches the critical value. The watering will be continued until the moisture content reaches the value of the minimum soil moisture capacity.

The transformation of climate scenarios into temporal sequences of irrigation demand also has some peculiarities. The temporal series must be presented for every plot and defined by its statistical characteristics. Because of the inadequacy of the climatic indices in the global warming scenarios, the following three-step technique is necessary. First, the model's sensitivity to various climatic indices, in particular air temperature, precipitation, radiation balance, and air moisture deficit, must be investigated. Second, the correlation between temperature and precipitation, on the one hand, and other climatic indices, on the other hand, must be estimated. These correlations are presented in the irrigation demand model, but are not included in climate scenarios. Third, climatic indices including their dependencies must be estimated and subsequently inserted into the equations of the Budagovsky model.

12.4.3 Joint modeling of river runoff and irrigation demand sequences

The simulated river runoff and irrigation demand sequences should match the given statistical characteristics of the data. Two stages are required to accomplish this task; the first stage simulates annual sequences, and the second stage simulates the intra-year distribution using the fragment method (Svanidze, 1980).

The initial data for the simulation algorithm include the average annual values of the river runoff and the irrigation demand, the deviations of these processes, the auto-correlation coefficients, and the cross-correlation coefficient matrix. The multidimensional stochastic process was simulated for the Terek River basin. For this simulation, the vector components include four river runoffs and six irrigation demands. The sequences were observed for 20 years. The intercorrelation coefficients for the annual inflows ranged between 0.5 and 0.8; the coefficients for the annual irrigation demands ranged between 0.5 and 0.9; the coefficients for the inflows and irrigation demands were negative, and ranged between −0.1 and −0.5. The auto-correlation coefficients for the annual runoffs at different sites ranged between 0.07 and 0.32, and the same indices for the irrigation demands ranged between −0.3 and +0.3. The value of C_s/C_v (coefficient of skewness/coefficient of variation) ranged between −2.5 and 11.5. The average accuracy of simulation for the multidimensional vector with 150-year components is equal to 0.2% for the expectation, 2.8% for the deviation, and 0.05% for the coefficients of cross-correlation.

12.4.4 Basin-wide WRS simulation

Elements of the WRS Simulation Model

The purpose of a simulation model is to investigate WRS performance and elaborate its operation rules in an algorithmic nature. Such a model incorporates equations balancing the input at sites and reservoirs, and involves algorithms for both reservoir control and water distribution at the sites.

The WRS simulation model is based on the aggregative approach developed by Buslenko (1977) and Emelyanov *et al.* (1988). This method divides a complex system into standard elements. Formal description of every element is based on the theory of piecewise linear Marcovian processes. This makes it possible to simulate complex structured systems, allowing for a broad range of operation algorithms. A river-basin simulation model based on the aggregative approach (Shnaydman, 1992) contains basic reservoir, site, generator, and control-device elements.

The reservoir element simulates functioning of a reservoir in the WRS. For every reservoir, the following is known: the live reservoir capacity, the reservoir losses, the supply regime, the coordinates of the rule curves, and the reliability of the output.

The site element simulates water distribution among water users within the system. For every site, the following is known: each user's schedule of water demand, including stochastic demand for irrigation; the users' reliabilities and priorities; and the rules of water distribution between users.

The control-device element simulates the algorithm of joint operation of reservoirs. The control algorithm is based on a two-step principle of WRS simulation. For any event (reaching the beginning or end of every time interval, filling to maximum storage or decreasing to "dead volume," or crossing a rule curve), the state of every reservoir and site has been computed at the first step. The state of a reservoir includes its storage, inflow, losses, and abundance of water or yield deficit. The state of a site includes the water demand deficit, if any. Output signals from every element are formed and arrive at the control device. The control algorithm recalculates releases from the reservoirs to ensure a minimum deficit of water resources for users in the WRS for that instant. New releases are adjusted for every reservoir and the model is run again (the model time does not advance). The control algorithm of joint operation of reservoirs was developed for deterministic water demands (Shnaydman, 1992) and for stochastic water demands (Shnaydman, 1993).

The generator produces a sequence of stochastic values of flow and demand discharges for the WRS inputs. The output element is the terminal element, and it receives the output flow.

Reservoir Control Method

Every reservoir in a WRS is controlled by means of rule curves. The rule curve method provides an algorithmic relation between the amount of water stored and the reservoir yield. A single reservoir has a target draft zone, a reduced draft zone, and an excess zone. The method has already been summarized for a single reservoir in the literature (Kritsky and Menkel, 1952).

The rule curve method has several peculiarities for the coordinated control of reservoirs in a WRS. The draft from each reservoir depends not only on the amount of stored water, but also on inflows, stored water, yields, and losses from other reservoirs, and water deficits at sites. Within this approach, the target zone of every reservoir is divided into a compensation zone, where additional water can be released downstream, and an indifferent zone, where only the target draft can be provided. The dividing line between the compensation zone and the indifferent zone can be obtained from simulation experiments (Shnaydman, 1992).

Each reservoir can be one of two types. The first type is an upstream donor reservoir. A reservoir qualifies as this type if it has abundant water and its storage level is in the compensation zone. This type of reservoir can release additional water downstream. The second type of reservoir is in an indifferent state or in a deficit state. This implies that its storage level is below the compensation line, and that it cannot donate additional water to downstream users.

A donation can be given to the downstream deficit sites from the nearest upstream higher priority reservoir. If the upstream reservoir is also in an indifferent state, water is donated from the nearest upstream, higher priority reservoir where water is abundant. If the donor reservoir's yield is not enough to compensate for the downstream deficit, the next higher priority reservoir with abundant water is selected.

12.4.5 Numerical evaluation of strategies

A numerical technique of gauging the effectiveness of WRS strategies must be developed. The estimation of the effectiveness is based on the difference between the best value $\Delta^{js}(+)$ of the WRS functioning index and the worst value $\Delta^{js}(-)$ (Borisov *et al.*, 1989). These can be calculated as follows:

$$\Delta^{js}(+) = \sqrt{\sum_{k} p_k^s [R_k^j - R_0^j(+)]^2} \ , \tag{12.1}$$

and

$$\Delta^{js}(-) = \sqrt{\sum_{k} p_k^s [R_k^j - R_0^j(-)]^2} \ , \tag{12.2}$$

where $R^j(+)$, $R^j(-)$, and R^j are values of the jth functioning index corresponding to best, worst, and estimated situations, respectively. The p represents the subjective probability. A similar approach has been developed (Kaczmarek, 1990) where the index $1/\Delta^{js}(+)$ is called the "robustness of the system." Values of R^j are assumed to be either an essential, most sensitive, or integral quality index (Kindler and Tyszewsky, 1989).

A change in the WRS operation strategy is equivalent to a change in parameters or operation rules. Such a decision can be made if the deviation from the normal value [computed with equation (12.1)] is more than the deviation from the worst value [computed with equation (12.2)]. More specifically, the decision is made if

$$\Delta^{js}(+) > \Delta^{js}(-) \ . \tag{12.3}$$

Table 12.1. Variation in water demand due to changes in climatic indices (Mozdok site).

Percentage of variation in water demand	Precipitation (±10%)	Soil moisture deficit (±10%)	Radiation balance (±10%)	Air temperature (±10%)
Average year	±5	±2	±13	±2
Moderately dry year	±3	±2	±15	±2

Table 12.2. Statistical parameters of irrigation demand sequences for various climate scenarios (Mozdok site).

Climate scenarios		Mathematical expectation		Standard deviation		Coefficient of	Auto-correlation
ΔT (°C)	Prec. (%)	mln. m^3	%	mln. m^3	%	variation	coefficient
0	0	883	100	196	100	0.19	0.02
+3	−20	1,030	116	200	103	0.22	0.02
+3	+20	875	99	226	115	0.26	0.00

12.5 Results and Discussion

12.5.1 Sensitivity of water demands to climate indices

For the Mozdok site, the sensitivity of the Budagovsky model to the following parameters was tested: precipitation, soil moisture deficit, radiation balance, and air temperature. The results of computations for the average year and the moderately dry year are presented in *Table 12.1.*

Table 12.1 shows that the most sensitive parameter of the model is radiation balance. The cross-correlation coefficients between the climatic indices are negligible, with the exception of the connection between air temperature (t) and radiation balance (R). The coefficient of correlation is 0.98, and the regression equation can be presented as R=4.6+0.3t.

Statistical characteristics of irrigation demand for a 230 × 10^3 hectare plot in the Mozdok site for different climate scenarios are shown in *Table 12.2.*

The probability curves for various climate scenarios are shown in *Figure 12.4*. Scenario 1 represents the current climate; Scenario 2 represents the decreased-precipitation, or unfavorable, scenario; and Scenario 3 represents the increased-precipitation, or favorable, scenario. *Figure 12.4* reveals that irrigation demand for an average year in the favorable scenario decreased by 1%. In the unfavorable scenario, irrigation demand increased by 16%.

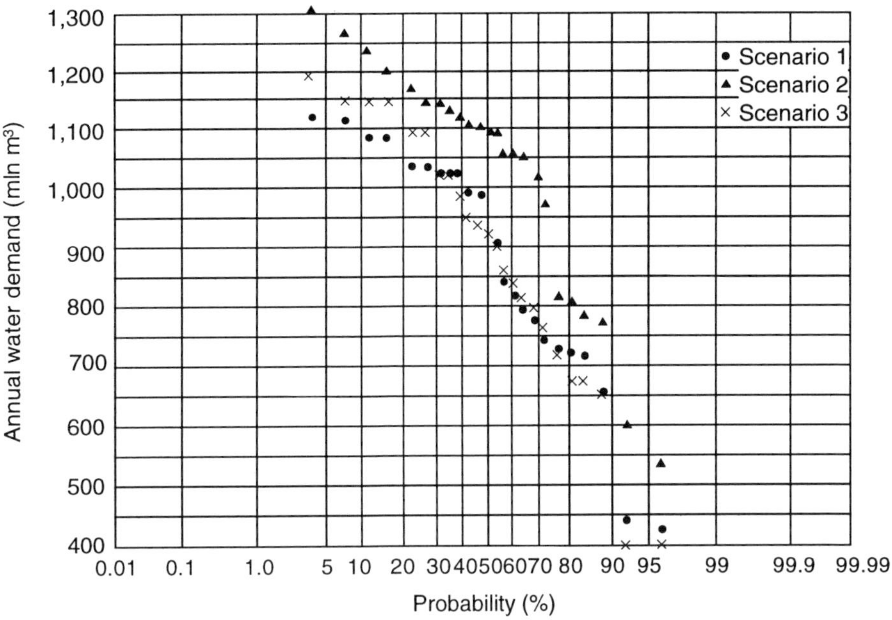

Figure 12.4. The irrigation demand reliability curves for various climate scenarios for the Mozdok site.

For a dry year (25% reliability), the irrigation demand increased by 5% and 13% for the favorable and unfavorable scenarios, respectively.

The increase of precipitation in the favorable scenario leads to a 15% drop in mean irrigation demand and to a 15% increase of standard deviation. Consequently, it is not possible to reduce the capacity of reservoirs in the favorable scenario because the capacity is required to maintain the current reliability of water supply.

12.5.2 Investigation of the sensitivity to climate variations of the Terek WRS

Two groups of experiments were run in which the river runoffs and irrigation demands were adjusted. In the first group (Variant A), the mean runoff was changed by –20% to +20% and the irrigation demands were increased by 0% to +20%. In the second group (Variant B), the mean runoff and irrigation demand were changed by the same percentages, and the runoff and irrigation demand deviations were increased by 25%.

Table 12.3. Variations of the reliability at the Mozdok site.

Irrigation demand	Change in runoff (%)				
change (%)	−20	−10	0	+10	+20
0	0.50/0.43	0.74/0.69	0.77/0.76	0.75/0.69	0.79/0.74
+10	0.41/0.40	0.72/0.68	0.79/0.69	0.74/0.74	0.75/0.66
+20	0.30/0.36	0.67/0.55	0.67/0.73	0.68/0.72	0.64/0.71

Table 12.4. Variations of the resiliency at the Mozdok site.

Irrigation demand	Change in runoff (%)				
change (%)	−20	−10	0	+10	+20
0	0.48/0.40	0.69/0.68	0.77/0.78	0.68/0.66	0.75/0.69
+10	0.35/0.31	0.70/0.65	0.84/0.72	0.68/0.65	0.86/0.67
+20	0.26/0.27	0.64/0.54	0.68/0.78	0.67/0.72	0.57/0.75

Table 12.5. Variations of normal annual volume of water resources deficit reduced to the basic variant in the Terek River basin.

Irrigation demand	Change in runoff (%)				
change (%)	−20	−10	0	+10	+20
0	4.98/6.81	1.77/2.31	1.00/1.52	0.85/1.32	0.79/1.21
+10	6.30/8.47	2.12/3.74	1.17/2.08	1.13/1.72	0.76/1.49
+20	7.69/10.47	2.91/4.21	1.64/2.86	1.49/2.12	1.18/1.81

The reliability, resilience, and normal annual volume of the water supply deficit have been computed for the Terek River sites in the WRS simulation model. Results for the Mozdok site and for the entire river basin are shown in *Tables 12.3* to *12.5*. Each table element shows the results of Variants A and B as the numerator and the denominator, respectively. An analysis of *Tables 12.3* to *12.5* shows that the reliability and resilience at Mozdok varied insignificantly when the runoffs and the irrigation demands varied from −10% to +20% with respect to the basic variant. However, a serious deterioration of reliability and resilience appeared when the runoff was reduced by 20%.

Changes in the annual volume of the water resources deficit are shown in *Figure 12.5* for different runoff and water demand values. When the runoff varies within −10% to +20%, the annual volume of the water resources deficit increases. It increases more than threefold if the sequence deviation equals the observed deviation, and more than fourfold if the sequence deviation is

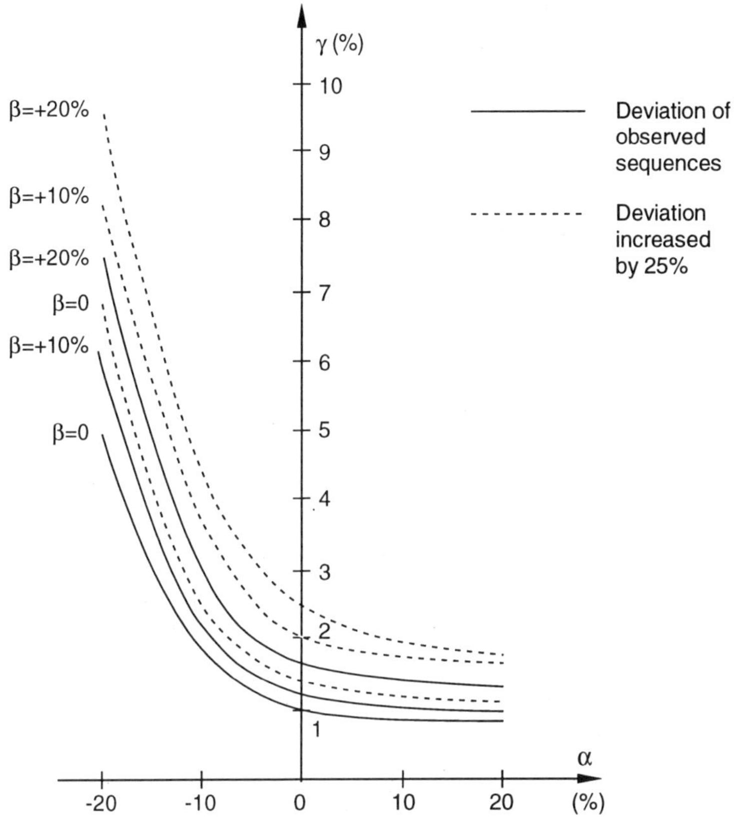

Figure 12.5. Dependence between the annual volume of water resources deficit (γ) (normalized to the basic variant without climate changes) and variations of runoff (α, %) and irrigation demand (β, %) in the Terek River basin.

increased by 25%. A tenfold increase in the normal annual volume of water resources deficit is observed for a 20% decrease of the mean runoff.

12.5.3 Strategy evaluation under climate change

For this analysis, the increased-precipitation and the decreased-precipitation climate scenarios were again employed. Subjective probabilities of various changes in river runoff and irrigation demand are shown in *Table 12.6*. The nonzero elements in the upper section of the table correspond to the decreased-precipitation scenario. The nonzero elements in the lower section of the table correspond to the increased-precipitation scenario.

Table 12.6. Subjective probability matrix for various conventional time levels.

Irrigation demand change (%)	Change in runoff (%) −20	−10	0	+10	+20
Decreased-precipitation scenario					
0	0/0	0.1/0.0	0.1/0.0	0/0	0/0
+10	0.2/0.0	0.2/0.0	0/0	0/0	0/0
+20	0.1/0.0	0.3/0.0	0/0	0/0	0/0
Increased-precipitation scenario					
0	0/0	0/0	0.0/0.2	0.2/0.4	0.0/0.1
+10	0/0	0/0	0/0	0/0	0.1/0.0
+20	0/0	0/0	0/0	0/0	0/0

The WRS functioning indices [$\Delta^{js}(+)$ and $\Delta^{js}(-)$] were computed for the normal annual deficit of water resources. Climate aridization tends to worsen WRS performance [$\Delta^{js}(+) = 7.4$ and $\Delta^{js}(-) = 3.2$]. The variations are not strong, because the subjective probabilities of river runoff and irrigation demand changes are small. In this case, equation (12.3) is not true and, therefore, the WRS performance strategy should not be changed. This conclusion is confirmed by computations for the next time-level scenario, where the WRS performance indices $\Delta^{js}(+)$ and $\Delta^{js}(-)$ equal 0.3 and 9.3, respectively. The upper limit of both of these indices equals 9.5. Therefore, increased precipitation tends to improve WRS performance compared with the previous case. These results show that the worsening of WRS performance is unimportant, because WRS performance will stabilize at the second time level (2030); therefore, no additional investments are necessary for the construction of reservoirs, the reduction of irrigated areas, etc.

12.6 Conclusions

Investigation of water resources strategies under climate change represents decision making under uncertainty. The solution is based on experiments with a WRS simulation model. Computations were made using a set of hypotheses about the river runoff and irrigation demand variations. The hypotheses were assigned by statistical characteristics corresponding to various climate scenarios. The trustworthiness of results under river runoff and irrigation demand variations is greatly affected by the subjective probabilities. The WRS strategy is evaluated with the help of two performance indices showing the deviations from the best and the worst computed values.

The WRS simulation model includes the algorithms for the operation of every reservoir based on the rule curve method; the distribution of water resources between different users at a site; joint operation of reservoirs in cascade under the stochastic water demand; irrigation demand computation; joint simulation of runoff and irrigation demand sequences; and decision making about the WRS performance strategy under climate changes.

The performance of the Terek River WRS was investigated for two climate scenarios, corresponding to decreased precipitation and increased precipitation. Computations showed that the worsening of the WRS performance indices for the first time level was insignificant and did not result in strategy variation; under the subsequent increased-precipitation scenario, the WRS performance stabilizes.

References

Borisov, A.N., Alekseev, A.V., Merkur'eva, G.V. *et al.*, 1989, *Treatment of Fuzzy Information in Decision-Making Systems*, Radio and Svijaz, [in Russian].

Budagovsky, A.I., and Dzhogan, L.Ja., 1980, Methods of iterative improvement of techniques to calculate irrigation rates, *Vodnye Resursy*, **6**(1):29–51 [in Russian].

Buslenko, N.P., 1977, *Modeling of Complex Systems*, Nauka, Moscow, Russia [in Russian].

Emelyanov, S.V. *et al.*, 1988, *Technology of System Modeling and Simulation*, Mashinostroenie, Verlag-Technik, Moscow-Berlin [in Russian and German].

Fiering, M.B., and Matalas, N.C., 1990, Decision making under uncertainty, in P.E. Waggoner, ed., *Climate Change and U.S. Water Resources*, John Wiley & Sons, New York, NY, USA.

Kaczmarek, Z., 1990, Impact of Climatic Variations on Storage Reservoir Systems, WP-90-020, International Institute for Applied Systems Analysis, Laxenburg, Austria.

Kindler, J., and Tyszewsky, S., 1989, Multicriteria evaluation of decision rules in the design of a storage reservoir, *Proceedings of the Baltimore Symposium on Closing the Gap Between the Theory and Practice*, International Association of Hydrological Sciences (IAHS) Publicaton No. 180, IAHS Press, Wallingford, UK.

Klemeš, V., 1985, Effects of hypothetical climate change on reliability of water supply from reservoirs, in *Sensitivity of Water Resources Systems*, World Meteorological Organization, WCP-98, Geneva, Switzerland.

Kritsky, S.N., and Menkel, M.F., 1952, *Water Resource Calculations*, Gidrometeoizdat, Leningrad, Russia [in Russian].

Larichev, O.I., 1979, *Science and Art in Decision-Making*, Nauka, Moscow, Russia [in Russian].

Nikonov, A., *et al.*, 1987, *The Effect of Climatic Variations on Agriculture in Semi-Arid Zone of the European USSR. A. The Stavropol Territory*, International Institute for Applied Systems Analysis and United Nations Environmental Program, Laxenburg, Austria.

Priazhinskaya, V.G., ed., 1992, *Modeling of Water Resource Systems (Economical and Ecological Aspects)*, Nauka, Moscow, Russia [in Russian].

Schwartz, H., 1977, Climatic Change and Water Supply: How Sensitive is the Northeast? - US NASA. Climate, Climatic Change and Water Supply, Washington, DC, USA.

Shnaydman, V.M., 1989, Study of water resource system operation taking account of climate change, *International Conference "Climate and Water,"* Vol. 2, Finland:383–393.

Shnaydman, V.M., 1992, The application of the aggregative approach in simulation modeling of water resources systems, *Water Resources Management*, **6**:135–148.

Shnaydman, V.M., 1993, The influence of climate variations on an irrigation water resources system performance strategy, *Water Resources Management*, **7**:39–56.

Svanidze, G.G., 1980, *Mathematical Modeling of Hydrological Series*, Water Resources Publications, Fort Collins, CO, USA.

TVA (Tennessee Valley Authority), 1983, Sensitivity of the Tennessee Valley Authority Reservoir System to Global Climate Change, Rep. 28-1-680, 101, Norris, TN, USA.

Zadeh, L.A., 1965, Fuzzy sets, *Information and Control*, **8**(N3):338–353.

Chapter 13

Climate Change and Streamflow and the Sizing of Flow Control Reservoirs: Two Catchments in Belgium

Daniel Gellens

13.1 Introduction

The purpose of this chapter is to follow up impact studies of CO_2-induced climate change that were carried out on river basins in Belgium (Bultot *et al.*, 1988a, 1988b; Gellens, 1991) and in Switzerland (Bultot *et al.*, 1992). A secondary purpose is to continue to investigate the impacts of changes in relative humidity under the changed CO_2 conditions (Bultot and Gellens, 1989).

A sensitivity study was carried out on two catchments in Belgium with sharply different geomorphological characteristics. The possible modifications of the streamflow regime and the water cycle were estimated by running the Institut Royal Météorologique de Belgique (IRMB) hydrological model in current conditions and in increased-CO_2 conditions.

Finally, possible impacts of climate change on the design of a flow control reservoir were investigated. A simplified flood control reservoir and a low-flow control reservoir were considered, and the sensitivity of their characteristics to increased atmospheric CO_2 was studied.

Z. Kaczmarek et al. (eds.),
Water Resources Management in the Face of Climatic/Hydrologic Uncertainties 339–355.

13.2 The IRMB Hydrological Model and the Drainage Basins

The IRMB model is a conceptual daily-step hydrological model (Bultot and Dupriez, 1976a, 1976b, 1985) developed to simulate the water balance of midsize catchments, that is, catchments for which the daily step is sufficient to correctly simulate the flood peaks. In practice this corresponds to a range of areas between 200 km^2 and 2,000 km^2. The model allows estimation of the flow at the outlet and the flow's components (base flow, interflow, surface flow); estimation of the various transfers between the different water phases (water interception by the different types of vegetation, throughfall, snowmelt, evaporation, runoff, infiltration, etc.); and estimation of the water content of the catchment (e.g., content of the aeration zone of the soil and of the saturated zone). Catchment particulars are taken into account by considering the areas covered by the various types of vegetation, their albedo, leaf area index, and water capacity of the aeration zone of the soil.

In this study, two catchments in Belgium have been investigated; their geographic positions are shown in *Figure 13.1.* They differ in their main geomorphological and phytogeographical features.

- The Dyle River basin, with its outlet at Saint Joris-Weert, has a flat drainage area of 645 km^2 with a thick aquifer in sandy loamy soil. The vegetative cover is mainly crops and meadows. The Dyle River is a tributary of the Rupel River, itself tributary of the Scheldt River.
- The Semois River basin, with its outlet at Membre, has a rock-floored basin of 1,235 km^2 with gently marked relief and small perched aquifers. It is mainly covered by meadows and forests and is a tributary of the Meuse River.

13.3 Current Climate Conditions

The Belgian climate is temperate and humid with a mean annual rainfall around 1,000 mm. Precipitation is almost uniformly distributed throughout the year (*Table 13.1*; reference period 1901 to 1986 = Scenario 0). The mean annual temperature is approximately 8°C; the mean winter temperature is just above 0°C and the mean summer temperature is about 15°C. *Table 13.1* shows that the climate conditions around the Dyle catchment are slightly drier and warmer than are those at the Semois catchment; this difference is mainly due to the difference in their altitudes. The catchments have

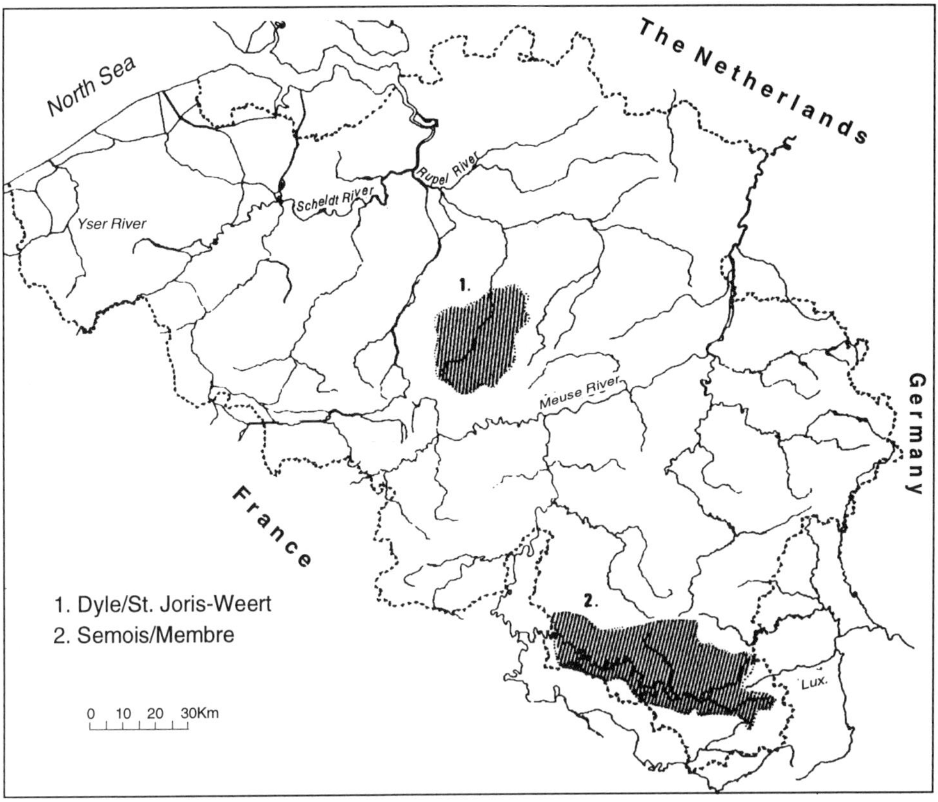

Figure 13.1. Location of the catchments studied.

similar levels of potential evapotranspiration (PET, estimated by a modified Penman method; Bultot *et al.*, 1983), about 550 mm per year (see *Table 13.3*). Due to the slightly drier climate conditions of the Dyle catchment, the effective evapotranspiration reaches only 85% of the PET, compared with 92% in the case of the Semois catchment.

13.4 Climate Change and Impact Study

The IRMB model has been used to simulate the potential impact of a CO_2-related climate change on the water cycle of catchments in Belgium (Bultot *et al.*, 1988a, 1988b; Gellens, 1991).

In this study a basic $2\times CO_2$ scenario, called Scenario 1, has been created using the zonal means of the GCM output. In *Table 13.2*, the rise in temperature and the rainfall modifications under $2\times CO_2$ conditions are

Table 13.1. Monthly regimes of the air temperature and precipitation under current conditions (Scenario 0).

	Jan.	Feb.	Mar.	Apr.	May	Jun.	Jul.	Aug.	Sep.	Oct.	Nov.	Dec.	Year
Air temperature (°C)													
Dyle catchment	2.1	2.5	5.3	8.2	12.4	14.9	16.7	16.4	14.0	10.1	5.6	3.0	9.3
Semois catchment	0.0	0.8	3.7	6.8	10.9	13.8	15.7	15.0	12.1	7.9	3.5	0.9	7.6
Precipitation (mm)													
Dyle catchment	69.8	55.3	60.1	57.2	65.9	67.2	73.0	76.5	64.2	71.9	76.6	77.1	814.8
Semois catchment	119.5	93.0	86.9	81.1	79.1	85.1	90.1	88.7	87.8	98.6	119.2	127.8	1,156.9

Table 13.2. Specific air temperature and precipitation changes under $2\times CO_2$ conditions (Scenario 1).

	Jan.	Feb.	Mar.	Apr.	May	Jun.	Jul.	Aug.	Sep.	Oct.	Nov.	Dec.	Year
Air temperature increments (K)[a]	3.1	3.4	3.4	3.1	2.8	2.7	2.5	2.3	2.3	2.7	2.8	3.2	2.9
Precipitation increments (mm)[b]	9.3	10.5	9.9	10.2	−1.2	−2.7	−1.6	−2.2	0.0	5.3	8.1	8.7	54.3

[a]Source: Manabe and Stouffer (1980).
[b]Sources: Manabe *et al.* (1981), and Washington and Meehl (1983).

compared with present conditions (Scenario 0). Changes in PET have been assessed by considering the potential changes in the different components of the energy budget. In particular, the parameters of the Brunt-Monteith formula for the calculation of infrared radiation balance have been modified (Bultot *et al.*, 1988a). The annual regime of PET under 2×CO_2 conditions is given in *Table 13.3.* This basic scenario, created a few years ago, is in agreement with current estimations of the impact of a CO_2 doubling on the climate (IPCC, 1990).

Scenario 1 was used to systematically modify the various climatological data needed to run the IRMB model (Bultot *et al.*, 1988b). Running the model under the 2×CO_2 conditions permitted a comparison of the current behavior of catchments in Belgium with their behavior under the 2×CO_2 climate conditions of Scenario 1 (Bultot *et al.*, 1988b).

The main finding of this study was that the sensitivity of the catchments to climate change depends greatly on local conditions, particularly on the geomorphological characteristics of the basins. Focusing on results related to modifications of the streamflow and its components, and to alterations of the flood and low-flow regimes, certain changes can be seen.

- In catchments with prevailing surface flow, like the Semois basin, the evolution would involve the following changes:
  - An increase in flood frequencies during the winter season combined with an increase of the extreme river stages, leading to greater flood risks.
  - A decrease in streamflow during the summer season and a consequent increased risk of water pollution.
  - Possible limits on water availability from local ground water storages in the summer and autumn.
- In catchments with high infiltration rates, like the Dyle basin, the impact would include the following changes:
  - An increase in ground water storage, effecting an increase in the base flow throughout the year, which in turn would increase flood risks.
  - A reduction of the number of low-stage occurrences in summer, resulting in reduced river pollution.
  - A possible increase in water availability from the aquifers.

Similar results have been highlighted in a larger set of catchments in the UK (Arnell *et al.*, 1990), using monthly streamflow simulations.

In this context, it was also necessary to introduce some additional hypotheses concerning variables needed for running the IRMB model and about which no clear information is yet available (Bultot *et al.*, 1988b). The main uncertainty is related to the evolution of the relative humidity in the $2\times CO_2$ conditions (Bultot and Gellens, 1989). Relative humidity together with the temperature determine the values of water vapor pressure and saturation deficit, which are used to estimate the PET (Bultot *et al.*, 1983) under the changed CO_2 conditions. Generally, relative humidity is assumed to remain unmodified by the CO_2 increase. This implies that as the temperature rises there is a simultaneous increase of the saturation deficit, and hence an increase in PET.

13.5 Role of the Relative Humidity Assumption

A previous study (Bultot and Gellens, 1989) has shown that the assumption that relative humidity will not change under $2\times CO_2$ conditions has significant consequences for the potential and effective evapotranspiration values. To assess the impact of this assumption, it has been assumed that the relative humidity in $2\times CO_2$ conditions has changed at fixed ratios of its present value.

In addition to the case of unmodified relative humidity (Scenario 1), three cases have been examined in which the relative humidity is equal to 95%, 105%, and 110% of its present value. The corresponding scenarios are called Scenario .95, Scenario 1.05, and Scenario 1.10, respectively. This assumption implies clear modifications of the estimated potential and effective evapotranspiration. The annual regimes of these two variables are given in *Table 13.3* for the two catchments studied in this chapter.

The impacts of a change in relative humidity are similar in both the basins and can be summarized as follows:

- Any increase (or decrease) in the relative humidity or evapotranspiration implies a regular decrease (or increase) in the PET.
- The effective evapotranspiration behaves the same way, except in the summer, from June through August, when it remains practically equal to its value under Scenario 1.
- During the autumn and the winter months, the potential and effective evapotranspiration values estimated under increased relative humidity are very close to, or slightly lower than, the present values.

Table 13.3. Monthly regimes of the potential and effective evapotranspiration under current conditions (Scenario 0), under 2×CO_2 conditions (Scenario 1), and under 2×CO_2 conditions of changed relative humidity values.

	Jan.	Feb.	Mar.	Apr.	May	Jun.	Jul.	Aug.	Sep.	Oct.	Nov.	Dec.	Year
Potential evapotranspiration (mm)													
Dyle catchment													
Scenario 0	8.5	14.0	32.3	59.7	89.5	91.9	88.7	74.5	56.9	29.6	10.7	8.3	564.6
Scenario .95	13.2	20.4	41.1	69.0	98.7	100.7	96.2	80.8	63.2	35.5	15.8	13.1	647.7
Scenario 1	10.3	17.9	39.0	66.8	96.3	100.1	95.5	80.1	59.7	32.4	13.2	9.9	621.2
Scenario 1.05	7.8	15.5	36.6	64.6	93.6	99.4	94.6	79.0	56.1	29.2	10.6	7.3	594.3
Scenario 1.10	5.6	13.3	34.4	62.3	90.9	98.7	93.7	78.1	52.8	26.3	8.6	5.1	569.8
Semois catchment													
Scenario 0	4.4	9.8	26.2	56.6	84.5	91.3	96.6	79.9	55.3	26.3	7.6	4.8	543.3
Scenario .95	7.2	15.5	33.2	64.0	91.2	98.1	102.4	84.5	60.0	30.9	11.4	7.6	606.0
Scenario 1	5.5	13.7	31.9	62.7	89.6	98.3	102.4	84.4	57.3	28.6	9.6	5.6	589.6
Scenario 1.05	4.0	12.1	30.5	61.3	87.8	98.2	102.1	84.0	54.6	26.3	7.9	3.9	572.7
Scenario 1.10	2.8	10.6	29.1	59.9	86.0	98.1	101.9	83.7	52.1	24.3	6.5	2.6	557.6
Effective evapotranspiration (mm)													
Dyle catchment													
Scenario 0	8.6	14.1	31.1	52.6	75.8	75.6	71.4	60.3	44.4	26.0	10.6	8.3	478.8
Scenario .95	13.2	20.2	38.5	59.4	80.5	78.5	73.1	61.7	46.6	30.1	15.3	12.9	530.0
Scenario 1	10.3	17.6	36.6	57.9	79.1	78.6	73.0	61.5	44.4	27.7	12.8	9.9	509.4
Scenario 1.05	7.8	15.4	34.7	56.2	77.6	78.6	72.7	61.1	42.2	25.3	10.4	7.3	489.3
Scenario 1.10	5.6	13.2	32.7	54.6	76.0	78.7	72.5	60.7	40.1	23.0	8.4	5.1	470.6
Semois catchment													
Scenario 0	4.4	9.7	26.0	54.7	80.0	83.7	85.0	70.4	49.2	24.9	7.6	4.8	500.4
Scenario .95	7.2	15.4	32.7	61.5	85.4	88.0	87.4	72.4	52.3	28.9	11.3	7.6	550.1
Scenario 1	5.5	13.6	31.5	60.4	84.1	88.3	87.5	72.4	50.1	26.8	9.5	5.6	535.3
Scenario 1.05	4.0	12.1	30.1	59.2	82.6	88.5	87.5	72.2	47.8	24.8	7.8	3.9	520.5
Scenario 1.10	2.8	10.5	28.8	57.9	81.1	88.7	87.5	72.1	45.8	22.9	6.5	2.6	507.2

Scenario .95 = 95% of present value; Scenario 1.05 = 105% of present value; Scenario 1.10 = 110% of present value.

To determine the related impacts on the streamflow, the IRMB model has been run completely in each scenario. The streamflow annual regimes have been calculated and are given in *Table 13.4*.

The first salient feature of the regimes is the very regular evolution of the streamflow as a function of the adopted rate of humidity change in the scenarios. For all months, the increments between Scenario .95 and Scenario 1 are nearly the same as those between Scenario 1 and Scenario 1.05. It must be pointed out that streamflow increments during the summer are not close to zero, even if the effective evapotranspiration remains fairly constant in the different scenarios.

In all the simulations the streamflow undergoes a clear increase during the winter, which confirms results of previous studies. This streamflow increase could jeopardize the safety conditions along the river courses, as mentioned in Section 13.4.

In all the scenarios, even with increased relative humidity under $2\times CO_2$ conditions, the streamflows of the Semois basin are lower than the present values from at least June through September. In Scenario .95, this period reaches seven months, i.e., from May to October. On the other hand, the streamflow of the Dyle catchment shows only a slight decrease in Scenario .95 (compared with Scenario 0, the maximum decrease reaches .4 mm/month in September). This indicates that, in terms of impact on the streamflow regime, knowledge of the exact rate of variation of the relative humidity is not necessary to foresee a streamflow reduction in the Semois catchment. For the Dyle catchment this is not so. This difference can be explained by the geomorphology of the basin. In the Semois catchment the total flow is predominantly constituted by surface runoff. Thus, the response of the flow at the outlet directly depends on the rainfall increments. Because the summer increments are negative in the rainfall scenario adopted in this study, it is understandable that the streamflow increments are also negative. On the other hand, in the case of the Dyle catchment the thick aquifer may sustain the streamflow during the dry season so that the rainfall decrease is hidden by the increase in the base flow. In Scenario .95 the increase in the evapotranspiration is such that the rainfall decrease characterizing this season induces a slight decrease in the total flow.

Additional information given by the IRMB model concerning the soil moisture regimes is shown in *Table 13.5*. Due to the increased evapotranspiration rate, the soil moisture is used in a larger measure and consequently the soil becomes slightly drier throughout the year. This loss could reach nearly 5% during the summer. In Scenarios 1.05 and 1.10, however, due to the increased relative humidity and the reduced PET, the soil remains

Table 13.4. Monthly regimes of the streamflow under current conditions (Scenario 0), under 2×CO_2 conditions (Scenario 1), and under 2×CO_2 conditions of changed relative humidity values.

	Jan.	Feb.	Mar.	Apr.	May	Jun.	Jul.	Aug.	Sep.	Oct.	Nov.	Dec.	Year
Streamflow (mm)													
Dyle catchment													
Scenario 0	21.3	19.1	20.2	18.3	17.7	15.3	15.2	14.8	14.6	16.5	17.6	20.3	210.9
Scenario .95	22.2	20.0	20.8	19.0	17.8	15.2	15.0	14.6	14.2	16.3	17.7	20.9	213.7
Scenario 1	23.4	21.1	22.0	20.1	18.9	16.3	15.9	15.4	15.1	17.3	18.7	22.1	226.3
Scenario 1.05	24.6	22.2	23.2	21.2	19.9	17.1	16.8	16.3	16.0	18.3	19.8	23.3	238.7
Scenario 1.10	25.7	23.2	24.2	22.2	20.9	17.9	17.7	17.1	16.9	19.3	20.8	24.4	250.3
Semois catchment													
Scenario 0	96.2	80.7	70.6	53.6	32.4	19.8	19.9	18.8	18.8	31.7	55.3	94.5	592.3
Scenario .95	106.5	86.6	70.5	55.4	30.9	16.8	16.8	15.5	15.5	28.8	55.2	101.3	599.8
Scenario 1	108.5	88.0	71.8	56.4	31.7	17.3	17.1	15.7	16.0	30.1	57.0	103.4	613.0
Scenario 1.05	110.1	89.4	73.0	57.5	32.6	17.8	17.4	16.0	16.6	31.4	58.8	105.4	626.0
Scenario 1.10	111.4	90.7	74.2	58.5	33.5	18.4	17.8	16.3	17.1	32.7	60.3	107.0	637.9

Scenario .95 = 95% of present value; Scenario 1.05 = 105% of present value; Scenario 1.10 = 110% of present value.

Table 13.5. Monthly regimes of the soil moisture under current conditions (Scenario 0), under 2×CO_2 conditions (Scenario 1), and under 2×CO_2 conditions of changed relative humidity values.

	Jan.	Feb.	Mar.	Apr.	May	Jun.	Jul.	Aug.	Sep.	Oct.	Nov.	Dec.	Year
Soil moisture (%)													
Dyle catchment													
Scenario 0	99.6	99.2	96.4	89.9	81.3	74.3	69.8	70.8	74.6	82.9	92.0	97.7	85.7
Scenario .95	99.3	98.7	95.3	88.9	78.8	69.7	64.4	65.1	69.4	78.7	89.7	96.9	82.9
Scenario 1	99.5	99.0	95.7	89.5	79.7	70.7	65.2	65.8	70.5	80.3	91.0	97.5	83.7
Scenario 1.05	99.7	99.3	96.1	90.1	80.7	71.7	66.1	66.7	71.8	82.0	92.2	98.0	84.5
Scenario 1.10	99.8	99.4	96.5	90.6	81.7	72.7	66.9	67.5	73.0	83.5	93.2	98.4	85.3
Semois catchment													
Scenario 0	99.9	99.8	98.6	93.9	86.8	79.1	73.9	72.6	77.4	87.7	96.6	99.1	88.8
Scenario .95	99.9	99.6	97.9	93.0	84.6	75.2	69.1	67.5	72.8	84.6	95.3	98.8	86.5
Scenario 1	99.9	99.7	98.1	93.3	85.2	75.7	69.4	67.8	73.4	85.4	95.8	99.0	86.9
Scenario 1.05	99.9	99.8	98.2	93.5	85.7	76.3	69.8	68.1	74.1	86.3	96.3	99.1	87.3
Scenario 1.10	100.0	99.8	98.4	93.8	86.3	76.9	70.2	68.5	74.8	87.1	96.6	99.2	87.6

Scenario .95 = 95% of present value; Scenario 1.05 = 105% of present value; Scenario 1.10 = 110% of present value.

slightly wetter during the cold season. This is particularly true in the case of the Dyle catchment, where the high infiltration rate compensates for the evaporation rise. No emphasis is given to this variable in this chapter.

13.6 Impact of a CO_2 Increase on a Simplified Flood Storage Reservoir

For the two catchments it was supposed that for safety reasons or to reduce the downstream damages streamflow must be limited at the outlet beyond a determined threshold. In the present case the threshold is a multiple of the annual mean module of the streamflow.

The flood control reservoir model used is very simple. A temporary storage is supplied with the daily outflows of the basin. The outlet of the reservoir is equipped with a bottom outlet or a system that allows flows lower than a chosen stage to outflow freely. When the input is greater than the threshold, the output is limited to this value and the surplus is stored in the reservoir. This surplus water is progressively released when the input becomes smaller than the threshold.

This oversimplified model allows the estimation of the maximum content of the reservoir and the mean number of days during which the reservoir is necessary to limit the outflow. The sensitivity of these two parameters to the CO_2-induced climate change and to the relative humidity hypothesis considered is investigated.

The thresholds beyond which the outflow must be limited are estimated for present conditions and for the climate change scenarios. For the Dyle catchment, a threshold equal to twice the annual module was chosen because the streamflow remains fairly quiet throughout the year. For the Semois catchment, characterized by large winter floods, a stage of three times the module was adopted to limit the extent of the reservoir capacity. These results are given in *Table 13.6*.

The case in which the reservoir has been built to exactly satisfy the present conditions is also studied. The maximum capacity is assumed to be fixed, and when the content of the reservoir reaches this value, the surplus flows out via a spillway system. The number of days when this occurs and the flood peak overflows are given in *Table 13.6*.

It appears that for both catchments the maximum contents of the dam, i.e., the greatest amount of water held in the reservoir, and the number of days during which the reservoir is necessary are greater in the $2{\times}CO_2$ conditions. The rate of change of the reservoir capacity is more important

Table 13.6. Characteristics of a flood control reservoir under current conditions (Scenario 0), under 2×CO_2 conditions (Scenario 1), and under 2×CO_2 conditions of changed relative humidity values.

	Scenario 0	Scenario .95	Scenario 1	Scenario 1.05	Scenario 1.10
Dyle catchment					
Threshold: 1.16 mm (two times annual module)					
Maximum content of the reservoir	3.51 mm	4.73 mm	5.78 mm	6.98 mm	8.07 mm
	$2.26\times10^6 m^3$	$3.05\times10^6 m^3$	$3.73\times10^6 m^3$	$4.50\times10^6 m^3$	$5.21\times10^6 m^3$
Mean annual number of days during which the reservoir is used	15.5	20.3	23.7	30.1	36.3
With maximum reservoir content estimated in present conditions (3.51 mm):					
Total number of days during which the reservoir overflowed (from 1901 to 1986)		14	32	53	71
Flood peak overflowing the reservoir (mm)		17.2	17.2	20.4	20.8
Semois catchment					
Threshold: 4.86 mm (3 times annual module)					
Maximum content of the reservoir	114.2 mm	161.9 mm	165.9 mm	169.4 mm	172.2 mm
	$141.0\times10^6 m^3$	$199.9\times10^6 m^3$	$204.9\times10^6 m^3$	$209.2\times10^6 m^3$	$212.7\times10^6 m^3$
Mean annual number of days during which the reservoir is used	39.2	47.7	49.9	51.9	53.4
With maximum reservoir content estimated in present conditions (114.2 mm):					
Total number of days during which the reservoir overflowed (from 1901 to 1986)		14	15	16	17
Flood peak overflowing the reservoir (mm)		18.81	18.81	18.94	19.06

Scenario .95 = 95% of present value; Scenario 1.05 = 105% of present value; Scenario 1.10 = 110% of present value.

for the Dyle catchment but the values are small. For the Semois River, an increase of at least 50×10^6 m^3 is estimated. The work necessary for modifying the reservoir would be extensive and perhaps difficult to realize in practice.

If the reservoir were built for the Semois catchment according to current conditions, it would overflow approximately 15 days during the 86-year period. The number of days of overflow is nearly the same in each scenario for the Semois catchment for the entire period. In the case of the Dyle River, the number of overflow days increases from about 15 days to about 70 days. This corresponds to an increased number of extreme events, such as floods, leading to reservoir overflows. This might be explained by the fact that in this type of catchment the total flow is sustained by the base flow; therefore, a number of separate events may exceed reservoir capacity.

13.7 Impact of CO_2 Increase on a Simplified Low-Flow Control Reservoir

A reservoir designed to supply water to ensure a streamflow at least equal to a given threshold during the dry season was investigated. This storage may be useful for maintaining a certain level of water quality or a minimum amount of water for downstream users. In this particular case, a water supply reservoir is studied where water demand is constant throughout the year and the water is not taken out of the river channel, but is used to enhance the outlet streamflow.

The mathematical model is very simple. The reservoir contains water available to sustain the flow. When the flow is lower than the threshold, the water needed to reach the threshold is taken out of the storage until the flow returns to normal levels. As soon as the streamflow becomes greater than the threshold, the surplus is collected in the reservoir to rebuild the reserve. This simplified model does not take into consideration characteristics of an actual reservoir (e.g., there are no limitations of the reservoir inflow due to the size of the inflow channel or to the power of the pump; there is no streamflow level beyond which no exploitation is possible for water quality reasons, etc.).

The reservoir is sized to suppress the low-flow episodes, i.e., periods when the flow does not meet or exceed the monthly median discharge. The low-flow thresholds are estimated for current conditions. The mass curve method (Domokos, 1987) has been used to size the reservoir. The simulated historic streamflow from 1901 to 1986, and the simulated streamflow series

Table 13.7. Characteristics of a low-flow control reservoir under current conditions (Scenario 0), under 2×CO_2 conditions (Scenario 1), and under 2×CO_2 conditions of changed relative humidity values.

	Scenario 0	Scenario .95	Scenario 1	Scenario 1.05	Scenario 1.10
Dyle catchment					
Threshold: 0.39 mm					
Minimum capacity of the reservoir	29.64 mm	44.73 mm	22.91 mm	16.25 mm	11.58 mm
	$19.12\times10^6 m^3$	$28.85\times10^6 m^3$	$14.78\times10^6 m^3$	$10.48\times10^6 m^3$	$7.47\times10^6 m^3$
Mean annual number of days during which the reservoir sustains the outflow	12.6	13.5	9.5	6.1	3.9
With reservoir capacity estimated in present conditions (3.51 mm):					
Total number of days during which the reservoir cannot sustain the outflow (from 1901 to 1986)		106	0	0	0
Volume reliability		0.998728	1.0	1.0	1.0
Semois catchment					
Threshold: .15 mm					
Minimum capacity of the reservoir	14.19 mm	15.41 mm	14.67 mm	13.96 mm	13.34 mm
	$17.52\times10^6 m^3$	$19.3\times10^6 m^3$	$18.12\times10^6 m^3$	$17.24\times10^6 m^3$	$16.47\times10^6 m^3$
Mean annual number of days during which the reservoir sustains the outflow	4.4	5.9	5.3	4.8	4.4
With reservoir capacity estimated in present conditions (14.19 mm):					
Total number of days during which the reservoir cannot sustain the outflow (from 1901 to 1986)		12	6	0	0
Volume reliability		0.999743	0.999900	1.0	1.0

Scenario .95 = 95% of present value; Scenario 1.05 = 105% of present value; and Scenario 1.10 = 110% of present value.

under the different climate change scenarios have been used to estimate the minimum reservoir capacity. These capacities are given in *Table 13.7*, along with the mean annual number of days during which the reservoir sustains the river flow. It is assumed that the reservoir has been built for use under current conditions. The reliability (defined as the ratio of the water supplied – with or without the help of the reservoir – and the demand) of this reservoir under the 2×CO_2 conditions is examined, and the number of days in the 86-year period that the reservoir failed to sustain the outflow is calculated. These results are also given in *Table 13.7*.

The minimum capacity of the reservoir is more sensitive to climate change for the Dyle River than for the Semois River. For the Dyle River the minimum capacity is equal to 19.12×10^6 m^3 under current conditions. Under a changed climate this value would range between 28.85×10^6 m^3 in Scenario .95 and 7.47×10^6 m^3 in the 2×CO_2 conditions of Scenario 1.10. For the Semois River, the current value is equal to 17.52×10^6 m^3; the corresponding changed-climate values range between 19.30×10^6 m^3 and 16.47×10^6 m^3. In both catchments the capacity becomes smaller than its present value if the relative humidity is assumed to increase. This implies that in these cases a reservoir built in accordance with current conditions would still have 100% reliability. In the Semois catchment, the size of the reservoir needed would increase in the first two scenarios (Scenario .95 and Scenario 1), but the range of variation of the capacity is very small. This implies that a reservoir built with a sufficient margin of safety (in this case around 10%) may still maintain the streamflow level in a worst-case scenario. This conclusion cannot be applied to the Dyle River. The capacity of the reservoir for the Dyle catchment ought to be 50% greater than its present value. At this stage it is not easy to explain the difference in the behavior of the two catchments; however, it seems that for basins similar to the Dyle River, a reservoir sized for current conditions could be inappropriate if the conditions in Scenario .95 occur.

13.8 Discussion and Conclusions

Results of this study show that for catchments with prevailing surface flow, such as the Semois River, the streamflow decreases in the summer for all scenarios. This directly reflects the rainfall modification adopted in the scenarios and has negative implications for low-flow stages, for water quality, and for water resources during the summer. For catchments where infiltration is important, such as the Dyle River with its thick aquifer, the streamflow shows only a slight reduction during the summer in Scenario .95. Significant negative effects would thus remain very limited in this catchment.

The total flow during the rest of the year, and in particular during the winter, is characterized by an increase in both catchments for all scenarios. Concerning the total flow regime, it seems that the winter increase, which is mainly due to rise in rainfall, may compromise the safety conditions along the river. As there is general agreement among GCM simulations regarding a winter precipitation increase (IPCC, 1990), it seems that this impact on streamflow is relatively well established and merits particular attention in the future.

The sensitivity of flow control reservoirs to climate change was investigated by considering two simplified types of single-purpose reservoirs. First, a flood control reservoir used to limit the flood peaks beyond a given threshold chosen in the present conditions was considered. In this case, a multiple of the annual module was used. For both catchments and for all scenarios, the minimum required capacity of the reservoir needed to eliminate the peaks is greater in the $2\times CO_2$ conditions, a consequence of the increased winter streamflow. The rise of the reservoir capacity is equal to nearly half of the present value for the Semois catchment and to a third of the present value for the Dyle catchment. If the reservoir is created for current conditions, the number of days during which water could overflow is rather stable in the Semois catchment, corresponding to the same flooding events in the different scenarios. This number is more variable in the Dyle catchment, indicating that floods could cause more damage here than in the Semois catchment. This shows the paradox that the climate change impact could be more significant in this kind of catchment than in the Semois basin, even if the change of reservoir size is more important for the Semois River.

Finally, a low-flow control reservoir was considered to sustain the streamflow above a given threshold, in particular the low-flow stage. For the two catchments the simulations show that if the relative humidity increases, the storage built under current conditions would be large enough to adequately sustain the flow. In the case of a decrease in the relative humidity the size of the storage ought to be larger than it is now. The estimated changes of the minimum size of the reservoir between the different scenarios are greater in the Dyle basin than in the Semois basin. Consequently, it appears that for the Semois River a reservoir built with a small safety range could still satisfy the output conditions; however, for the Dyle River significant modifications could be necessary.

The purpose of the set of sensitivity studies run in this chapter was to show how the catchment particulars affect the sensitivity of simple parameters. This kind of approach should be generalized to a larger number of catchments before any conclusion applicable for all of Belgium can be drawn. Projects of this nature are currently in progress; the results will certainly help to identify the impacts in relation to the characteristics of the catchments in the country as a whole.

Acknowledgments

This research has been carried out, in part, in fulfillment of the contract GC/34/022 of the Global Change Impulse Project of the Federal Office for Scientific, Technical and Cultural Affairs, Brussels, Belgium.

References

Arnell, N.W., Brown, R.P.C., and Reynard, N.S., 1990, Impact of Climatic Variability and Change on River Flow Regimes in the UK, Report No. 107, Institute of Hydrology, Wallingford, UK.

Bultot, F., and Dupriez, G.L., 1976a, Conceptual hydrological model for an average-sized catchment area, I. Concepts and relationships, *Journal of Hydrology*, **29**(3/4):251–272.

Bultot, F., and Dupriez, G.L., 1976b, Conceptual hydrological model for an average-sized catchment area, II. Estimate of parameters, validity of model, applications, *Journal of Hydrology*, **29**(3/4):273–292.

Bultot, F., and Dupriez, G.L., 1985, Daily effective evapotranspiration from a river basin, in Casebook on Operational Assessment of Areal Evaporation, *Oper. Hydrol. Rep.*, **22** (WMO-No. 635):80–105.

Bultot, F., and Gellens, D., 1989, Simulation of the impact of atmospheric CO_2 doubling on precipitation and evapotranspiration: Study of the sensitivity to various hypotheses, *Proceedings of the Conference on Climate and Water*, Government Printing Center, Helsinki, Finland, 11–15 September, Vol. I:73–92.

Bultot, F., Coppens, A., and Dupriez, G.L., 1983, Estimation de l'évapotranspiration potentielle en Belgique (Procédure Révisée), Institut Royal Métérologique Publications, Série A, No. 112, Uccle-Bruxelles, Belgium.

Bultot, F., Dupriez, G.L., and Gellens, D., 1988a, Estimated regime of energy-balance components, evapotranspiration and soil moisture for a drainage basin in the case of a CO_2 doubling, *Climatic Change*, **12**(1):39–56.

Bultot, F., Coppens, A., Dupriez, G.L., Gellens, D., and Meulenberghs, F., 1988b, Repercussions of a CO_2 doubling on the water cycle and on the water balance: A case study for Belgium, *Journal of Hydrology*, **99**:319–347.

Bultot, F., Gellens, D., Spreafico, M., and Schädler, B., 1992, Repercussions of a CO_2 doubling on the water balance: A case study in Switzerland, *Journal of Hydrology*, **137**:199–208.

Domokos, M., 1987, Hydrology of reservoirs, in Ö. Starosolszky, ed., *Applied Surface Hydrology*, Water Resources Publications, Littleton, CO, USA.

Gellens, D., 1991, Impact of a CO_2-induced climatic change on river flow variability in three rivers in Belgium, *Earth Surface Processes and Landforms*, **16**:619–625.

IPCC (Intergovernmental Panel on Climate Change), 1990, *Climate Change: The IPCC Scientific Assessment*, J.T. Houghton, G.J. Jenkins, and J.J. Ephraums, eds., Cambridge University Press, Cambridge, UK.

Manabe, S., and Stouffer, R.J., 1980, Sensitivity of a global climate model to an increase of CO_2 concentration in the atmosphere, *Journal of Geophysical Research*, **85**(C10):5529–5554.

Manabe, S., Wetherald, R.T., and Stouffer, R.J., 1981, Summer dryness due to an increase of atmospheric CO_2 concentration, *Climatic Change*, **3**(4):347–385.

Washington, W.M., and Meehl, G.A., 1983, General circulation model experiments on the climatic effects due to a doubling and quadrupling of carbon dioxide concentration, *Journal of Geophysical Research*, **88**(C11):6600–6610.

Chapter 14

National Assessment: Poland

Zdzislaw Kaczmarek and Janusz Kindler

14.1 Introduction

During an average year Poland's rainfall and snowfall yield about 197 km^3 of water. Of this, 142 km^3 go back into the atmosphere due to evaporation, and 55 km^3 feed the Baltic Sea through the river systems. The spatial distribution of surface runoff is presented in *Figure 14.1*. *Figure 14.2* shows annual runoff for the years 1951 to 1990, ranging between 34.4 km^3 in 1952 and 79.5 km^3 in 1981. An additional 5 km^3 of river flow enters Poland from neighboring countries. Together this gives an annual per capita rate of about 1,500 m^3 – the lowest value in Europe. Taking into account the interannual and intra-annual stochastic variability of climatic and hydrologic processes, it has been estimated (Kaczmarek, 1978) that reliable water resources, available 95% of time, are equal to about 22 km^3 unevenly distributed throughout the country. Due to environmental constraints only 30% to 40% of these surface water resources can be effectively used for agricultural, industrial, or residential needs.

Water can become a barrier to social and economic development due to several mutually dependent factors:

- Natural water scarcity depending on the relation between regional water supply and demand.
- Pollution of rivers, lakes, and ground water aquifers.
- Technological and economic shortcomings.
- Institutional impediments and low public awareness.

Z. Kaczmarek et al. (eds.),
Water Resources Management in the Face of Climatic/Hydrologic Uncertainties 356–377.

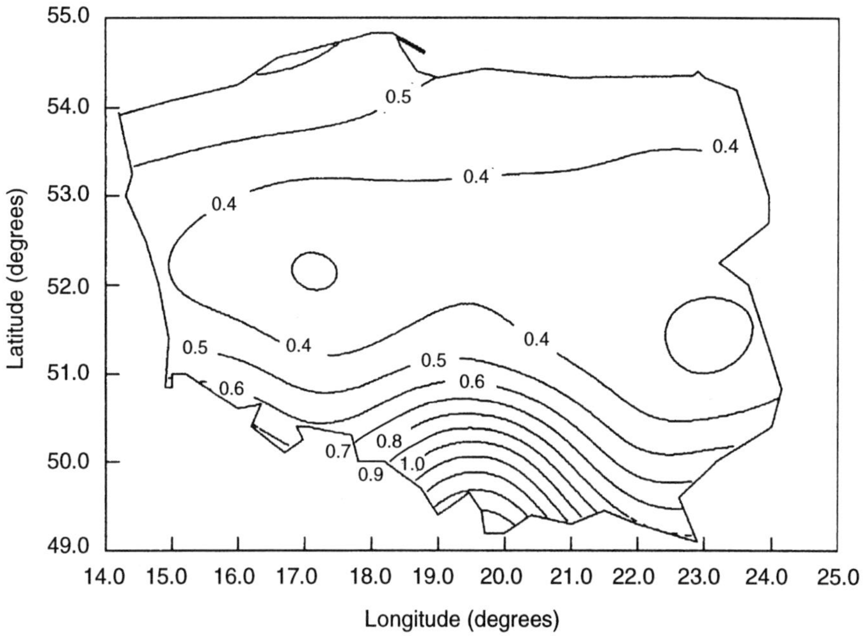

Figure 14.1. Spatial distribution of annual runoff (mm/day) in Poland.

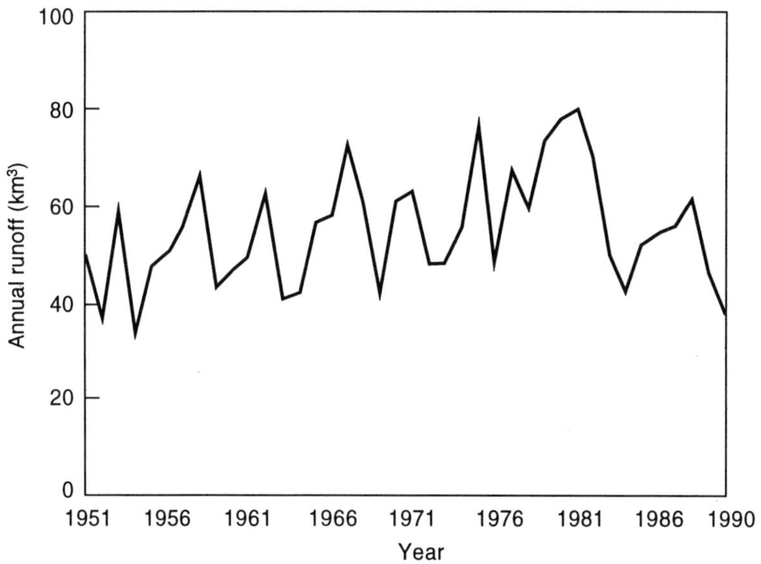

Figure 14.2. Variability of annual runoff in the years from 1951 to 1990.

Only the first two factors are sensitive to climate change. The others are subject to policy decisions, which if rationally applied may help to adapt water resources systems to non-stationarity of geophysical processes.

Despite natural water scarcity, Poland's economy is water intensive. According to recent data, about 170 m^3 of water are used to produce US$1,000 of GNP; in Germany only 62 m^3 of water is used to produce the same amount of GNP. The water shortages observed in some years in several regions of the country are deeply rooted not only in natural scarcity but also in inefficient industrial water use and in the high level of water pollution.

When a nation's long-term economic future is considered, the issue of climate change should not be neglected and various scenarios of possible trends must be investigated. Science is responsible for developing a methodology for climate impact assessment and for providing the rational basis necessary to undertake adaptive measures. However, such assessment must be undertaken with the understanding that the main indicators of water economy projected over the next century will be influenced not only by climate, but also by population processes, economic growth, and technological progress. As mentioned earlier, water as a barrier to economic growth is usually the result of many different factors, some of which are quite far removed from water resources per se. All of these factors are interrelated and their cumulative impacts must be taken into account when improvement programs are undertaken.

14.2 Modeling of Climatic Scenarios for Poland

A climate change scenario should be a feasible and internally consistent quantitative estimate of a possible future climate. As suggested in another part of this book, general circulation models (GCMs) are the most sophisticated tools for evaluating the likely effects of the increasing concentration of greenhouse gases on global climate. There are two main forms of GCM results:

- The equilibrium-type response to a certain level of change (usually a doubling) in the concentration of greenhouse gases.
- The transient reaction of the climate system to time-dependent changes of the chemical composition of the atmosphere.

Most of the water resources impact studies reported in the literature are based on equilibrium climate scenarios. The usual procedure is to add the differences (ΔT) of model results for $2\times CO_2$ and $1\times CO_2$ conditions to measured temperature data. Observed precipitation values are multiplied by the ratios (rP) of GCM rainfall for the conditions described above. A similar approach has been used in this case study to evaluate changes in Poland's water resources. Very little is known about possible changes of other climatic variables, for example the relative air humidity, which influence catchment processes.

For a water resources impact study by the Institute of Geophysics of the Polish Academy of Sciences, five GCM-based climate scenarios obtained from the National Center for Atmospheric Research in Boulder, CO, USA, were analyzed:

- The Canadian Climate Center (CCC) equilibrium model.
- The Geophysical Fluid Dynamics Laboratory (GFDL) equilibrium model – version R15.
- The GFDL – R30 equilibrium model.
- The Goddard Institute for Space Studies (GISS) equilibrium model (1982 version).
- The GFDL transient model.

The model output results for $1\times CO_2$ conditions were compared with observed climatic data for the years 1951 to 1990 at 41 stations distributed throughout the country. For temperature patterns, the best agreement was observed for the CCC and GFDL-R15 models; for precipitation, all models gave an equally unsatisfactory picture of the present climate.

It was decided to implement the water resources impact assessment on the basis of the GFDL-R15 and GISS models. For the $2\times CO_2$ conditions, the first may be characterized as "warm-dry" and the latter as "warm-wet" scenarios. In both cases data from the model output were interpolated to a grid with a resolution of one degree latitude by one degree longitude, covering the central European region between 10° E to 24° E longitude and 49° N to 55° N latitude. To illustrate possible changes in Poland's climate, the mean yearly values of temperature increments (ΔT) and precipitation ratios (rP) for the GFDL-R15 model are given in *Figures 14.3* and *14.4*. The monthly distributions of temperature disturbances and rainfall for a location in central Poland (18.67 E, 52.20 N) are shown in *Figure 14.5* and *14.6*. In both scenarios the relative air humidity and the duration of sunshine were assumed to remain unchanged for the $2\times CO_2$ conditions.

It must again be stressed that two main problems are encountered when GCM data are used as input variables in hydrological assessment studies. First, the spatial scale of current global atmospheric models is much coarser than is required for water resources analysis. Second, regional precipitation

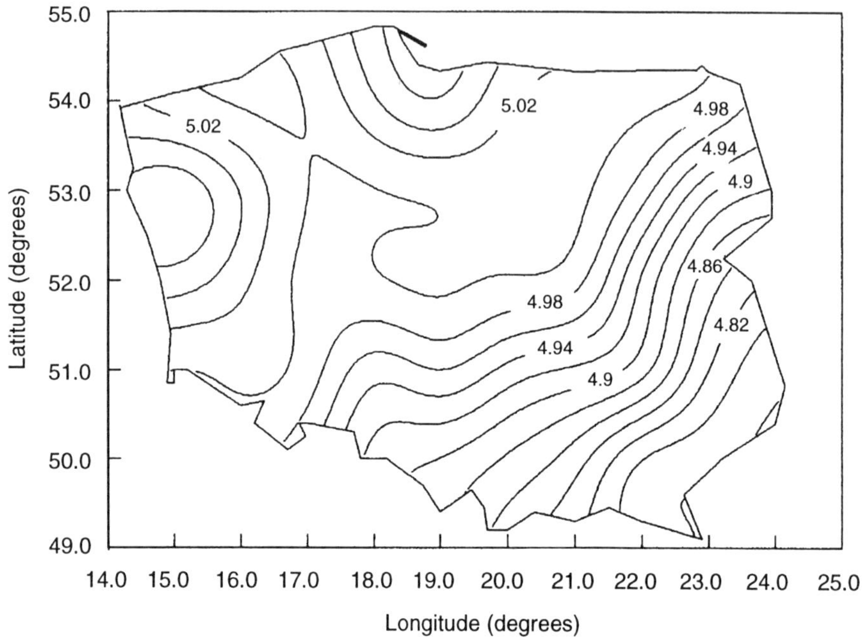

Figure 14.3. Increments of annual temperature calculated by GFDL-R15 model.

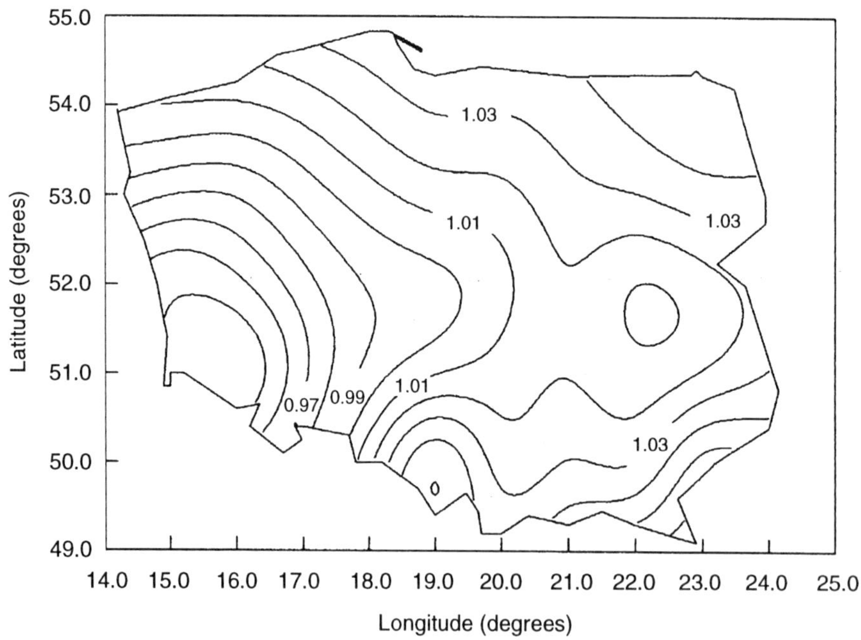

Figure 14.4. The $2\times CO_2$:$1\times CO_2$ precipitation ratio calculated by GFDL-R15 model.

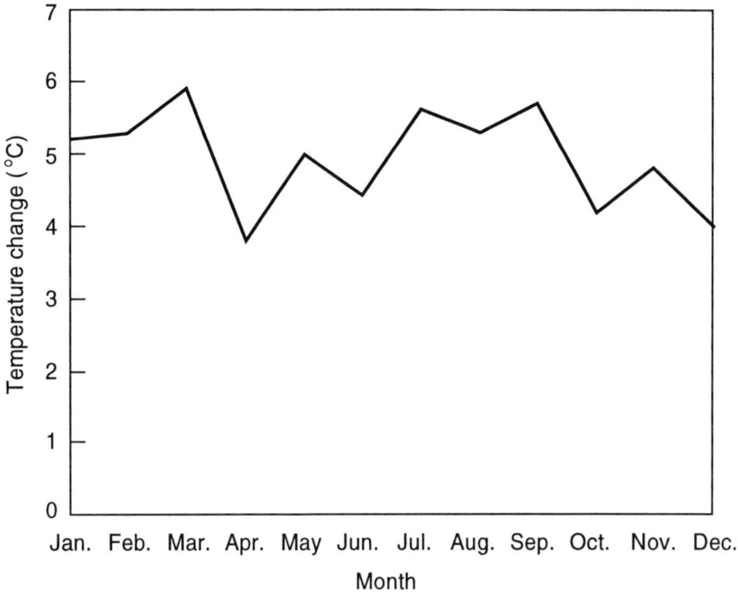

Figure 14.5. Temperature changes in central Poland related to a doubling of CO_2.

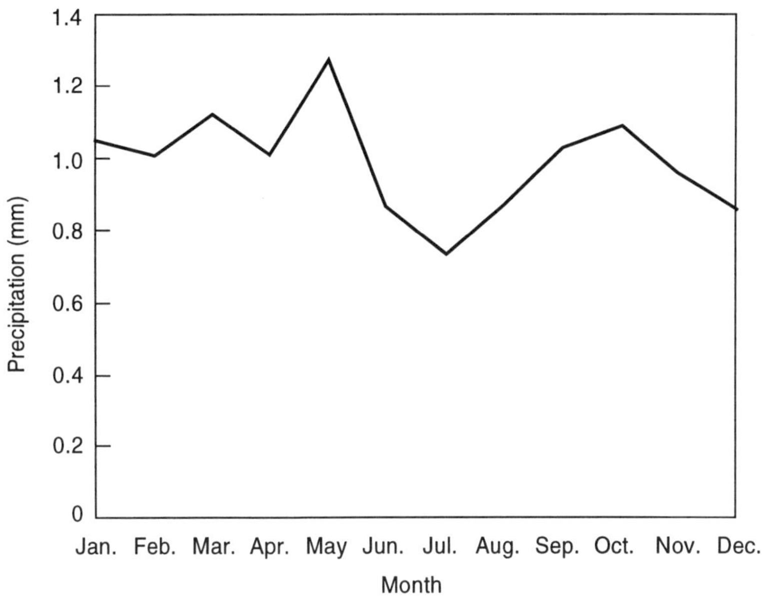

Figure 14.6. Precipitation changes in central Poland related to a doubling of CO_2.

estimates obtained by means of GCMs are currently highly uncertain. The regional climate change scenarios based on global atmospheric models therefore should not be interpreted as a prediction of future climate conditions. For water resources impact studies, several different scenarios should be used to assess the possible range of hydrologic implications and to evaluate various water resource strategies to cope with changed climate conditions (Carter *et al.*, 1992).

14.3 Modeling Hydrological Processes

To assess the impact of climate on water resources, some type of model of the hydrological processes is needed. There is a range of possible model types, from simple empirical relationships to complex conceptual models with a great number of parameters. Empirical models are "black-box" representations of the hydrological processes, and are usually based on regression functions between climate inputs and some indices of hydrological response. Examples are the Langbein (1949) diagrams, and the Turc formula (Turc, 1954), which permit annual runoff to be calculated from mean annual precipitation and average annual temperature representative for a given catchment. Conceptual models are simplified representations of the physical processes involved in the hydrological cycle, based on a particular concept of how a river catchment works. They usually include a number of parameters that must be either identified by means of a calibration procedure or estimated from empirical relationships with measurable catchment properties.

For the assessment of the sensitivity of Poland's water resources to climate change, a conceptual water balance model called CLIRUN3 (Climate/Runoff Model – 3rd version) has been applied (Kaczmarek, 1993). The model differs from previous approaches in two respects:

- It is time continuous, i.e., the water balance components vary as continuous functions of time within certain assumed time intervals, e.g., months. In the present version of CLIRUN, the input climatic data are averaged monthly values; however, in principle they may also be treated as time dependent.
- The stochastic properties of water balance components are expressed either as simulated time series, or by a set of probabilistic matrices based on stochastic storage theory (Kaczmarek, 1974; Klemeš, 1981).

The water balance equation that connects precipitation, evapotranspiration, runoff, and catchment storage was first formulated by Penck (1986) and has since been applied to many hydrological problems. It was found to be an efficient tool for climate/water resources impact studies. In the case of CLIRUN3, the water balance equation for the mth time interval (month, season) is written in the following form:

$$S_{\max}\frac{dz(t)}{dt} = P_m - R_s[z(t),\Omega_m] - R_g[z(t)] - R_b - E[z(t),\Omega_m] \quad , \quad (14.1)$$

where $S_{\max}$ is the catchment water-holding capacity (mm); t is time; $z(t) = S(t)/S_{\max}$, the relative catchment storage; P_m is the effective precipitation (mm/day), calculated from the rainfall data and snow budget estimates; $R_s[z(t),\Omega_m]$ represents immediate (surface) runoff (mm/day); $R_g[z(t)]$ represents delayed runoff (mm/day); R_b is a base flow independent of seasonal and multi-annual climate fluctuations (mm/day); $E[z(t),\Omega_m]$ is actual evapotranspiration (mm/day); and Ω_m is a set of climatic elements influencing water balance. All input and output variables are taken to be uniformly distributed throughout the river catchment.

Conceptualization of the water balance equation consists in assuming functional relationships describing components of equation (14.1). It is difficult to prove that one approach is better than another, and we should agree with Gburek (1971) that "a model system is merely a researcher's idea of how a physical system interacts and behaves, and in the case of watershed research, models are usually extremely simplified mathematical descriptions of complex situations." Details of the CLIRUN conceptualization and the methods used for solving equation (14.1) are described in the literature (Gburek, 1971).

Computer programs for both versions of the stochastic water balance model have been written for the IBM-compatible personal computer. The input variables are either a time series (for a simulation model) or statistical parameters (for a matrix-type model) of effective precipitation and potential evapotranspiration. The user may choose among various methods available for calculating these variables.

The CLIRUN model depends on three parameters, $S_{\max}$, α, and ε, determined by calibration and based on minimization of the mean square error of catchment outflows. In the case of the simulation model, the time series of recorded runoff data for a certain part of the observation period may be used for calibration; the remainder of the records allows the validation of the results. For the stochastic-matrices version, optimization of model parameters is based on minimization of mean square differences between calculated and observed average monthly runoff values. Future studies should look for empirical relationships between the model parameters and physical characteristics of catchments.

Finally, it should be stressed that the terms *sensitivity analysis* and *impact studies* used in this study are sometimes used interchangeably in the literature, which may lead to confusion and misunderstandings. To avoid this, these two concepts should be clearly distinguished. According to the definitions used in Chapter 1, sensitivity refers to the reaction of a hydrological system to disturbances in the input variables. To find the sensitivity and elasticity parameters of the water balance, alternative climate scenarios need not be assumed. On the other hand, a climate change impact on a hydrological variable is understood as the difference between the numerical values of that variable obtained for an alternative climate scenario and the values under current climatic conditions. An impact of climatic forcing Ω may be then characterized by

$$\Delta y = y(\Omega_{\text{alternative}}) - y(\Omega_{\text{present}}) \quad . \tag{14.2}$$

This means that to perform an impact analysis, one or more alternative climate scenarios must be specified and applied to the model serving as the transfer function between climatic and hydrological phenomena. The case study discussed in this chapter is impact oriented. It should be added, however, that in the context of limited knowledge of possible regional climate change, some authors have expressed the opinion that sensitivity analysis may be more informative for decision makers than are the results of impact studies.

14.4 Possible Changes in Poland's Water Resources

The CLIRUN3 model was used to calculate water balance components (runoff, evapotranspiration, and storage values) for 31 river catchments and 60 grid cells with a resolution of one degree latitude by one degree longitude covering Poland.

The study used climatic and hydrological data measured from 1951 to 1990 at 41 meteorological stations for 31 river catchments. Also, GFDL and GISS temperature increments and precipitation ratios were used to develop climate scenarios for $2\times CO_2$ conditions.

Model parameters were identified for river catchments and were then interpolated to the grid. Input variables are gridded mean values and coefficients of the variation of monthly precipitation (adjusted with regard to snow accumulation processes), as well as gridded mean characteristics of potential evapotranspiration calculated by the energy budget method (Zubenok, 1977). The output variables are mean values, standard deviations, and coefficients of the correlation of key water balance elements – runoff, actual evapotranspiration, and active catchment (or grid-box) storage. As a result, monthly characteristics of these variables were collected for each grid cell for current climate conditions (based on observations) and for selected alternative equilibrium climate scenarios.

The output data permit the estimation of the probability distributions of a grid cell's water balance elements for each climate scenario under consideration. The data also permit the analysis of changes in hydrological characteristics for $2\times CO_2$ conditions. Some of the results are shown in *Figures 14.7* through *14.14* in the form of isolines representing percentage change of selected water balance variables. The results differ greatly depending on the climate model used (GFDL or GISS). For both scenarios there is a tendency for runoff and soil moisture to decrease during the drought period, which in Poland is usually in August and September. Detailed analysis of output files also shows a shift in the occurrence of floods from March and April to January and February, caused by changes in snow accumulation and snowmelt processes. Changes in monthly water balance components for GFDL-R15 and GISS scenarios for one grid box in central Poland are shown in *Figure 14.15*. The analysis of output data suggests that the highest sensitivity of runoff to climate change has been observed in those regions with currently low runoff coefficients.

Among various physical characteristics of surface water resources, the temperature regime of lakes and rivers plays an important role. It may influence biochemical processes in freshwater ecosystems, consequently influencing the quality parameters of water bodies. Water temperature is also a decisive factor in ice formation. Several studies have been undertaken recently in Poland to assess the possible effects of climate change on thermal regime of rivers, on stratified and unstratified lakes, and on free surface evaporation. Calculations were based on models developed by Jurak and Wisniewski (1989) and by Jurak (1992). Some of the results obtained for the GFDL climate scenario are presented in *Figures 14.16* to *14.18*. The rise of the temperature of the Vistula River at points 185 km and 380 km downstream is due to thermal pollution caused by power plants.

The various impact studies described above should be treated as preliminary assessments of possible changes of hydrological regime in Poland, caused by expected climate change of anthropogenic origin. Further investigations are needed to evaluate a range of expected disturbances in

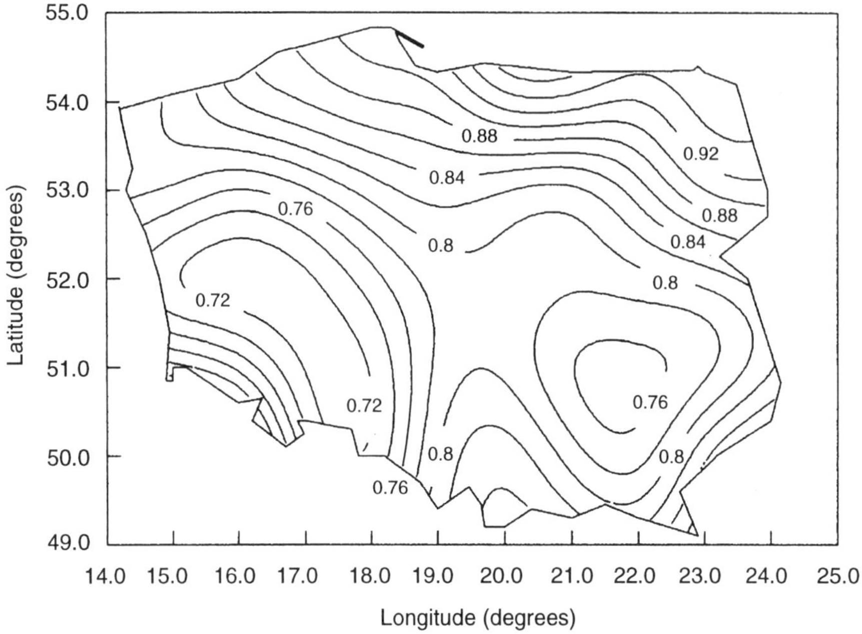

Figure 14.7. Change in annual runoff (mm/day) under $2\times CO_2$ conditions (GFDL model).

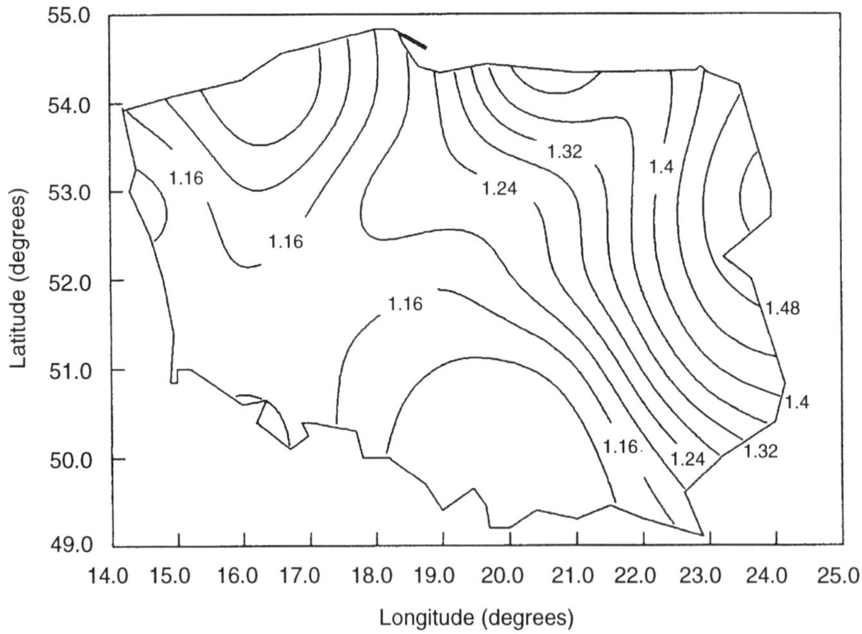

Figure 14.8. Change in annual runoff (mm/day) under $2\times CO_2$ conditions (GISS model).

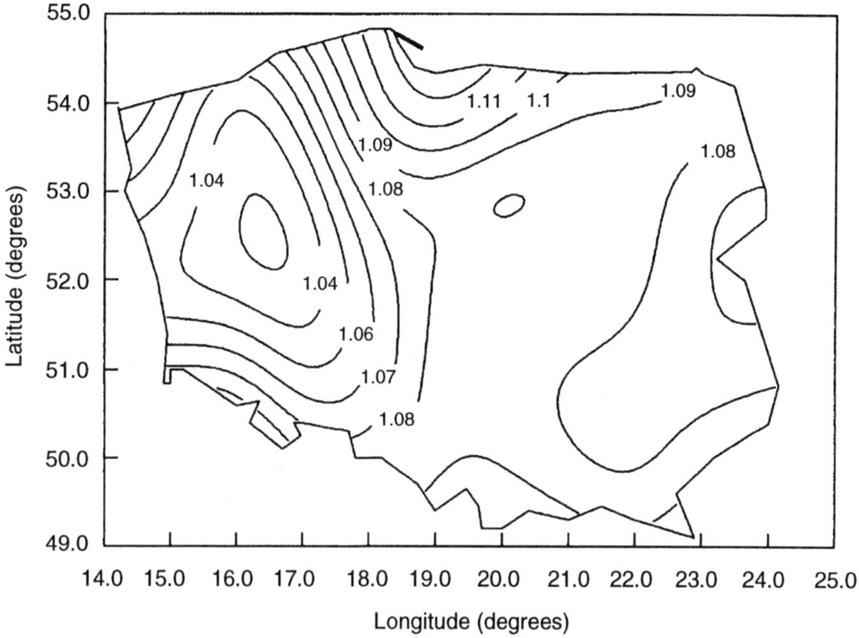

Figure 14.9. Change in evapotranspiration (mm/day) under $2{\times}CO_2$ conditions (GFDL model).

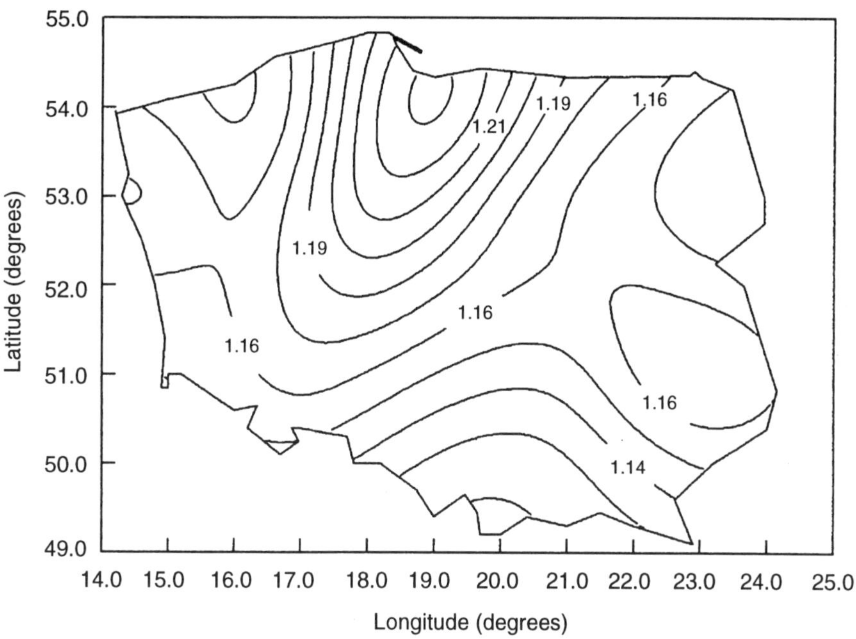

Figure 14.10. Change in evapotranspiration (mm/day) under $2{\times}CO_2$ conditions (GISS model).

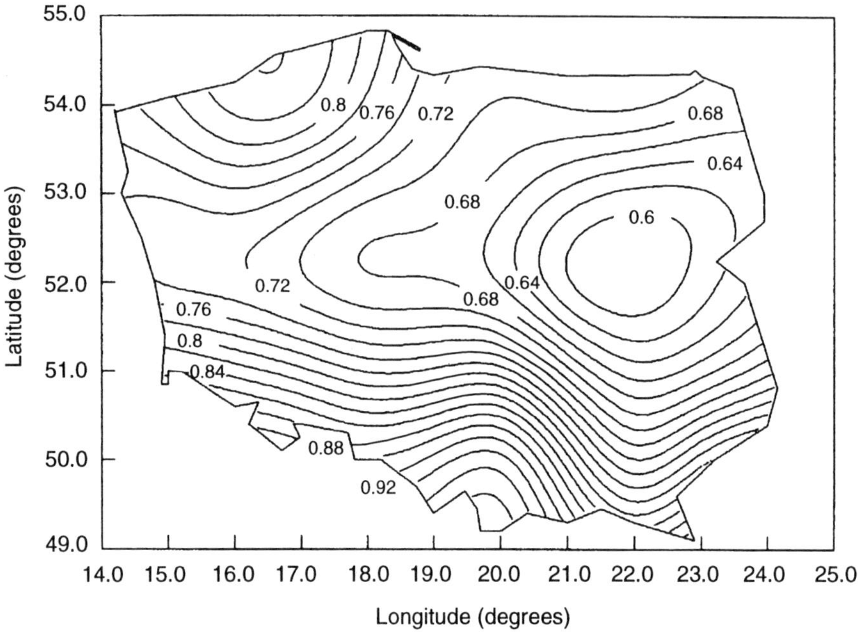

Figure 14.11. Change in September's runoff (mm/day) under 2×CO_2 conditions (GFDL model).

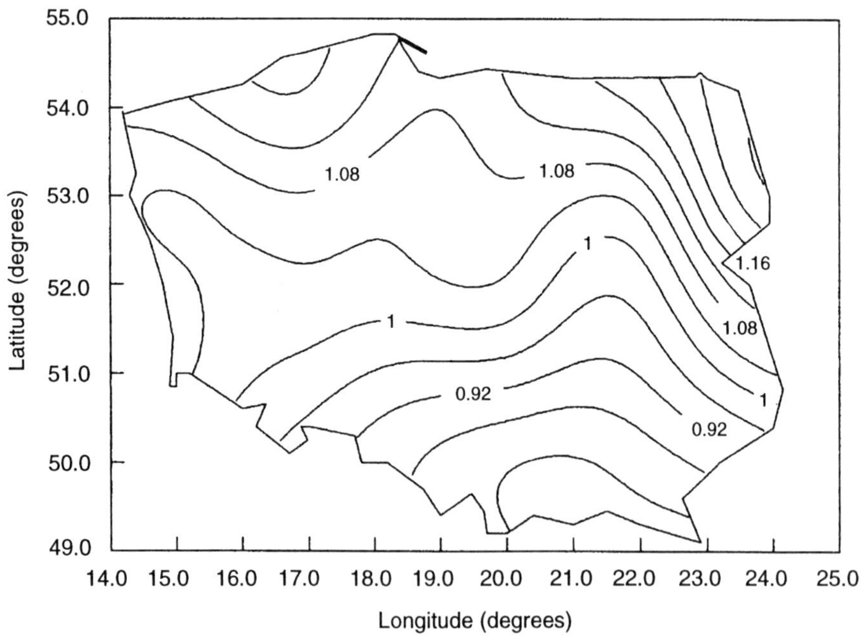

Figure 14.12. Change in September's runoff (mm/day) under 2×CO_2 conditions (GISS model).

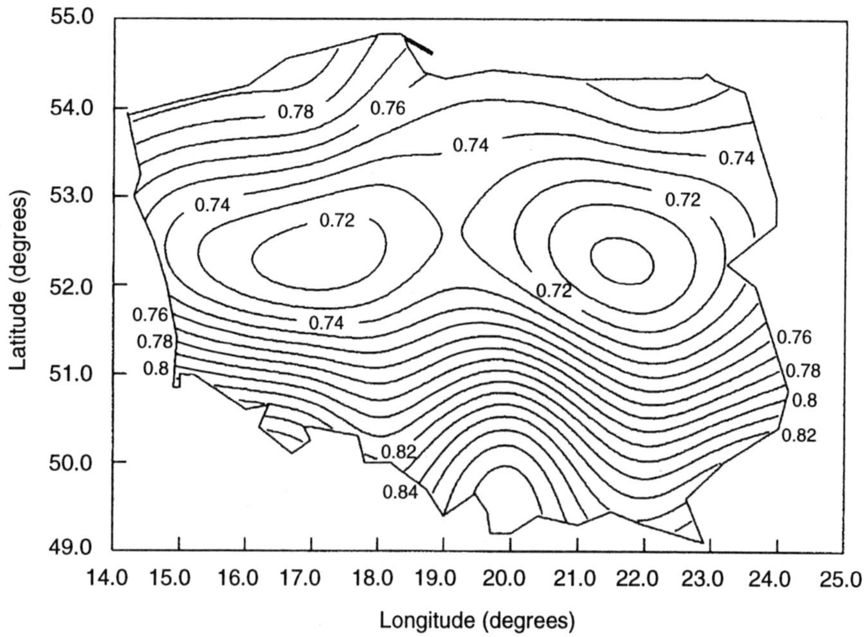

Figure 14.13. Change in September's runoff (mm/day) under $2\times CO_2$ conditions (GFDL model).

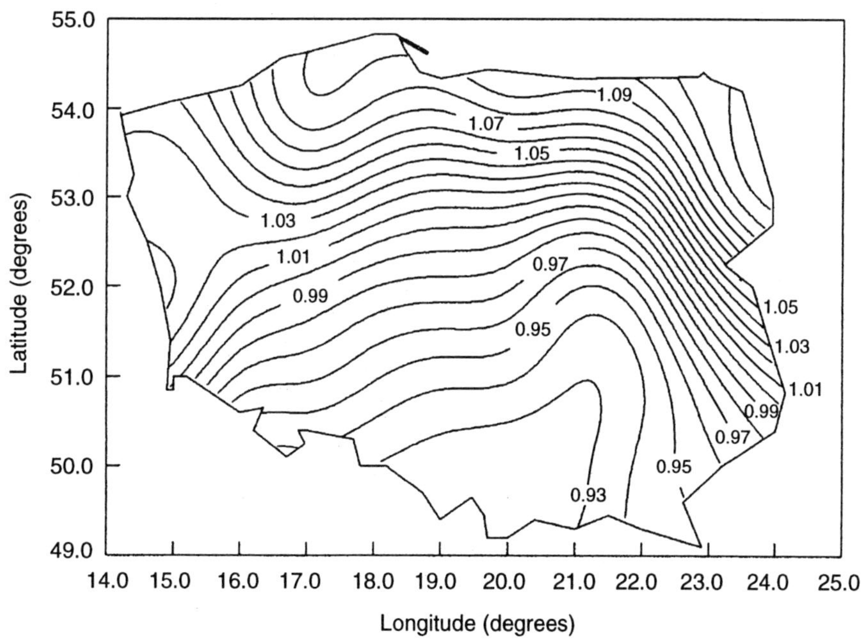

Figure 14.14. Change in September's runoff (mm/day) under $2\times CO_2$ conditions (GISS model).

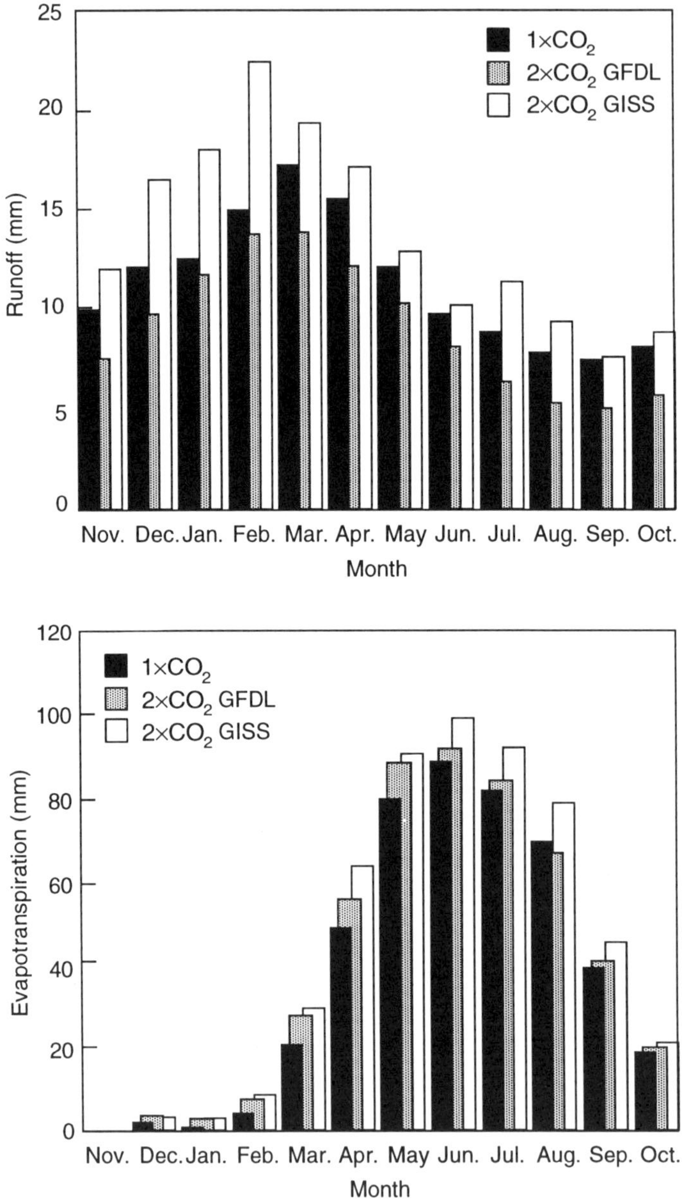

Figure 14.15. Changes in water balance for a grid box (19.5 E, 52.5 N) in central Poland.

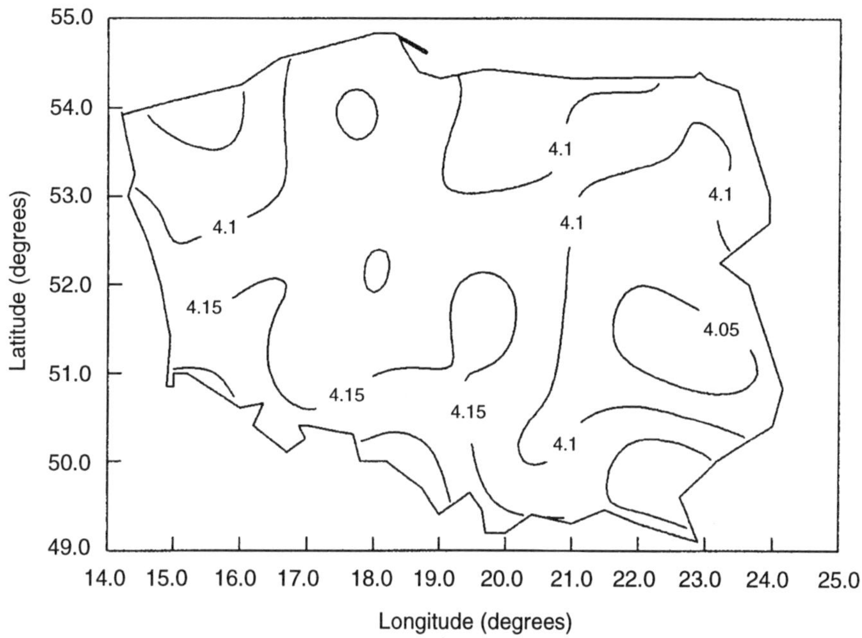

Figure 14.16. Change in summer temperature (°C) of shallow lakes (May to October) under $2 \times CO_2$ conditions (GFDL model).

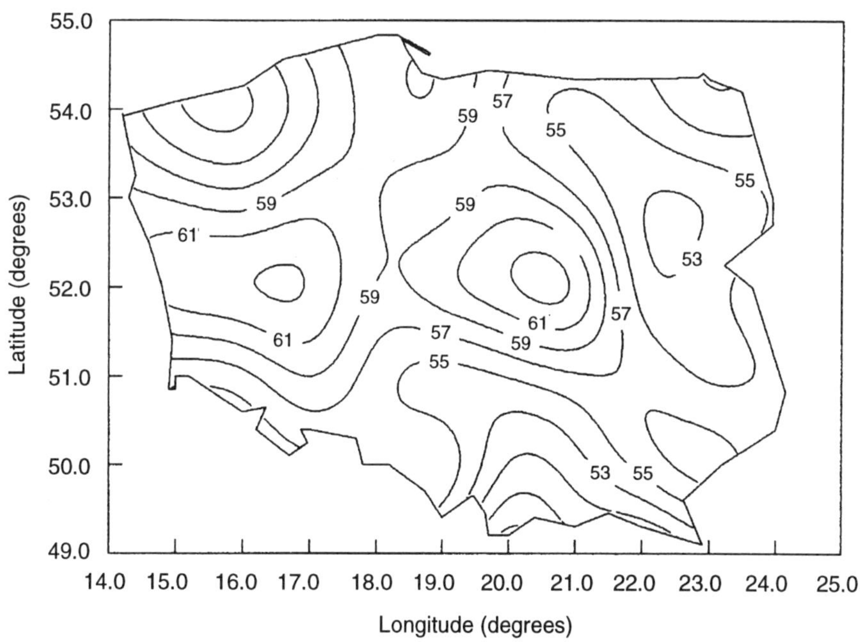

Figure 14.17. Change in summer evaporation (mm) from shallow lakes (May to October) under $2 \times CO_2$ conditions (GFDL model).

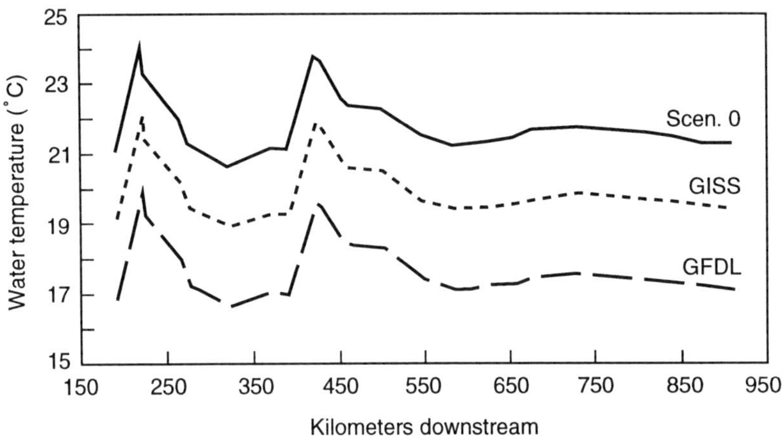

Figure 14.18. Average summer water temperature (May to October) for Vistula River in 1973.

water resources characteristics, including thermal regime and water quality processes, for a set of climate scenarios based on updated versions of atmospheric models.

14.5 Water Resources Management in an Uncertain Environment

The fundamental problem of responding to possible consequences of global warming resulting from human activity is deciding what adaptive measures should be undertaken in the face of highly uncertain climatic threats. The choices must be made on the basis of incomplete knowledge of future water supply and demand, and the policy alternatives must be analyzed with respect to the risks of assuming incorrect future scenarios. During the last century, water resources planning has been based on the assumption that variability of hydrological phenomena is governed by stationary stochastic processes. In the era of global environmental change this concept has become questionable.

To date, trends in water demand caused by demographic and socioeconomic factors are identified without reference to possible changes in environmental conditions, including climate. The experience of water resources management agencies in various countries shows that socioeconomic processes influencing water use in the long run cannot be accurately predicted.

In most of the past studies done in Poland to formulate the nations' long-term water resources strategies, future demands were highly overestimated. Even more difficult to assess are possible implications of climate change on future water requirements.

Water withdrawal in Poland in 1990 was equal to 14.2 km^3, with the following distribution among sectors: industry, 2.9 km^3; energy (used for cooling), 6.6 km^3; agriculture, 1.7 km^3; and domestic use, 3.0 km^3. According to our estimates, water demand in the middle of the next century may increase by 40% to 60% due to non-climatic factors. There is very little information on the effect of possible temperature and precipitation changes on water requirements in various sectors of Poland's economy. Preliminary estimates, based mostly on the literature, lead to the conclusion that a temperature increase may have only a moderate impact on industrial and domestic water use. The unknown factor is Poland's agricultural sector, which at present uses a relatively small amount of water for irrigation. According to the dry-warm climate scenario, however, the situation may change dramatically because a threshold between irrigated and nonirrigated agriculture may be surpassed in most of Poland's lowland territory. In this case, a demand for spray irrigation may increase substantially.

Engelman and LeRoy (1993), with reference to Falkenmark and Widstrand (1992), introduced a "water-stress index" by assuming that countries with less than 1,700 m^3 per person per year may occasionally suffer local water difficulties, and that countries with less than 1,000 m^3 per person per year will experience general water scarcity. This last benchmark has been accepted as a water-scarcity indicator by the World Bank. For the 1990 population level of 38.5 million inhabitants, the "water stress" and "water scarcity" thresholds for Poland are 65.4 km^3 and 38.5 km^3 of available water supply, respectively. Assuming a 5% increase of the per capita indices for $2\times CO_2$ climate conditions and projecting a population of 46.9 million, the threshold values for Poland in 2050 would be 83.7 km^3 of available water in the case of "water stress" and 49.2 km^3 of available water in the case of "water scarcity." Taking into account the interannual variability of runoff, the probability of "water scarcity," "water stress," and "water abundance" for present socioeconomic and climatic conditions (the year 1990) and for $2\times CO_2$ scenarios (the year 2050) were calculated. Results are shown on *Figure 14.19.* It can be seen that the GFDL (warm-dry) and GISS (warm-wet) scenarios project two very different pictures of Poland's water resources management problems.

There are three different approaches to adapting water resources systems to climate changes. First, decisions can be postponed until more

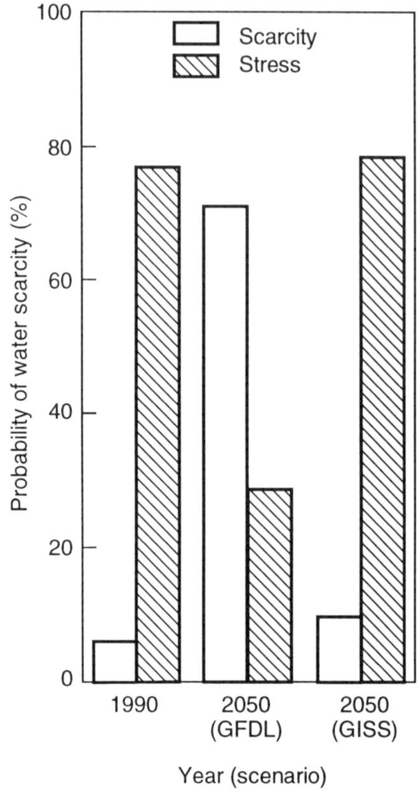

Figure 14.19. Probability of water scarcity in Poland in the present and by the year 2050.

reliable information on global processes becomes available. The existing water schemes remain unchanged, and new ones are designed in accordance with current procedures. The Commission for Hydrology of the World Meteorological Organization has adopted a "CHy Statement on the Hydrological and Water Resources Impacts of Global Climate Change" (Commission for Hydrology, 1988), which reads: "...given the added burden of uncertainty about climate change it is certainly inappropriate at this time to discard available analytical procedures or to engage in expensive alterations to build facilities." However, this approach may lead to decisions being made too late to protect water resources systems from the negative consequences of climate change.

Second, there is the "minimum regret" approach, where decisions are made to solve current problems in the best possible way while preparing water resources systems for potential surprises and shocks. Waggoner and Revelle (1990) describe this policy by saying, "So long as the future remains unsure, and that seems a long time ahead, rational people will make decisions that solve present problems and make water supply robust, resilient and flexible for any future."

Finally, in the third approach it is assumed that certain optimality rules are applied to a range of climate and water resources scenarios. Decisions should be made by comparing costs, benefits, losses, and risks for each scenario, partly based on subjective interpretation of the scenarios and results of analysis. Application of Bayesian theory in the decision-making process in an environment of uncertainty is advocated by some analysts. However, this requires assignment of prior probabilities to assumed climate scenarios, which at present may only reflect different degrees of subjective belief about the accuracy of climate models.

In Poland's present economic situation the second policy approach, that of undertaking preventive measures against possible negative consequences of climate change, seems to be the most rational. Not much can be done in Poland to reduce natural resource scarcity. Probably one of the best ways to cope with the consequences of climate change is to reduce the current variability of runoff by means of increased storage capacity. Another way of improving the supply–demand balance is to implement demand-management and water-conservation strategies. There is plenty of evidence that central regulation alone is insufficient and that it must be supported by appropriate economic incentives.

If climatic and hydrologic conditions lead to temporal or regional water stress, it is crucial to protect available resources from contamination. The present level of pollution of water resources in Poland is very high, which is the second reason for the possible emergence of water as a barrier to growth. Above all, the program of water quality improvement in Poland provides for the construction of new wastewater treatment plants and for more effective use of the existing facilities.

There is a need for increased public awareness of Poland's water resources problems and for awareness of the possible negative consequences of changing environmental conditions. The present lack of awareness is caused not so much by the population's lack of sensitivity to water issues, but rather by the lack of belief in the cumulative effects of individual undertakings. The notion of "our water" must be given practical meaning, which is closely related to the ongoing process of establishing real self-governing regional authorities.

Finally, improvements in institutional arrangements are crucial if water is not to become a barrier to growth. In Poland, water authority would be gradually transferred from the 49 regional administrations to the newly established river-basin authorities. These important institutional changes are now being introduced within the framework of the process leading to more decentralized decision making in the entire country.

All of the measures discussed above – flow regulation, water quality improvement, demand management based on rational economic principles, public participation, and more effective institutional arrangements – should be applied to solve current national water resources problems. They may at the same time help to prepare water resources systems to cope with possible shocks and threats caused by global changes of environmental processes.

References

Carter, T.R., Parry, M.L., Nishioka, S., and Harasawa, H., 1992, *Preliminary Guidelines for Assessing Impacts of Climate Change*, Intergovernmental Panel on Climate Change (IPCC), World Meteorological Organization (WMO), and United Nations Environmental Program (UNEP), Geneva, Switzerland.

Commission for Hydrology, Abridged Final Report of the Eighth Session, October 1988, World Meteorological Organization, No. 175, Geneva, Switzerland.

Engelman, R., and LeRoy, P., 1993, *Sustaining Water: Population and the Future of Renewable Water Supplies*, Population Action International, Washington, DC, USA.

Falkenmark, M., and Widstrand, C., 1992, Population and Water Resources: A Delicate Balance, *Population Bulletin*, Population Reference Bureau.

Gburek, W.J., 1971, Discussion of hydrologic consequences of rainfall augmentation, *American Society of Civil Engineers Journal of Hydraulic Div.*, **97**(HY12):2114–2115.

Jurak, D., 1992, Water temperature and evaporation in the light of various climate scenarios it Przeglad Geofizyczny, **37**(3–4):149–159 [in Polish].

Jurak, D., and Wisniewski, A., 1989, Simple Model of Thermal Pollution in Rivers, WP-89-066, International Institute for Applied Systems Analysis, Laxenburg, Austria.

Kaczmarek, Z., 1974, Some problems of stochastic storage with correlated inflows, International Association of Hydrological Sciences (IAHS) Publication No. 100, IAHS Press, Wallingford, UK.

Kaczmarek, Z., 1978, Water resources of Poland and principles for their rational use, *The Review of the Polish Academy of Sciences*, No. 4:35–45.

Kaczmarek, Z., 1993, Water balance model for climate impact analysis, *Acta Geophysica Polonica*, **41**(4):423–437.

Klemeš, V., 1981, Applied stochastic theory of storage in evolution, *Advances in Hydrosciences*, **12**:79–141.

Langbein, W.B., 1949, Annual runoff in the United States, USGS Circular 5, US Department of Interior, Washington, DC, USA.

Penck, A., 1986, Untersuchungen über Verdunstung und Abfluß von größeren Landflächen, Bd. 5, No. 5., Geogr. Abt., Vienna, Austria.

Turc, L., 1954, Water balance of soils: Relationship between precipitation, evapotranspiration, and runoff, *Annales Agronomiques*, **5**:491–595 and **6**:5–131 [in French].

Waggoner, P.E., ed., 1990, *Climate Change and U.S. Water Resources*, John Wiley & Sons, New York, NY, USA.

Waggoner, P.E., and Revelle, R.R., 1990, Summary, in P.E. Waggoner, ed., *Climate Change and U.S. Water Resources*, John Wiley & Sons, New York, NY, USA.

Computer Software Reference

Kaczmarek, Z., 1993, *CLIRUN: Climate/Runoff Model Version 3*, Institute of Geophysics, Warsaw, Poland.

Chapter 15

Water Resources Management Under Climatic Variability*

Zdzislaw Kaczmarek, Kenneth M. Strzepek, and László Somlyódy

15.1 Introduction

The material presented in the first six chapters provides an overview of the issues and methodologies concerning the analysis of water resources under climatic uncertainty. Chapters 8 to 14 present seven case studies of the impact of climate change on a variety of water resources systems of various sizes. Although much information is presented and knowledge can be gained from studying these chapters, the unfortunate fact is that there is no single lesson or message that can be drawn, save that the uncertainties involved in climate change assessment dominate the quantitative results produced. How then do we proceed with this lack of a clear scientific direction? As part of the IPCC process a group of 22 international water resources experts – 6 lead authors and 16 collaborating authors – made a considerable effort to answer this very question. The discussion that follows draws heavily on the consensus of this group of experts. Although the focus of the discussion is broader than that of the material contained in this book, the editors feel that the insights are important and enhance the book's overall contribution to the literature.

*This chapter draws heavily from the work of the Intergovernmental Panel on Climate Change (IPCC) Working Group II B.6 on Water Resources Management, Z. Kaczmarek, Convening Lead Author, with N. Arnell, K. Hanaki, G. Mailu, L. Somlyódy, and E. Stakhiv, Lead Authors.

Z. Kaczmarek et al. (eds.),
Water Resources Management in the Face of Climatic/Hydrologic Uncertainties 378–387.

15.2 The Management Process

Assessments of climate change impacts on water resources management attempt to show how the range of possible physical changes in temperature, precipitation, and runoff is likely to affect the range of water uses and their socioeconomic implications and possible response options. Unlike the previous IPCC reports (1990a, 1990b, 1992), which combined hydrologic effects and water resources impacts, this chapter focuses primarily on the ability of water resources management to cope with anticipated climate changes. Because water resources management inherently reflects a broad range of management strategies designed to adapt to contemporary climate variability, much of the discussion will be devoted to showing how these practices will also serve to ameliorate alternative climate change scenarios.

The previous IPCC assessments outlined the extensive difficulties of conducting meaningful analyses of climate change impacts on hydrology and water resources (Shiklomanov *et al.*, 1990; Stakhiv *et al.*, 1992). To date, little new information has been added to change the conclusions of the earlier reports. This book also reflects the state of the art in 1992. The cascading uncertainties of climate change impact analysis are simply too great to identify regional or watershed-specific impacts. It has been noted by many analysts (Rogers, 1993a, 1993b; Klemeš, 1993; Stakhiv, 1993) that water management practices that serve the present range of climate variability may also adequately ameliorate the range of perturbations expected to accompany climate change. This does not imply that water resources impacts might not be severe in some regions – there simply is no way of pinpointing either the magnitude or the incidence of those impacts (Lettenmaier *et al.*, 1994a, 1994b). The case studies above demonstrate this fact, reporting a wide range of impacts in the same location due to slight changes in the climate scenario analyzed. Furthermore, it is doubtful whether the current fashion of conducting "worst case" analyses, wherein the most extreme scenario of a given GCM is used to develop hydrologic responses, is useful for a critical appraisal of regional sensitivities to climate change. If anything, this type of analysis skews the evaluation and deflects the search for pragmatic responses.

Most water managers, whether in agencies dealing with large, multiple reservoir systems or in small utilities dependent on ground water, constantly concern themselves with three types of issues and designs: those dealing with new investments for capacity expansion, those dealing with the operation and maintenance of existing systems, and those dealing with demand modifications (Rogers, 1993a). There is continuous adaptation, driven by new hydrologic information, ecological constraints, water quality standards, and shifts in demands and preferences. This situation is a reality for managed water systems, but this is less common in systems that are not managed (e.g., wetlands) dependent entirely on either river flow, ground water level, or precipitation. These points related to water quality are well documented in Chapter 8 on the Delaware River basin and in Chapter 10 on California's Central Valley.

Transformation of the structure and characteristics of water supply and demand due to climatic and non-climatic factors may create serious social and political problems. Socioeconomic factors greatly influence the ability to solve these problems; without necessary capital and technology solutions cannot be implemented. The need for external financial help may arise in cases of unfavorable trends in hydrologic and atmospheric processes.

15.3 Considerations for Response Strategies

In general, most countries and civilizations have faced water shortages due either to natural climate variability or overexploitation and pollution of the resource base. Water resources management inherently deals with mitigating the effects of hydrologic extremes and providing a greater degree of reliability in the delivery of water-related services. However, no enterprise is risk free. A society decides what level of risk it will bear through its acceptance of certain levels of risk and reliability, as expressed by standards and criteria. The marginal costs of reducing each additional increment of risk usually rise rapidly as reliability approaches 100%. Hence, water managers typically deal with 90%, 95%, and 99% levels of reliability as useful performance measures of the available quality and quantity of water. The same is true in the mitigation of other natural hazards, especially in traditional approaches to drought mitigation, flood control, and damage mitigation. Given some of the preliminary results of GCM experiments regarding potential changes in rainfall intensity and frequency (Gordon *et al.*, 1992; Whetton *et al.*, 1993), it appears that flood-related consequences of climate change may be as serious and widely distributed as the adverse impacts of droughts. This should raise concerns over dam safety and levee design criteria and cause us to rethink our floodplain management policies. The results on reservoir design in Belgium reported in Chapter 13 illustrate this point well. The recent devastating floods of 1993 in the upper Mississippi River basin have resulted in a US Interagency Floodplain Management Review Committee report (1994) on just such policy issues. This is another example of the reality of constant adaptation in the field of water resources management.

The first IPCC reports (1990a, 1990b) contained a discussion on the philosophy of adaptation and a list of adaptation options suited to the range of water management problems expected under climate change. Based on a review of the most recent literature, no new water management actions or strategies unique to climate change can be added to the list. However, it should be noted that many nations have pledged to implement action plans for sustainable water resources management as part of their obligation toward AGENDA 21. In this respect, the principles laid out in that document would serve as a useful guide for developing a strategy to enable nations, river basin authorities, and water utilities to prepare for and partially accommodate the uncertain hydrologic effects that might accompany global warming.

There are many possibilities for individual adaptation measures or actions. An overview of water supply and demand management options is presented by Frederick (1992) as part of an attempt to develop approaches to handling increasing water scarcity. A long-term strategy requires a series of plausible development scenarios formulated based on different combinations of population growth assumptions and economic, social, and environmental objectives (Carter *et al.*, 1992). After these scenarios are established, taking into account the possibility of climate change, a set of alternative long-term strategies for water management must be formulated consisting of different combinations of water management measures, policy instruments, or institutional changes designed to best meet the objectives of a particular growth and development scenario and its consequent CO_2 emissions rate. The response strategies must then be compared and appraised according to their levels of service reliability, costs, and environmental and socioeconomic impacts. Chapter 11 on the Lake Kariba operation and Chapter 12 on the Terek River basin both demonstrate the use of optimization approaches to determine an "optimal" response strategy to climate change impacts. Some strategies will be better suited to dealing with climate change uncertainty, i.e., more robust and resilient. Others will focus on environmental sustainability, as discussed in Chapters 8, 9, and 10. Some are likely to emphasize reliability of supply. The reality is that the selection of an "optimal" path is a decision based on social preferences and political realities more so than on engineering design criteria. Engineering design criteria, however, also evolve over time, and are updated as new meteorological and hydrologic records are extended and the performance of water management systems is tested under varying conditions. This is discussed in Chapter 13.

All major institutions that deal with water resources planning and management agree that future water management strategies should include various cost-effective combinations of the following management measures:

- Direct measures to control water use (regulatory, technological).
- Indirect measures that affect behavior (incentives, taxes).
- Institutional changes for improved management of resources.
- Improvement in the operation of water management systems.
- Direct measures that increase availability of supply (reservoirs, pipelines).

Different strategies apply to different circumstances, as the case studies illustrate. Watersheds that have little or no control of natural flows and are largely dependent on precipitation must implement a different set of water management strategies than river basins with a high degree of control in the form of reservoirs, canals, levees, etc. Chapter 14 points out that Polish rivers have limited storage potential and require a different approach to adaptation that is suited to river basins without a high degree of control of natural flows. Similarly, rapidly urbanizing areas will require different responses than will agricultural regions. There is no standard, prescribed approach. However, if a rational management strategy is undertaken to deal with reasonably foreseeable regional needs in the absence of climate change, then according to the principles espoused in AGENDA 21 that strategy will also serve to offset much of the range of possible adverse consequences of climate change.

15.4 Concluding Remarks and Research Needs

Understanding of the implications of climate change for water resources has advanced significantly over the last few years with the completion of several regional case studies (although very few such studies were carried out in the developing countries). The level of uncertainty remains high, depending on time discretization, spatial scale, and climatic variables of interest. Most existing hydraulic structures were designed to mitigate negative consequences of climate and hydrological variability. The addition of one more factor – climate change – complicates planning techniques; but water analysts are not unprepared to deal with various types of uncertainties. Most existing operational tools can be applied for changed climatic conditions as they are now used in very different climatic zones. For the decision-making process the critical question is what practical conclusions can be drawn from a set of calculations based on uncertain future scenarios. If the range of possible actions is not too broad, then even subjective expert judgement may help. If, however, the alternative solutions (concerning, for example, the required capacity of a reservoir) are highly dispersed, then the decision-making process may involve serious controversies. Traditional water resources performance criteria, such as reliability, dam safety, resilience, and robustness, may help in the assessment of the impact of climate change on water resources systems. The Bayesian approach and other similar techniques have been advocated for finding a rational solution, but their practical application still requires methodological improvements. A wide variety of methodologies are presented in the case studies.

Uncertainties are propagated through the numerous levels of analysis as one moves from multiple CO_2 emission scenarios, through the comparison of different GCM outputs, to transference of precipitation data to hydrologic runoff and basin water budgets, to impacts on each water user sector, to water management decisions, and finally to the socioeconomic and environmental impacts of response measures and strategies. All case studies and assessments available for this book are based on earlier IPCC scenarios and steady-state GCM output. Although the case studies cover different areas and regions, few new insights are offered. A fundamental gap in hydrological analysis, and hence in water management decision making, is the complete absence of vital information on future climate variability. Most water management decisions are based on some knowledge about climate and hydrologic variability, and GCMs are not capable of providing such information. Attempts have been made to replicate contemporary variability in the form of specific weather patterns associated with atmospheric circulation conditions, as measured by changes in atmospheric pressure (Bardossy and Plate, 1992). This is based on the fact that, of all the variables that are forecast by the GCMs, atmospheric pressure is thought to be the most accurate (Matyasovszky *et al.*, 1993). Although this method brings us closer to providing a time series of hydrological events explicitly linked to transient scenarios generated by GCMs, it is based on the assumption that the stochastic characteristics of weather patterns of contemporary climate will be similar for future climate scenarios.

A more fundamental evaluation problem is a methodological flaw that is characteristic of most climate change impact studies to date, including those focusing on water resources. It is the failure to distinguish the incremental impacts due exclusively to climate change over the course of an extended 50- to 100-year time horizon from those changes that would occur in the absence of climate change. Socioeconomic variables depend on geopolitics, religious pressure, epidemics, etc., and are determined quite differently from the physical variables that govern climate change. Economic assessments may not be reliable for more than a few years. Therefore, socioeconomic projections are rarely consistent with climate change projections, which are generally valid for longer time periods. This point is fundamental to the establishment of an adaptive versus an anticipatory strategy. Indeed, it is very difficult to establish scenarios that reflect future changes and actions that will occur in a region in the absence of climate change. Yet, that is what is required according to most accepted impact analysis procedures, including those promoted by the IPCC (Carter *et al.*, 1992). In the case of water resources management, in particular, adaptation will occur continuously and primarily in response to population growth and shifts in demand and sectoral water use. Chapters 8 and 12, with their focus on river basins, take into account a more comprehensive approach than do most climate impact assessments.

The emphasis in the next decade is likely to be on demand management, regulatory controls, legal and institutional changes, and economic instruments. Furthermore, the choice of options for adaptation is inherently localized and dependent on the institutions and culture of water use. The only significant difference that is likely to result from climate change is in the timing of a project's implementation. Instead of a reactive, incremental adaptation strategy, water managers might follow an anticipatory one. In effect, such knowledge would serve to reduce, but not eliminate, the adverse environmental and socioeconomic impacts likely to be associated with future climate change perturbations. This is because, despite the best planning efforts, water management systems are not designed to eliminate all risk from natural hazards. They only eliminate a significant portion of risk, which results in an improvement of net social well-being based on a formal risk–cost tradeoff analysis (Stakhiv, 1994).

It is difficult to generalize water resources problems on a global scale; most of the difficult problems arise at, and should be solved at, the national or regional level. Chapter 9 provides a global perspective on lake stratification by globally integrating a number of local cases. Chapter 14 presents an assessment of national runoff by extrapolating from modeling of representative catchments that account for one-third of the total area of Poland. Many nations are confronted with difficulties in supplying clean water for domestic use, industry, and agriculture, and may face increased stress due to a number of climatic and non-climatic factors, as discussed in Chapters 8, 9, and 10. Today, problems should be solved in a way that makes water supply more robust and flexible for an uncertain future. Local conditions determine the importance of incremental climate change impacts on water resources; however, arid and semiarid regions, as well as countries with high population growth, may be particularly vulnerable. In most cases, regulated water systems are better prepared to cope with negative consequences of climate change than are systems that are not managed. The choice of adaptation strategies in managed systems should be based on cost-effectiveness analysis. Because most hydraulic structures require rather long implementation times, any water resources impact study should use appropriate time horizons and areas. Under the condition of growing uncertainties, solutions with shorter implementation times (e.g., several small-scale reservoirs instead of a single large one) should be given preference, if such options are available.

Regarding the impact of climate change on thermal pollution, no drastic water quality changes are anticipated due to likely climate alteration. This is particularly true for clean water bodies. For contaminated water bodies, characterized by complex interactions of many different pollution problems, climate change adds an incremental impact that can be estimated only with significant uncertainty at present. The analysis of the incremental impact should be incorporated into future sensitivity studies; existing hydrothermal and water quality models can be used for this purpose. In countries such as the USA, where waste load allocation methods are routinely used within the framework of permit procedures, the calculations should be updated to see the possible water quality management consequences of changed climate. Water quality management has no specific means of coping with the impacts of climate change; therefore, some traditional tools should be applied to adapt water systems to changed conditions. Changes are anticipated to take place slowly; time is available for monitoring the consequences and for planning measures for the reconstruction of existing control facilities or the establishment of new ones, should they be necessary.

More research is needed on possible changes in sectoral water demands, particularly for irrigation water requirements; the impact of changes in atmospheric processes on variability of hydrologic phenomena, including flood frequency characteristics; the effects of climate on quality of aquatic systems and soil salinization; the impact of climate change on water resources thresholds; and the impact on possible regional and international conflicts to guarantee access to water resources.

References

Bardossy, A., and Plate, E., 1992, Space-time model for daily rainfall using atmospheric circulation patterns, *Water Resources Research*, **28**(5):1247–1259.

Carter, T., Parry, M., Nishioka, S., and Harasawa, H., eds., 1992, *Preliminary Guidelines for Assessing Impacts for Climate Change*, Intergovernmental Panel on Climate Change (IPCC), Geneva, Switzerland.

Frederick, K., 1992, *Balancing Water Demands with Supplies: The Role of Management in a World of Increasing Scarcity*, Technical Paper No. 189, The World Bank, Washington, DC, USA.

Gordon, H. *et al.*, 1992, Simulated changes in daily rainfall intensity due to the enhanced greenhouse effect: Implications for extreme rainfall events, *Climate Dynamics*, **8**:83–102.

IPCC (Intergovernmental Panel on Climate Change), 1990a, *Climate Change: The IPCC Scientific Assessment*, J.T. Houghton, G.J. Jenkins and J.J. Ephraums, eds., Cambridge University Press, Cambridge, UK.

IPCC (Intergovernmental Panel on Climate Change), 1990b, *Climate Change: The IPCC Impacts Assessment*, W.J.McG. Tegart, G.W. Sheldon, and D.C. Griffith, eds., Australian Government Publishing Service, Canberra, Australia.

IPCC (Intergovernmental Panel on Climate Change), 1992, *The Supplementary Report to the IPCC Impacts Assessment*, W.J.McG. Tegart, G.W. Sheldon, and J.H. Hellyer, eds., Australian Government Publishing Service, Canberra, Australia.

Klemeš, V., 1993, Design implications of climate change, in T. Ballentine and E. Stakhiv, eds., *Proceedings of Conference on Climate Change and Water Resources Management*, US Army Institute for Water Resources, Fort Belvoir, VA, USA.

Lettenmaier, D., Wood, E., and Wallis, J., 1994a, Hydro-climatological trends in the continental United States: 1948–88, *Journal of the Climate*, **7**:586–607.

Lettenmaier, D. *et al.*, 1994b, Water Management Impacts of Global Warming: Draft Reports: Tacoma, Washington, Savannah River system, Boston, Massachusetts, Missouri River system, Columbia River system, US Army Institute for Water Resources, Fort Belvoir, VA, USA.

Matyasovszky, I., Bogardi, I., Bardossy, A., and Duckstein, L., 1993, Space-time precipitation reflecting climate change, *Hydrological Sciences Journal*, **38**:539–558.

Rogers, P., 1993a, *America's Water: Federal Roles and Responsibilities*, The MIT Press, Cambridge, MA, USA.

Rogers, P., 1993b, What water managers and planners need to know about climate change and water resources management, in T. Ballentine and E. Stakhiv, eds., *Proceedings of Conference on Climate Change and Water Resources Management*, US Army Institute for Water Resources, Fort Belvoir, VA, USA: I/1–I/14.

Shiklomanov, I., Lins, H., and Stakhiv, E., 1990, Hydrology and water resources, in W.J.McG. Tegart, G. Sheldon, and D. Griffiths, eds., *The IPCC Impacts Assessment*, Australian Government Publishing Service, Canberra, Australia.

Stakhiv, E., 1993, Water resources planning and management under climate uncertainty, in T. Ballentine and E. Stakhiv, eds., *Proceedings of Conference on*

Climate Change and Water Resources Management, US Army Institute for Water Resources, Fort Belvoir, VA, USA: IV/20-IV/36.

Stakhiv, E., 1994, Managing water resources for adaptation to climate change, in L. Duckstein and E. Parent, eds., *Engineering Risk and Reliability in a Changing Physical Environment*, Proc. NATO ASI, Deauville, Kluwer, Dordrecht, The Netherlands.

Stakhiv, E., Lins, H., and Shiklomanov, I., 1992, Hydrology and water resources, in W.J.McG. Tegart and G. Sheldon, eds., *The Supplementary Report to the IPCC Impacts Assessments*, Australian Government Publishing Service, Canberra, Australia.

US Interagency Floodplain Management Review Committee, 1994, *Sharing the Challenge: Floodplain Management into the 21st Century*, Report to the Administration Floodplain Management Task Force, US Government Printing Office, Washington, DC, USA.

Whetton, P., Fowler, A., Haylock, M., and Pittock, A., 1993, Implications of climate change due to the enhanced greenhouse effects on flood and droughts in Australia, *Climatic Change*, **25**:289–317.

Index